GEOMETRY

new edition

Ray C. Jurgensen

Richard G. Brown

Alice M. King

Editorial Adviser
Albert E. Meder, Jr.

Teacher Consultant
Anita M. Cooper

HOUGHTON MIFFLIN COMPANY · BOSTON
Atlanta   Dallas   Geneva, Ill.   Hopewell, N.J.   Palo Alto   Toronto

THE AUTHORS

Ray C. Jurgensen, formerly Chairman of the Mathematics Department and holder of the Eppley Chair of Mathematics, Culver Academies, Culver, Indiana.

Richard G. Brown, Mathematics Teacher, Phillips Exeter Academy, Exeter, New Hampshire.

Alice M. King, Professor of Mathematics, California State Polytechnic University, Pomona, California.

EDITORIAL ADVISER

Albert E. Meder, Jr., Dean and Vice Provost and Professor of Mathematics, Emeritus, Rutgers, The State University of New Jersey.

TEACHER CONSULTANT

Anita M. Cooper, former Mathematics Teacher, Walnut Hills High School, Cincinnati, Ohio.

CONTENTS

12 Transformations 423

Symbols

adj. ∠	adjacent angles (p. 17)
alt. int. ∠	alternate interior angles (p. 43)
∠, ∠	angle(s) (p. 8)
a	apothem (p. 335)
≈	is approximately equal to (p. 221)
$\overset{\frown}{BC}$	arc with endpoints B and C (p. 253)
A	area (p. 322)
B	area of base (p. 360)
b	length of base (p. 322)
$\odot O$	circle with center O (p. 245)
C	circumference (p. 339)
comp. ∠	complementary angles (p. 18)
cm	centimeter(s) (p. 67)
≅	congruent, is congruent to (p. 82)
↔	corresponds to (p. 82)
corr. ∠	corresponding angles (p. 43)
cos	cosine (p. 226)
cm^3	cubic centimeter(s) (p. 361)
m^3	cubic meter(s) (p. 361)
°	degrees (p. 9)
$D_{O,k}$	dilation with center O and scale factor k (p. 488)
d	diameter; distance; length of diagonal (p. 340; p. 388; p. 326)
H_O	half turn about point O (p. 433)
h	height, length of altitude (p. 322, p. 359)
T^{-1}	inverse of transformation T (p. 445)
I	identity transformation (p. 445)
km	kilometer(s) (p. 209)
L.A.	lateral area (p. 360)
JL	length of $\overline{JL}$, distance between points J and L (p. 5)
$\overleftrightarrow{AB}$	line containing points A and B (p. 2)
$S: A \to A'$	S maps point A to point A'. (p. 424)
m	meter(s) (p. 168)
mm	millimeter(s) (p. 228)

(x, y)	ordered pair (p. 387)
∥	parallel, is parallel to (p. 41)
▱	parallelogram (p. 123)
p	perimeter (p. 335)
⊥	perpendicular, is perpendicular to (p. 17)
π	pi (p. 339)
n-gon	polygon with n sides (p. 61)
$M_j \circ M_k$	product of M_j and M_k (p. 440)
quad.	quadrilateral (p. 69)
r	radius (p. 339)
$\frac{a}{b}$, $a:b$	ratio of a to b (p. 167, p. 168)
$\overrightarrow{AB}$	ray with endpoint A, passing through point B (p. 5)
rect.	rectangle (p. 135)
M_j	reflection in line j (p. 427)
rt. ∠	right angle (p. 10)
rt. △	right triangle (p. 133)
$R_{O,90}$	rotation about point O through $90°$ (p. 432)
s-s. int. ∠	same-side interior angles (p. 43)
$\overline{AB}$	segment with endpoints A and B (p. 5)
s	length of a side of a regular polygon (p. 322, p. 335)
~	similar, is similar to (p. 174)
sin	sine (p. 226)
l	slant height (p. 364)
m	slope (p. 398)
cm^2	square centimeter(s) (p. 323)
m^2	square meter(s) (p. 323)
$\sqrt{x}$	positive square root of x (p. 203)
supp. ∠	supplementary angles (p. 18)
T.A.	total area (p. 360)
tan	tangent (p. 220)
trans.	transversal(s) (p. 127)
trap.	trapezoid (p. 241)
△, △	triangle(s) (p. 56)
vert. ∠	vertical angles (p. 18)
V	volume (p. 361)

1

Points, Lines, Planes, and Angles

Undefined Terms and Basic Definitions

Objectives
1. Understand *point, line,* and *plane;* draw representations of them.
2. Use undefined terms to define some basic terms in geometry.
3. Use symbols for lines, segments, rays, and distances.
4. Name angles and find their measures.
5. Apply postulates from algebra.

1-1 *Points, Lines, and Planes*

The simplest figure studied in geometry is a *point.* All other figures are made of points.

• B

• A

Two points   A line is a   A circle is a
             set of points.   set of points.

A point is pictured by drawing a dot. In order to be seen, the dot must have some size, but the point represented by the dot has no size at all. Likewise, a picture of a line has some thickness, whereas the line itself has no thickness.

A point is named by a capital letter, for example, *point A*. Whenever we refer to "two points," we shall mean two different points. "Three points" will mean three different points, and so on. This will also be our custom with regard to lines and other geometric figures.

Our term *line* will always mean a straight line. Often a line is referred to by a single lowercase letter, such as *line m*. If you know that a line contains the points *A* and *B*, you can also call it *line AB* ($\overleftrightarrow{AB}$) or *line BA* ($\overleftrightarrow{BA}$). The arrowheads in the drawing suggest that the line extends in both directions without ending.

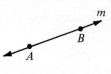

A floor, wall, or tabletop suggests a geometric *plane*. A plane, unlike a tabletop, extends without ending and has no thickness. Although a plane has no edges, we usually picture a plane by drawing a four-sided figure as shown below. We label a plane with a capital letter.

Plane *M*

Plane *N*

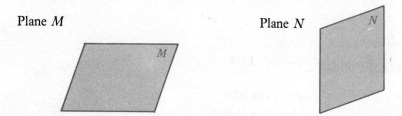

Some expressions commonly used to describe relationships between points, lines, and planes follow. In these expressions, *intersects* means "meets" or "cuts." The **intersection** of two figures is the set of points that are in both figures. Dashes in the diagrams indicate parts hidden from view in figures in space.

A is in *l*.
A is on *l*.
l contains *A*.
l passes through *A*.

l and *h* intersect in *O*.
O is the intersection of *l* and *h*.

k and *P* are in *M*.
M contains *k* and *P*.
j intersects *M* in *P*.
P is the intersection of *j* and *M*.

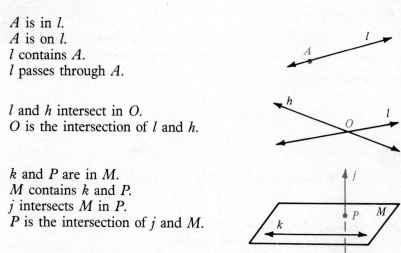

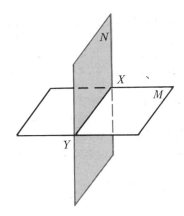

M and N intersect in $\overleftrightarrow{XY}$.

$\overleftrightarrow{XY}$ is the intersection of M and N.

$\overleftrightarrow{XY}$ is in M and N.

M and N contain $\overleftrightarrow{XY}$.

Classroom Exercises

1. Suppose there is a line AB. Does $\overleftrightarrow{AB}$ begin at point A?
2. Does a plane have edges?
3. Suppose point P is in plane Q. Are there any points in Q that are a million kilometers from P?
4. Is there any limit to the number of points in a line?
5. Is there any limit to the number of points in a plane?
6. Is there any limit to the number of lines in a plane?
7. Can a given point be in two lines? in ten lines?
8. Can a given line be in two planes? in ten planes?
9. Is it possible for two planes not to intersect at all?

Written Exercises

Classify each statement as true or false.

A
1. l is in Q.
2. Q contains P.
3. k is in Q.
4. l passes through P.
5. P is on k.
6. P is the intersection of l and k.
7. Q passes through k at P.
8. k and l do not intersect.

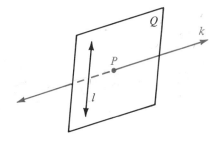

Classify each statement as true or false.

9. Plane R intersects plane S in line AB.

10. $\overleftrightarrow{AB}$ is in plane R.

11. S contains $\overleftrightarrow{AB}$.

12. Point C is in R and S.

13. R and S contain D.

14. D is in line h.

15. h is in R.

16. h is in S.

Sketch and label the figures described. Use dashes for parts hidden from view.

B 17. Line m passes through points E and F. 18. Lines k and n intersect in point O.

19. Point P is in plane R. 20. $\overleftrightarrow{AB}$ is in plane T.

21. Line l intersects plane M in only one point, D.

22. Lines k and n cut plane R at point O.

23. Point A lies outside plane Q. Two lines, h and j, both contain A and intersect Q.

24. Plane V intersects plane W in $\overleftrightarrow{XY}$.

C 25. Three planes, X, Y, and Z, intersect in line l.

26. Three planes, A, B, and C, intersect in only one point, P.

27. Two nonintersecting planes, T and V, are both intersected by line k.

28. Lines h and j intersect in point E. Plane B contains h but does not contain j.

29. Planes F and G intersect in line a, planes F and H intersect in line b, and planes G and H intersect in line c.

1-2 *Definitions*

If you try to describe the diagram at the right, you will probably use the words "point" and "line." If you then try to define these words, you will have a difficult job. In geometry, *point*, *line*, and *plane* are not defined. We use such *undefined terms* as a basis for defining other terms. Here are our first definitions.

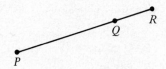

Collinear points are points all in one line.

Collinear points

Noncollinear points

Coplanar points are points all in one plane.

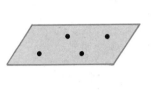 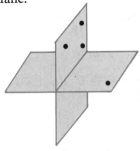

Coplanar points   Noncoplanar points

Space is the set of all points.

Segment *AB* ($\overline{AB}$) consists of points *A* and *B* and all points in $\overleftrightarrow{AB}$ which are between *A* and *B*. Points *A* and *B* are called the **endpoints** of $\overline{AB}$.

Ray *AB* ($\overrightarrow{AB}$) is the part of $\overleftrightarrow{AB}$ which starts at point *A* and extends without ending through point *B*. The endpoint of $\overrightarrow{AB}$ is *A*, the point named first.

$\overrightarrow{SR}$ and $\overrightarrow{ST}$ are called **opposite rays** if *S* is in $\overleftrightarrow{RT}$ and between *R* and *T*.

On a *number line* every point is paired with a number and every number is paired with a point. Below, point *J* is paired with -3, the *coordinate* of *J*.

J		K		L		M

$$-4 \quad -3 \quad -2 \quad -1 \quad 0 \quad 1 \quad 2 \quad 3 \quad 4$$

The **length** of $\overline{JL}$ is the distance between point *J* and point *L*. We use the symbol *JL* to represent this distance. We can find *JL* by counting the number of units between *J* and *L* or by subtracting the smaller coordinate (-3) from the larger coordinate (2).

$$JL = 2 - (-3) = 5$$
$$KL = 2 - 0 = 2$$
$$ML = 4 - 2 = 2$$

Equal segments are segments that have equal lengths. We have shown that $\overline{KL}$ and $\overline{ML}$ both have length 2. We indicate that $\overline{KL}$ and $\overline{ML}$ are equal segments when we write $KL = ML$, which states that their lengths are equal.

The **midpoint of a segment** is the point that divides the segment into two equal segments. In the figure, $AP = PB$. P is the midpoint of $\overline{AB}$.

A **bisector of a segment** is a line, segment, ray, or plane that intersects the segment at its midpoint. Line l is a bisector of $\overline{AB}$. $\overrightarrow{PQ}$ and plane X also *bisect* $\overline{AB}$.

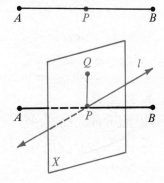

Classroom Exercises

1. Are points A, B, and C collinear?

2. Are points D, B, and E collinear?

3. Are points A, B, and C coplanar?

4. Are points A, B, C, and D coplanar?

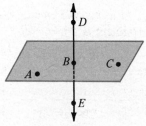

What does each symbol represent?

5. $\overline{PQ}$ **6.** $\overrightarrow{PQ}$ **7.** $\overleftrightarrow{PQ}$ **8.** PQ

9. How many endpoints does a segment have?

10. How many endpoints does a ray have?

11. How many endpoints does a line have?

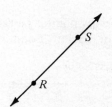

12. Is $\overline{RS}$ the same as $\overline{SR}$?

13. Is $\overrightarrow{RS}$ the same as $\overrightarrow{SR}$?

14. Is $\overleftrightarrow{RS}$ the same as $\overleftrightarrow{SR}$?

Use the number line for Exercises 15–24.

15. What is the coordinate of P? of R?

16. Name the point with coordinate 2.

17. Find each distance: (a) RS (b) RQ (c) PT

18. Name the midpoint of $\overline{RT}$.

19. What number is halfway between 1 and 2?

20. What is the coordinate of the midpoint of $\overline{ST}$?

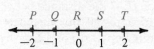

21. Name three other numbers between 1 and 2.

22. Can you list all the numbers between 1 and 2?

23. Is there a point on the number line for every number between 1 and 2?

24. Is there any limit to the number of points on the number line between S and T?

Written Exercises

Are the points collinear?

A 1. A, S, B

2. A, B, C

3. R, A, B

4. D, T, B

Are the points coplanar?

5. D, T, B, C

6. A, B, C, E

7. R, A, S, D

8. R, A, S, C

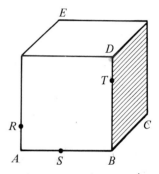

In the diagram, M is the midpoint of $\overline{GH}$. Classify each statement as true or false.

9. $GM = MH$

10. $\overleftrightarrow{KT}$ is a bisector of $\overline{GH}$.

11. $\overline{MT}$ bisects $\overline{GH}$.

12. $HM = \frac{1}{2}(HG)$

13. $\overline{GH}$ contains only three points.

14. $G, M,$ and T are collinear.

15. $\overrightarrow{MT}$ and $\overrightarrow{TM}$ are opposite rays.

16. $\overrightarrow{MT}$ and $\overrightarrow{MK}$ are opposite rays.

17. $\overrightarrow{KT}$ is the same as $\overrightarrow{KM}$.

18. $\overleftrightarrow{KT}$ is the same as $\overleftrightarrow{KM}$.

19. $\overline{KT}$ is the same as $\overline{KM}$.

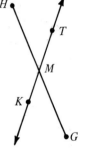

In the number line for Exercises 20–24, M is the midpoint of $\overline{UP}$, and $BU = PY$.

B 20. Find the length of $\overline{UP}$.

21. Name the coordinate of M.

22. Find BU.

23. Name the coordinate of Y.

24. Name a segment equal to $\overline{MB}$.

In Exercises 25–29, draw $\overline{CD}$ and $\overline{RS}$ so that the conditions are satisfied.

25. $\overline{CD}$ and $\overline{RS}$ are in the same line.

26. $\overline{CD}$ and $\overline{RS}$ intersect, but neither segment bisects the other.

27. $\overline{CD}$ and $\overline{RS}$ bisect each other.

28. $\overline{CD}$ bisects $\overline{RS}$, but $\overline{RS}$ does not bisect $\overline{CD}$.

29. $\overline{CD}$ and $\overline{RS}$ do not intersect, but $\overrightarrow{CD}$ and $\overrightarrow{RS}$ do intersect.

In Exercises 30–33, *P* is the midpoint of $\overline{NO}$; *Q* is the midpoint of $\overline{NP}$; *R* is the midpoint of $\overline{NQ}$; *S* is the midpoint of $\overline{NR}$; *T* is the midpoint of $\overline{NS}$; and *U* is the midpoint of $\overline{NT}$.

C **30.** If *NO* = 64, *NT* = _?_.   **31.** If *NO* = 56, *NT* = _?_.

   32. If *NO* = 2.4, *PR* = _?_.   **33.** If *US* = 12, *SP* = _?_.

34. Suppose that *D*, *U*, *N*, *K* are four points such that $\overleftrightarrow{DU}$ contains *N*, and $\overline{NK}$ contains *U*. Which of the following statements *must* be true?
   a. *D*, *U*, *N*, and *K* are collinear.   **b.** *N* is between *D* and *U*.
   c. $\overrightarrow{KN}$ contains *U*.   **d.** $\overrightarrow{UK}$ and $\overrightarrow{UD}$ are opposite rays.

35. State how many segments can be drawn between the points in each figure. No three points are collinear.

   a.   **b.**   **c.**   **d.**

   3 points   4 points   5 points   6 points
   _?_ segments   _?_ segments   _?_ segments   _?_ segments

   e. Without making a drawing, predict how many segments can be drawn between seven points, no three of which are collinear.

1-3 *Angles*

An **angle** (∠) is a figure formed by two rays that have the same endpoint. The two rays are called the **sides** of the angle and their common endpoint is the **vertex** of the angle.

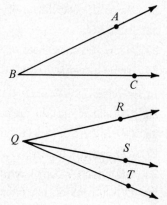

   The sides of the angle shown are $\overrightarrow{BA}$ and $\overrightarrow{BC}$. The vertex is point *B*. The angle can be named ∠ *B*, ∠ *ABC*, or ∠ *CBA*. If three letters are used to name an angle, the middle letter names the vertex.

   When there is no possibility of confusion, an angle can be named by just its vertex, as ∠ *B* above. On the other hand, it would be incorrect to refer to ∠ *Q* in the diagram at the right, since point *Q* is the vertex of three angles: ∠ *RQT*, ∠ *RQS*, and ∠ *SQT*.

You have probably measured angles in degrees with a protractor like the one shown below. Using the inner scale, you see that the measure of ∠*COD* is 35°. Some books use the notation "*m*∠*COD* = 35" to mean "the degree measure of ∠*COD* is 35." This notation emphasizes the distinction between the measure of an angle, which is a number, and the angle itself, which is a set of points. For convenience, we shall write this fact simply as ∠*COD* = 35°, since it is clear that we are talking about the measure of the angle rather than the angle as a set of points.

Using the outer scale, you see that ∠*FOE* = 35°. **Equal angles** are angles that have equal measure. ∠*COD* and ∠*FOE* are equal angles since their measures are both 35°.

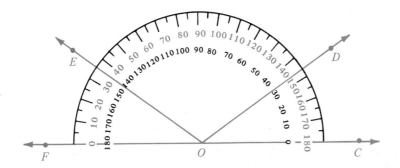

Sometimes angles are named by numbers. For example, in the diagram below, ∠1 is the same as ∠*TEM* and ∠2 is the same as ∠*MER*.

The diagram can be used to illustrate the addition of angle measures. We write "∠1 + ∠2 = ∠*TER*" as an abbreviation for "the measure of ∠1 plus the measure of ∠2 equals the measure of ∠*TER*." Note that ∠1 + ∠2 + ∠3 = 180°.

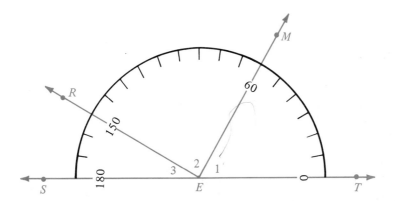

Without moving the protractor above, you can find the measure of ∠2 by subtracting angle measures: ∠2 = ∠*TER* − ∠1. Thus ∠2 = 150° − 60° = 90°.

Angles are classified according to their measures.

Acute angle: Measure between 0° and 90°
Right angle: Measure 90°
Obtuse angle: Measure between 90° and 180°
Straight angle: Measure 180°

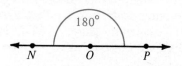

The small square indicates a right angle (rt. ∠).

∠*NOP* is a straight angle.

Classroom Exercises

Exercises 1–20 refer to the diagram.

1. What is the vertex of ∠4?

2. Name the sides of ∠4.

State another name for each angle.

3. ∠*ACD*   **4.** ∠*ABD*   **5.** ∠*EDC*

6. ∠6   **7.** ∠3   **8.** ∠5

9. Why is it confusing to refer to ∠*B*?

10. Name three angles which have *B* as the vertex.

11. How many angles have *D* as the vertex?

Classify each angle as acute, right, obtuse, or straight.

12. ∠1   **13.** ∠3   **14.** ∠*EDB*

15. ∠*CDB*   **16.** ∠*ADC*   **17.** ∠*ADE*

18. What is the measure of ∠*ABC*?

19. What is the measure of ∠2?

20. Complete: ∠5 + ∠6 + ∠7 = <u> ? </u>°

State the measure of each angle.

21. ∠*DOC*   **22.** ∠*HOG*

23. ∠*FOG*   **24.** ∠*FOE*

25. ∠*DOG*   **26.** ∠*HOC*

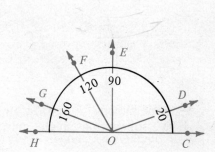

Estimate the measure of each angle.

27.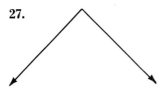

28.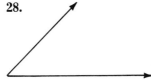

29.

Written Exercises

Exercises 1–14 refer to the diagram.

A **1.** Name the vertex of ∠5.

2. Name the sides of ∠4.

State another name for each angle.

3. ∠1   **4.** ∠3   **5.** ∠5

6. ∠ALD   **7.** ∠AST   **8.** ∠LES

Classify each angle as acute, right, obtuse, or straight.

9. ∠2   **10.** ∠LAS   **11.** ∠ATL

12. ∠S   **13.** ∠LTS   **14.** ∠EDT

Complete the statements.

15.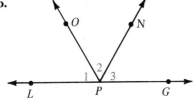

    a. ∠1 + ∠2 = ∠ ?

    b. ∠2 + ∠3 = ∠ ?

    c. ∠1 + ∠2 + ∠3 = ? °

16.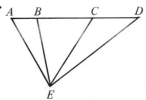

    a. ∠AED − ∠BED = ∠ ?

    b. ∠BED − ∠CED = ∠ ?

    c. ∠ACD − ∠ECD = ∠ ?

17. Using a straightedge, draw a large triangle. Then use a protractor to find the approximate measure of each angle and compute the sum of the three measures. Repeat this exercise for a triangle with different shape. Did you get the same result?

Without measuring, sketch each angle. Then use a protractor to check your accuracy.

18. 90° angle   **19.** 45° angle   **20.** 150° angle   **21.** 10° angle

Exercises 22–25 refer to the diagram.

B **22.** Name four right angles.

23. Name nine acute angles.

24. Name three obtuse angles and give their measures.

25. Name a pair of equal obtuse angles.

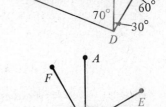

26. $\angle FBE$ is a right angle, $\angle ABD$ is a right angle, and $\angle EBD = 30°$. Find the measures of $\angle ABE$, $\angle ABF$, and $\angle FBC$.

27. Repeat Exercise 26, but suppose $\angle EBD = 33°$. What do you notice about $\angle EBD$ and $\angle ABF$?

The measures of the angles in the diagram are given in terms of x. Find the value of x.

28. $\angle 1 = 2x°$, $\angle 2 = 3x°$, $\angle 3 = (4x + 7)°$

29. $\angle 1 = 30°$, $\angle 2 = 4x°$, $\angle 3 = 6x°$

30. $\angle 1 = (60 - x)°$, $\angle 2 = x°$, $\angle 3 = 3x°$

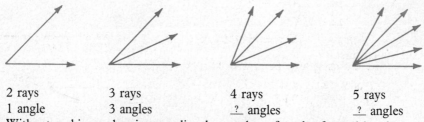

C **31. a.** Complete.

2 rays	3 rays	4 rays	5 rays
1 angle	3 angles	_?_ angles	_?_ angles

b. Without making a drawing, predict the number of angles formed by six noncollinear rays that have the same endpoint.

c. Which of the expressions below gives the number of angles formed by n noncollinear rays that have the same endpoint?

$n - 1$ $2n - 3$ $n^2 - 3$ $\dfrac{n(n - 1)}{2}$

Challenge

The two blocks of wood are duplicates of each other. It is possible to cut a hole in one block in such a way that you can pass the other block completely through the hole. How?

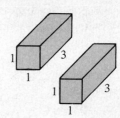

1-4 *Postulates from Algebra*

Since the length of a segment and the measure of an angle are given in terms of real numbers, the facts about real numbers and equality that you learned in algebra will be used in your study of geometry. These facts, which we shall accept without proof, are called *postulates*. The postulates that we shall use most often are listed below.

Addition Postulate If $a = b$ and $c = d$, then $a + c = b + d$.

Example 1: In the diagram, segments marked alike have equal lengths: $QR = PR$ and $RT = RS$. Therefore, $QR + RT = PR + RS$, or
$$QT = PS.$$

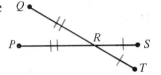

Subtraction Postulate If $a = b$ and $c = d$, then $a - c = b - d$.

Example 2: In the diagram, angles marked alike have equal measures: $\angle OLK = \angle NKL$ and $\angle 1 = \angle 2$. Therefore,
$\angle OLK - \angle 1 = \angle NKL - \angle 2$, or
$$\angle 3 = \angle 4.$$

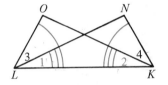

Multiplication Postulate If $a = b$, then $ca = cb$.

Example 3: In the diagram, M is the midpoint of $\overline{PQ}$, N is the midpoint of $\overline{RS}$, and $PM = RN$. Therefore, $2(PM) = 2(RN)$, or $PQ = RS$.

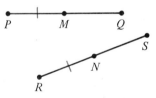

Division Postulate If $a = b$ and $c \neq 0$, then $\dfrac{a}{c} = \dfrac{b}{c}$.

We shall also use the following properties of equality as postulates in our work.

Reflexive Property $a = a$
Symmetric Property   If $a = b$, then $b = a$.
Transitive Property   If $a = b$ and $b = c$, then $a = c$.
Substitution Property   If $a = b$, then b may be substituted for a in any equation or inequality.

Example 4: In the diagram, $\angle 1 + \angle 2 = 90°$ and $\angle 1 = \angle 3$. We can substitute $\angle 3$ for $\angle 1$ to obtain $\angle 3 + \angle 2 = 90°$.

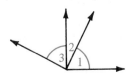

The other properties of algebra, such as the Distributive Property, are all available for our use in geometry. Now we'll review how the postulates are used in solving an equation.

1. $4x - 10 = \frac{2x}{3}$	1. Given equation
2. $12x - 30 = 2x$	2. Multiplication Postulate
3. $10x - 30 = 0$	3. Subtraction Postulate
4. $10x = 30$	4. Addition Postulate
5. $x = 3$	5. Division Postulate

Classroom Exercises

Some of the postulates from algebra are stated below in terms of angle measures. Complete each statement.

1. Reflexive Property: $\angle A = \underline{\ ?\ }$

2. Symmetric Property: If $\angle A = \angle B$, then $\angle B = \underline{\ ?\ }$.

3. Transitive Property: If $\angle A = \angle B$ and $\angle B = \angle C$, then $\angle A = \underline{\ ?\ }$.

4. Substitution Property: If $\angle A = \angle B$, then $\angle B$ may be substituted for $\underline{\ ?\ }$ in any equation or inequality.

In Exercises 5–15, justify each statement with a postulate from algebra.

5. If $x + 7 = 20$, then $x = 13$.

6. If $3x = 18$, then $x = 6$.

7. If $x - 8 = 6$, then $x = 14$.

8. If $\frac{x - 2}{5} = 10$, then $x - 2 = 50$.

9. If $2x + y = 70$ and $y = 3x$, then $2x + 3x = 70$.

10. If $\angle 1 + \angle 2 = 180°$, and $180° = \angle 3 + \angle 4$, then $\angle 1 + \angle 2 = \angle 3 + \angle 4$.

11. If $\angle 5 = \angle 6$ and $\angle 7 = \angle 8$, then $\angle ABC = \angle DEF$.

12. If $\angle ABC = \angle DEF$ and $\angle 7 = \angle 8$, then $\angle 5 = \angle 6$.

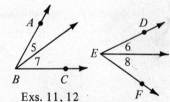

Exs. 11, 12

13. If $WO = PZ$ and $OX = YP$, then $WX = YZ$.

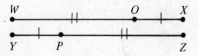

14. If $FL = AT$, then $FA = LT$.

15. If $FA = LT$, then $FL = AT$.

16. Suppose M is the midpoint of $\overline{AB}$ and N is the midpoint of $\overline{CD}$. What segments must be equal? Why?

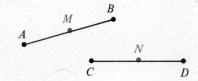

Written Exercises

In Exercises 1–17, justify each statement with a postulate from algebra.

A   **1.** If $x - 9 = 11$, then $x = 20$.   **2.** If $8x = 56$, then $x = 7$.

   3. a. If $9x + 5 = 50$, then $9x = 45$.

     b. If $9x = 45$, then $x = 5$.

   4. a. If $\dfrac{2x}{3} = 24$, then $2x = 72$.

     b. If $2x = 72$, then $x = 36$.

   5. a. If $12 - 20x = 15x + 7$, then $12 = 35x + 7$.

     b. If $12 = 35x + 7$, then $5 = 35x$.

     c. If $5 = 35x$, then $\frac{1}{7} = x$.

   6. a. If $\dfrac{x + 3}{2} = \dfrac{4 + x}{3}$, then $3(x + 3) = 2(4 + x)$.

     b. If $3(x + 3) = 2(4 + x)$, then $3x + 9 = 8 + 2x$.

     c. If $3x + 9 = 8 + 2x$, then $x = -1$.

   7. If $AB = CD$, then $AC = BD$.

   8. If $AC = BD$, then $AB = CD$.

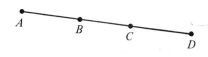

   9. If $\angle 1 = \angle 3$, then $\angle JOL = \angle KOM$.

   10. If $\angle JOL = \angle KOM$, then $\angle 1 = \angle 3$.

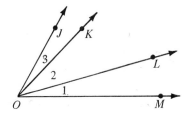

B   **11.** If $\angle 1 = \angle 2$ and $\angle 3 = \angle 4$, then $\angle PQR = \angle PRQ$.

   12. If $\angle PQR = \angle PRQ$ and $\angle 1 = \angle 2$, then $\angle 3 = \angle 4$.

   13. If $PT = PS$ and $TQ = SR$, then $PQ = PR$.

   14. If $QE = RE$ and $QS = RT$, then $SE = TE$.

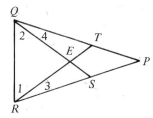

   15. If $\angle 1 = \angle 2$ and $\angle 2 = \angle 3$, then $\angle 1 = \angle 3$.

   16. If $\angle 1 = \angle 2$, $\angle 2 = \angle 3$, and $\angle 3 = \angle 4$, then $\angle 1 = \angle 4$.

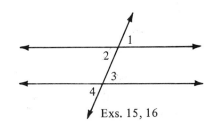

Exs. 15, 16

17. If $\angle ABC = 90°$, $\angle BCD = 90°$, and $\angle 2 = \angle 4$, then $\angle 1 = \angle 3$.

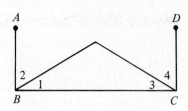

C 18. Consider the following statements:

Reflexive Property: Robot A is as rusty as itself.

Symmetric Property: If Robot A is as rusty as Robot B, then Robot B is as rusty as Robot A.

Transitive Property: If Robot A is as rusty as Robot B and Robot B is as rusty as Robot C, then Robot A is as rusty as Robot C.

A relation that is reflexive, symmetric, and transitive is an equivalence relation. The relation "is as rusty as" is an equivalence relation. Which of the following are equivalence relations?

a. is rustier than   **b.** has the same length as

c. is opposite (for rays)   **d.** is coplanar with (for lines)

SELF-TEST

Classify each statement as true or false.

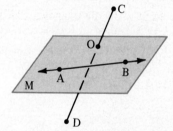

1. $\overline{CD}$ intersects plane M in point O.
2. $\overline{CD}$ and $\overleftrightarrow{AB}$ intersect.
3. A, B, and O are collinear.
4. A, B, and O are coplanar.
5. $\overrightarrow{AB}$ is in plane M.
6. If plane M bisects $\overline{CD}$, then $CO = DO$.

7. State which of the following are undefined terms: point, segment, ray, line, angle, midpoint, plane, coplanar points.

8. On a number line, the coordinates of points J and L are 6 and 10. $JL = \underline{\ ?\ }$

Complete each statement.

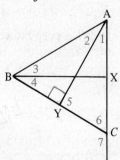

9. The vertex of $\angle 4$ is $\underline{\ ?\ }$.
10. $\angle 5 = \underline{\ ?\ }°$
11. $\angle BAC - \angle 2 = \angle\underline{\ ?\ }$
12. $\angle 6 + \angle 7 = \underline{\ ?\ }°$
13. If $\angle 6$ is an acute angle, $\angle 7$ is a(n) $\underline{\ ?\ }$ angle.

Justify each statement with a postulate from algebra.

14. If $AX = BY$ and $XC = YC$, then $AC = BC$.
15. If $\angle BAC = \angle ABC$ and $\angle 1 = \angle 3$, then $\angle 2 = \angle 4$.
16. If $\angle 1 = \angle 2$ and $\angle 2 = \angle 3$, then $\angle 1 = \angle 3$.

Special Pairs of Angles

Objectives

1. Define adjacent angles, bisector of an angle, and perpendicular lines.
2. Define complementary, supplementary, and vertical angles.
3. State and apply the theorems about special pairs of angles.

1-5 *Special Pairs of Angles*

Adjacent angles (adj. ∠s) are two angles in a plane that have a common vertex and a common side but no common interior points.

∠1 and ∠2 are adjacent angles.   ∠3 and ∠4 are *not* adjacent angles.

  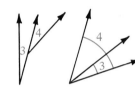

The **bisector of an angle** is a ray that divides the angle into two equal adjacent angles. In the diagram, $\overrightarrow{BX}$ bisects ∠ABC, and we can conclude three things:

$\angle ABX = \angle XBC$

$\angle ABX = \frac{1}{2}\angle ABC$

$\angle XBC = \frac{1}{2}\angle ABC$

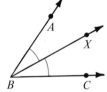

Perpendicular lines (⊥ lines) are two lines that form equal adjacent (90°) angles.

Here are three situations where this definition can be used.

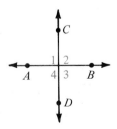

1. If $\overleftrightarrow{AB}$ is perpendicular to $\overleftrightarrow{CD}$ ($\overleftrightarrow{AB} \perp \overleftrightarrow{CD}$), then all the numbered angles are equal and 90°.
2. If any two of the adjacent angles are equal, then $\overleftrightarrow{AB} \perp \overleftrightarrow{CD}$.
3. If any of the numbered angles is 90°, then $\overleftrightarrow{AB} \perp \overleftrightarrow{CD}$.

The word "perpendicular" is also used for rays and segments which are parts of perpendicular lines. For example, in the diagram above, we can say $\overrightarrow{CD} \perp \overline{AB}$.

The diagram at the right shows two things:

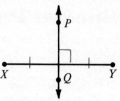

$$\overleftrightarrow{PQ} \text{ is perpendicular to } \overline{XY};$$
$$\overleftrightarrow{PQ} \text{ bisects } \overline{XY}.$$

Therefore, $\overleftrightarrow{PQ}$ is called a **perpendicular bisector** of $\overline{XY}$.

Complementary angles (comp. ∠) are two angles whose measures have the sum 90°. Each angle is called a *complement* of the other.

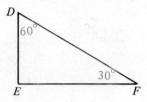
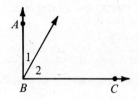

$\angle D$ and $\angle F$ are complementary.
$\angle D$ is a complement of $\angle F$.

If $\overrightarrow{BA} \perp \overrightarrow{BC}$, $\angle 1$ and $\angle 2$ are complementary.

Supplementary angles (supp. ∠) are two angles whose measures have the sum 180°. Each angle is called a *supplement* of the other.

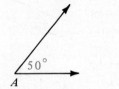
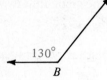
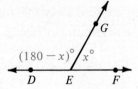

$\angle A$ and $\angle B$ are supplementary.
$\angle A$ is a supplement of $\angle B$.

$\angle DEG$ and $\angle GEF$ are supplementary.

Example: What is the measure of an angle if its supplement is three times as large as its complement?

Solution: Let the measure of the angle be $x°$. Then the supplement is $(180 - x)°$ and the complement is $(90 - x)°$.

Supplement → $180 - x = 3(90 - x)$ ← $3 \times$ Complement
$$180 - x = 270 - 3x$$
$$2x = 90$$
$$x = 45$$

The measure of the angle is 45°.

Vertical angles (vert. ∠) are two angles whose sides form two pairs of opposite rays. When two lines intersect, they form two pairs of vertical angles.

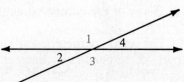

$\angle 1$ and $\angle 3$ are vertical angles.
$\angle 2$ and $\angle 4$ are vertical angles.

Classroom Exercises

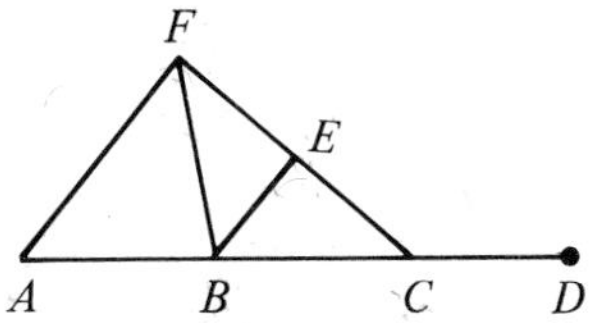

Exercises 1–3 refer to the diagram.

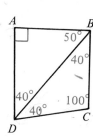

1. Name four pairs of adjacent angles in the diagram.

2. If $\overrightarrow{BE}$ bisects $\angle FBC$, what two angles are equal?

3. Name three pairs of supplementary angles.

Find the measures of a complement and supplement of $\angle R$.

4. $\angle R = 20°$ **5.** $\angle R = 55°$ **6.** $\angle R = 89°$ **7.** $\angle R = x°$

Exercises 8–13 refer to the diagram.

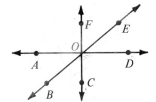

8. Name two right angles.

9. Name two adjacent complementary angles.

10. Name two complementary angles that are not adjacent.

11. Name a supplement of $\angle ADC$.

12. Name another pair of supplementary angles.

13. What angle is bisected?

Complete the pairs of vertical angles.

14. $\angle AOB$ and $\angle\,\underline{\ ?\ }$

15. $\angle AOF$ and $\angle\,\underline{\ ?\ }$

16. $\angle FOD$ and $\angle\,\underline{\ ?\ }$

17. $\angle EOA$ and $\angle\,\underline{\ ?\ }$

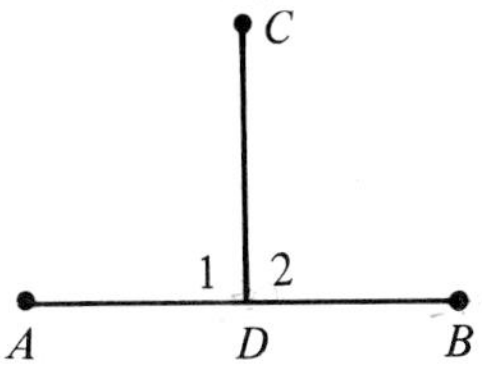

In the diagram, $\angle 1 = \angle 2$ and $AD = BD$. Classify each statement as true or false.

18. $\overline{CD} \perp \overline{DB}$ **19.** $\overline{AD} \perp \overline{CD}$ **20.** $\overline{BD} \perp \overline{DA}$

21. $\angle 2 = 90°$ **22.** $\angle CDA = 90°$ **23.** $\angle ADB = 90°$

24. $\angle CDB$ is a right angle.

25. $\overline{AB}$ is a perpendicular bisector of $\overline{CD}$.

26. $\overline{CD}$ is a perpendicular bisector of $\overline{AB}$.

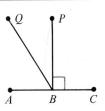

27. Two angles are equal and complementary. What is the measure of each?

28. Suppose two angles are equal. Do you think their supplements must also be equal? Briefly explain.

Written Exercises

In the diagram $\overline{PB} \perp \overline{AC}$.

A **1.** Name two right angles.

2. Name an obtuse angle.

3. Name a straight angle.

4. Name two complementary angles.

5. Name two supplementary angles that are not right angles.

Exercises 6–12 refer to the diagram.

6. Name two right angles.

7. Name an obtuse angle.

8. Name the bisector of ∠CBA.

9. Name a pair of supplementary angles.

10. Name a pair of adjacent complementary angles.

11. Name a pair of complementary angles that are not adjacent.

12. Name two pairs of perpendicular segments.

Find the measures of the complement and the supplement of ∠M.

13. ∠M = 80°   **14.** ∠M = 62°   **15.** ∠M = 88°   **16.** ∠M = y°

17. If a right angle is bisected, what is the measure of each angle formed?

18. **a.** Show how to form a 45° angle by folding a rectangular piece of paper.

   b. Can you form a $22\frac{1}{2}$° angle by folding?

B 19. ∠1 and ∠2 are complements.
   ∠3 and ∠4 are complements.
   a. If ∠1 = ∠3 = 27°, find the measures of ∠2 and ∠4.

   b. If ∠1 = ∠3 = x°, find the measures of ∠2 and ∠4.

   c. If two angles are equal, must their complements also be equal?

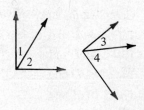

20. Complete the table.

∠1	∠2	∠3	∠4
30°	?	?	?
35°	?	?	?
x°	?	?	?

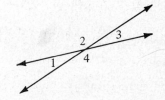

Are the vertical angles equal?

21. The supplement of an angle is twice as large as the angle. Find the measure of the angle.

22. The complement of an angle is four times as large as the angle. Find the measure of the angle.

23. The supplement of an angle is six times as large as the complement of the angle. Find the measure of the angle.

24. The supplement of an angle is 20° larger than the angle. Find the measure of the angle.

25. If an acute angle is bisected, the adjacent angles formed are also acute. What kind of adjacent angles are formed when an obtuse angle is bisected?

26. Fold down a corner of a rectangular sheet of paper as in Figure 1. Then fold the next corner so that the edges touch as in Figure 2.

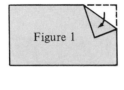

Figure 1

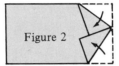

Figure 2

Open the paper and measure the angle between the fold lines. Repeat with another sheet of paper, folding the corner at a different angle. Explain why the measured angle is the same in both cases.

C 27. Explain why the complement of an angle can never be exactly half the supplement of the angle.

28. How many angles have measure equal to the difference between the supplement of the angle and twice the complement of the angle?

Biographical Note Benjamin Banneker

Benjamin Banneker (1731–1806), a noted American scholar, developed a keen interest in mathematics and science at an early age. Without formal instruction, he mastered a number of borrowed books on mathematics and often discovered errors in the authors' calculations. He spent many nights studying the stars and in 1789 demonstrated his proficiency in astronomy by accurately predicting a solar eclipse.

From 1791 until his death, he published almanacs containing information on astronomy, tide tables, medicinal products, and insect life. Banneker's almanacs contained many concepts and ideas far ahead of his time; including the formation of the Department of the Interior and the United Nations.

An accomplished surveyor as well as astronomer, Banneker was a member of the commission which defined the boundary line and laid out the streets of the District of Columbia.

1-6 *Our First Theorems*

Statements that are accepted without proof are called **axioms** or **postulates.** Statements that are proved are called **theorems.** The first theorems of geometry are proved from previously stated definitions and postulates. A proof of our first theorem is given below in *two-column form* with statements on the left and a reason for each statement on the right.

Theorem 1-1   If two angles are complements of equal angles (or of the same angle), then the two angles are equal.

Given: ∠1 is a complement of ∠2;
       ∠3 is a complement of ∠4;
       ∠2 = ∠4

Prove: ∠1 = ∠3

Proof:

Statements	Reasons
1. ∠1 is a complement of ∠2.	1. This is given information.
2. ∠1 + ∠2 = 90°	2. Definition of comp. ⚞
3. ∠3 is a complement of ∠4.	3. Given
4. ∠3 + ∠4 = 90°	4. Definition of comp. ⚞
5. ∠1 + ∠2 = ∠3 + ∠4	5. Substitution Property
6. ∠2 = ∠4	6. Given
7. ∠1 = ∠3	7. Subtraction Postulate

Example 1:   ∠POR and ∠QOS are right angles. What other two angles are equal?

Solution: ∠POQ = ∠ROS because both angles are complements of ∠QOR.

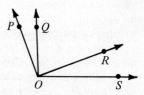

Theorem 1-2   If two angles are supplements of equal angles (or of the same angle), then the two angles are equal.

Theorem 1-2 is very much like Theorem 1-1. Proof of Theorem 1-2 is left as Exercise 32. We can use Theorem 1-2 to prove that vertical angles are equal. This is our next theorem.

Theorem 1-3   Vertical angles are equal.

Given: $\angle 1$ and $\angle 2$ are vertical angles.

Prove: $\angle 1 = \angle 2$

Proof:

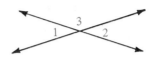

Statements	Reasons
1. $\angle 1 + \angle 3 = 180°$; $\quad \angle 2 + \angle 3 = 180°$	1. Def. of a straight $\angle$
2. $\angle 1$ is supp. to $\angle 3$; $\quad \angle 2$ is supp. to $\angle 3$	2. Def. of supp. $\angle$s
3. $\angle 1 = \angle 2$	3. Supplements of the same angle are equal. (Theorem 1-2)

In Steps 1 and 2 of the proof above, notice how statements can be written in pairs supported by the same reason.

Example 2:   In the diagram, $\angle 4 = \angle 5$.
Name two other angles equal to $\angle 5$.

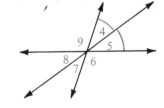

Solution: (1) $\angle 8 = \angle 5$ (Vert. $\angle$s are =.)

$\angle 7 = \angle 4$

$\angle 4 = \angle 5$

(2) $\angle 7 = \angle 5$ (Transitive Property or Substitution Property)

Classroom Exercises

1. Given: $\angle 1$ is a complement of $\angle 3$;
$\qquad \angle 2$ is a complement of $\angle 3$
State the theorem that allows you to conclude that $\angle 1 = \angle 2$.

2. Given: $\angle 4$ is a complement of $\angle 5$;
$\qquad \angle 6$ is a complement of $\angle 7$;
$\qquad \angle 5 = \angle 7$
State the theorem that allows you to conclude that $\angle 4 = \angle 6$.

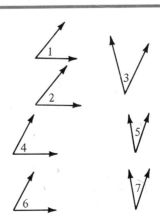

3. Given: ∠1 is a supplement of ∠3;
∠2 is a supplement of ∠3
State the theorem that allows you to
conclude that ∠1 = ∠2.

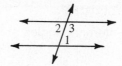

4. Given: ∠4 is a supplement of ∠5;
∠6 is a supplement of ∠7;
∠5 = ∠7
State the theorem that allows you to
conclude that ∠4 = ∠6.

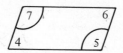

Classify each statement as true or false.

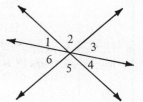

5. ∠1 = ∠4

6. ∠2 = ∠5

7. ∠3 = ∠6

8. ∠1 + ∠2 = ∠5

9. ∠2 + ∠3 = ∠5 + ∠6

10. ∠5 + ∠6 = 180°

11. ∠1 + ∠2 + ∠3 = 180°

12. ∠1 + ∠5 + ∠3 = 180°

Written Exercises

A **1.** Name a supplement of ∠2.

2. Name a supplement of ∠3.

3. If ∠2 = ∠3, write the theorem that
allows you to conclude that ∠1 = ∠4.

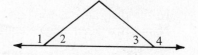

In the diagram, $\overline{LM} \perp \overline{MN}$ and $\overline{KN} \perp \overline{MN}$.

4. Name a complement of ∠2.

5. Name a complement of ∠3.

6. If ∠2 = ∠3, write the theorem that
allows you to conclude that ∠1 = ∠4.

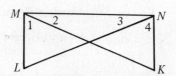

In the diagram, ∠3 is a supplement of ∠2.

7. ∠3 = ∠_?_

8. ∠4 = ∠_?_

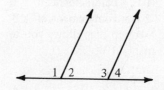

Find the measure.

9. ∠3

10. ∠1 + ∠2 + ∠3

11. ∠1 + ∠2

12. ∠1

13. ∠2

14. ∠4

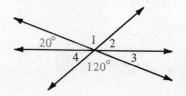

Find the value of x for each diagram.

15.

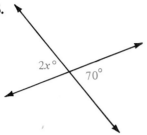

16.

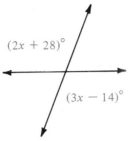

17.

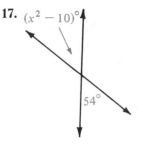

18. In the diagram, $\angle 1$ is supplementary to $\angle 3$.
Name two pairs of equal angles.

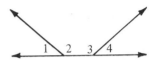

19. In the diagram, $j \perp \overrightarrow{PN}$ and $\overrightarrow{PI} \perp \overrightarrow{PK}$.
Name two pairs of equal acute angles.

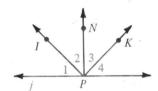

In the diagram, $\overleftrightarrow{FC} \perp \overleftrightarrow{LK}$ and $\angle AOC = 32°$.
Find the measure of each angle.

B **20.** $\angle FOS$ **21.** $\angle KOS$ **22.** $\angle FOA$

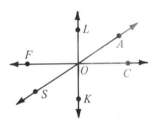

Find the values of x and y for each diagram.

23.

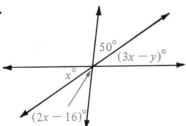

24.

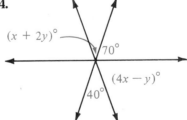

In the diagram, $\angle 4$ and $\angle 5$ are supplementary.
25. Name three angles equal to $\angle 5$.
26. Name three angles equal to $\angle 4$.

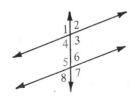

Complete Exercises 27–30, given: $\overline{AC} \perp \overline{BC}$;
$\angle 4$ is a complement of $\angle 1$; and
$\angle 3$ is a complement of $\angle 2$.

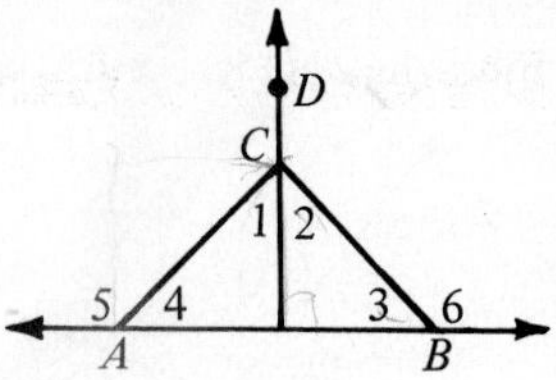

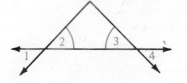

C 27. $\angle 4 = \angle\underline{\ ?\ }$　　　　**28.** $\angle 3 = \angle\underline{\ ?\ }$

29. $\angle DCA = \angle\underline{\ ?\ }$　　　**30.** $\angle DCB = \angle\underline{\ ?\ }$

31. Copy everything shown. Complete the proof.

Given: $\angle 2 = \angle 3$

Prove: $\angle 1 = \angle 4$

Proof:

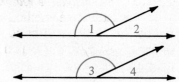

Statements	Reasons
1. $\angle 1 = \angle 2$	1. ___?___
2. $\angle 2 = \angle 3$	2. ___?___
3. $\angle 3 = \angle 4$	3. ___?___
4. $\underline{\ ?\ } = \underline{\ ?\ }$	4. Transitive Property

32. Prove Theorem 1–2: If two angles are supplements of equal angles, then the two angles are equal. Set up your proof like that for Theorem 1–1. The beginning is shown below.

Given: $\angle 1$ and $\angle 2$ are supplementary;
　　　　$\angle 3$ and $\angle 4$ are supplementary;
　　　　$\angle 1 = \angle 3$

Prove: $\angle 2 = \angle 4$

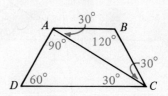

SELF-TEST

1. Name two perpendicular segments.

2. Name the bisector of $\angle DCB$.

3. Name two angles supplementary to $\angle D$.

4. Name two angles complementary to $\angle BAC$.

5. Name a pair of adjacent angles.

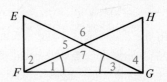

In the diagram, $\overline{EF} \perp \overline{FG}, \overline{HG} \perp \overline{FG},$ and $\angle 1 = \angle 3$. Write the theorem that allows you to conclude each statement.

6. $\angle 2 = \angle 4$　　　　　**7.** $\angle 6 = \angle 7$

8. Any angle supplementary to $\angle 6$ must equal $\angle 5$.

Proof in Geometry

Objectives

1. Understand the deductive method of proof.
2. Know four kinds of reasons used in proofs.
3. Understand the relationships described in the postulates and theorems of Section 1-8.
4. Write proofs in two-column form.

1-7 *Deductive Reasoning and Proof*

The word "geometry" literally means earth measurement. The early Egyptians had knowledge of practical rules of geometry which enabled them to construct irrigation systems, great temples, and pyramids.

Our study of geometry, however, is based on work that has come down to us from early Greek scholars interested in proving properties of geometric figures by logical reasoning rather than by measurement. Around 300 B.C. Euclid organized the Greek knowledge of geometry into a thirteen-volume work called *The Elements*, in which geometric facts are arrived at by *deductive reasoning*.

In the deductive method of proof we reason from undefined terms, definitions, and a few basic assumptions to conclusions. Statements deduced (proved) in this way are theorems that can be used to deduce still more theorems.

We have already begun the deductive process in this book.

Our starting point

Undefined terms: point, line, plane

Definitions: segment, ray, angle, complement, and so on

Postulates from algebra

Our first theorems

1-1 If two angles are complements of equal angles (or of the same angle), then the two angles are equal.

1-2 If two angles are supplements of equal angles (or of the same angle), then the two angles are equal.

1-3 Vertical angles are equal.

Now let us practice proving statements by reasoning deductively.

Example 1:

Given: $\angle 2 = \angle 3$

Prove: $\angle 1 = \angle 4$

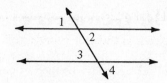

First we must plan the proof. We ask, How are $\angle 1$ and $\angle 4$ related to the angles which are given as equal? We see in the diagram that vertical angles are involved. A chain of equal angles may be set up, leading to the fact that $\angle 1 = \angle 4$ by the Transitive Property as follows:

Proof:

Statements	Reasons
1. $\angle 1 = \angle 2$	1. Vertical angles are equal. (*theorem*)
2. $\angle 2 = \angle 3$	2. Given
3. $\angle 3 = \angle 4$	3. Vertical angles are equal. (*theorem*)
4. $\angle 1 = \angle 4$	4. Transitive Property (*postulate*)

Example 2:

Given: $\overline{AC} \perp \overline{BC}$; $\angle 3$ is complementary to $\angle 1$.

Prove: $\angle 3 = \angle 2$

Proof:

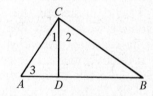

Statements	Reasons
1. $\overline{AC} \perp \overline{BC}$	1. Given
2. $\angle 2 + \angle 1 = 90°$	2. Definition of $\perp$ lines
3. $\angle 2$ is comp. to $\angle 1$.	3. Definition of comp. $\angle s$
4. $\angle 3$ is comp. to $\angle 1$.	4. Given
5. $\angle 3 = \angle 2$	5. Complements of the same angle are equal. (*theorem*)

Look back at the reasons given in the two examples of deductive proof.

Reasons Used in Proofs

Given information
Definitions
Postulates
Theorems that have already been proved

Classroom Exercises

Classify each term as defined or undefined.

1. line **2.** ray **3.** equal angles

4. segment **5.** angle **6.** perpendicular lines

7. point **8.** plane **9.** vertical angles

10. Supply the reasons for this proof that all right angles are equal.

Given: $\angle 1$ and $\angle 2$ are rt. $\angle$s.

Prove: $\angle 1 = \angle 2$

Proof:

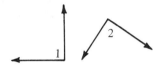

Statements	Reasons
1. $\angle 1$ is a rt. $\angle$; $\angle 2$ is a rt. $\angle$.	1. __?__
2. $\angle 1 = 90°$; $\angle 2 = 90°$	2. __?__
3. $\angle 1 = \angle 2$	3. __?__

11. Explain your plan for proving the following.

Given: $\angle 2 = \angle 3$;

$\angle 4 = \angle 5$

Prove: $\angle 1$ is supp. to $\angle 6$.

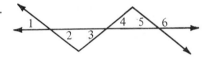

Written Exercises

Write the definition, postulate, or theorem that supports the statement about the diagram.

A **1.** If $\overleftrightarrow{CE} \perp \overleftrightarrow{AB}$, then $\angle ADE$ and $\angle BDE$ are equal.

2. If $\angle CDB$ is a right angle, then $\angle CDB = 90°$.

3. If $\angle 1 + \angle 2 = 90°$, then $\angle 1$ is complementary to $\angle 2$.

4. If $\overrightarrow{DF}$ bisects $\angle CDB$, then $\angle 1 = \angle 2$.

5. If $\angle ADC + \angle 1 = 135°$ and $\angle 1 = \angle 2$, then $\angle ADC + \angle 2 = 135°$.

6. If D is the midpoint of $\overline{AB}$, then $AD = DB$.

7. If $AD = CD$ and $DB = DE$, then $AB = CE$.

8. $\angle 1 = \angle 3$

9. If $\angle 2$ is complementary to $\angle 1$, and $\angle 4$ is complementary to $\angle 1$, then $\angle 2 = \angle 4$.

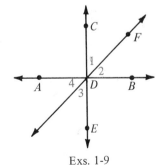

Exs. 1-9

Supply the reasons in each proof.

10. Given: $\angle 2 = \angle 3$;
$\qquad \angle 4 = \angle 5$

Prove: $\angle 1$ is supp. to $\angle 6$.

Proof:

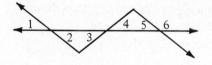

Statements	Reasons
1. $\angle 1 = \angle 2$	1. ?
2. $\angle 2 = \angle 3$	2. ?
3. $\angle 3 = \angle 4$	3. ?
4. $\angle 4 = \angle 5$	4. ?
5. $\angle 1 = \angle 5$	5. ?
6. $\angle 5 + \angle 6 = 180°$	6. ?
7. $\angle 1 + \angle 6 = 180°$	7. ?
8. $\angle 1$ is supp. to $\angle 6$.	8. ?

11. Given: $\overline{PQ} \perp \overline{QR}$;
$\qquad \overline{PS} \perp \overline{SR}$;
$\qquad \angle 1 = \angle 4$

Prove: $\angle 2 = \angle 5$

Proof:

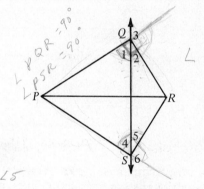

Statements	Reasons
1. $\overline{PQ} \perp \overline{QR}$; $\overline{PS} \perp \overline{SR}$	1. ?
2. $\angle 2 + \angle 1 = 90°$; $\angle 5 + \angle 4 = 90°$	2. ?
3. $\angle 2$ is comp. to $\angle 1$; $\angle 5$ is comp. to $\angle 4$.	3. ?
4. $\angle 1 = \angle 4$	4. ?
5. $\angle 2 = \angle 5$	5. ?

12. Explain your plan for proving that $\angle 3 = \angle 6$ in the diagram above now that you have proved that $\angle 2 = \angle 5$.

Copy the diagram. Write proofs in two-column form.

B 13. Given: $\angle 1 = \angle 2$; $\angle 3 = \angle 4$
$\qquad$ Prove: $\overrightarrow{OC} \perp \overleftrightarrow{AE}$

14. Given: $\overrightarrow{OC} \perp \overleftrightarrow{AE}$; $\overrightarrow{OC}$ bisects $\angle BOD$.
$\qquad$ Prove: $\angle 3 = \angle 4$

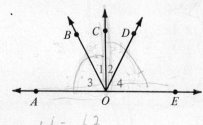

15. Explain your plan for proving the following.
Given: $\angle 2 = 120°$; $\angle 8 = 60°$
Prove: $\angle 4$ is supp. to $\angle 7$.

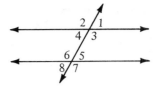

16. Draw any $\angle AOB$ and its bisector $\overrightarrow{OX}$. Now extend the rays $\overrightarrow{OA}$, $\overrightarrow{OB}$ and $\overrightarrow{OX}$ in the *opposite* direction. What can you conclude? Can you prove it?

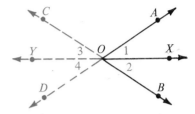

C 17. Make a diagram showing $\angle PQR$ bisected by $\overrightarrow{QX}$. Choose a point Y on the ray opposite $\overrightarrow{QX}$. Prove: $\angle PQY = \angle RQY$

18. Given: $\angle DBA = 45°$;
        $\angle DEB = 45°$
    Prove: $\angle DBC = \angle FEB$

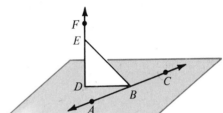

1-8 *Postulates of Points, Lines, and Planes*

To prove other statements about geometric figures, we must make some basic assumptions relating to points, lines, and planes. These assumptions are the postulates of geometry.

Postulate 1 A line contains at least two points; a plane contains at least three points not all in one line; space contains at least four points not all in one plane.

Postulate 2 Through any two points there is exactly one line.

Postulate 3 Through any three points there is at least one plane, and through any three noncollinear points there is exactly one plane.

Postulate 4 If two points are in a plane, then the line joining the points is in that plane.

Postulate 5 If two planes intersect, then their intersection is a line.

To help understand the relationships described in Postulates 1–5, illustrate each with a sketch or with cardboard and pencils for planes and lines. Remember that these are postulates. They are not proved.

Postulate 6   The points on a line can be paired with the real numbers in such a way that:

    a. Any two desired points can have coordinates 0 and 1.
    b. The distance between any two points equals the absolute value of the difference of their coordinates. (Ruler Postulate)

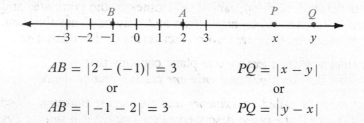

$$AB = |2 - (-1)| = 3 \qquad PQ = |x - y|$$
$$\text{or} \qquad\qquad\qquad \text{or}$$
$$AB = |-1 - 2| = 3 \qquad PQ = |y - x|$$

Postulate 7   Select a line AB and choose any point O between A and B. Consider the rays $\overrightarrow{OA}$ and $\overrightarrow{OB}$ and all the rays that can be drawn from O on one side of $\overleftrightarrow{AB}$. These rays can be paired with the real numbers from 0 to 180 in such a way that:

    a. $\overrightarrow{OA}$ is paired with 0, and $\overrightarrow{OB}$ with 180.
    b. If $\overrightarrow{OP}$ is paired with x, and $\overrightarrow{OQ}$ with y,
        then $\angle POQ = |x - y|°$. (Protractor Postulate)

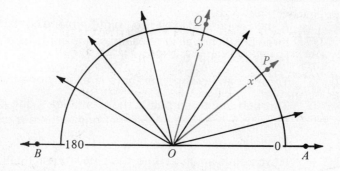

From these postulates it is possible to prove the following theorems. You will not study their proofs, but you may use the theorems.

Theorem 1-4	If two lines intersect, then they intersect in exactly one point.
Theorem 1-5	If there is a line and a point not in the line, then exactly one plane contains them.
Theorem 1-6	If two lines intersect, then exactly one plane contains them.
Theorem 1-7	Every segment has exactly one midpoint.
Theorem 1-8	Every angle has exactly one bisector.

The phrase "exactly one" appears several times in the postulates and theorems of this section. The phrase "one and only one" has the same meaning. For example, here are two correct forms of Theorem 1-6:

If two lines intersect, then *exactly one* plane contains them.
If two lines intersect, then *one and only one* plane contains them.

Either statement is called an *existence and uniqueness statement*. The theorem states that (1) the plane described exists and (2) it is unique (no more than one such plane exists).

Classroom Exercises

Classify each statement as true or false.

1. Three given collinear points can lie in more than one plane.
2. A given triangle can lie in more than one plane.
3. Two planes always intersect.
4. Two planes can intersect in only one point.
5. Two lines can intersect in two points.
6. Given any real number, a greater number can be named.
7. Between any two real numbers there is another real number.
8. There are more real numbers than there are points in a line.
9. There are more points in a line than there are real numbers.

In the diagrams below, what is the measure of ∠*CDE*?

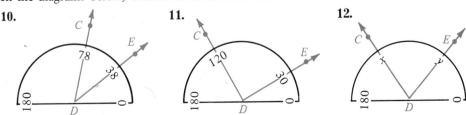

10.   **11.**   **12.**

Written Exercises

A 1. The picture suggests what would happen if we tried to show two "lines" drawn through two points. Write the postulate that makes such a situation impossible.

2. Is there any limit to the number of planes that can contain a given line?

3. How many planes contain a given line and also a point not in the line?

4. How many planes contain a given acute angle?

5. Plane *M* and plane *N* both contain point *P.* Do the planes have any other points in common? Write the postulate which answers this question.

6. Is it possible for there to be three points in space which are not in the same plane?

7. Is it possible for there to be four points in space which are not in the same plane?

8. A carpenter might check the flatness of a board by laying a straightedge across the board in several directions. Explain how this procedure relates to Postulate 4.

In the diagrams below, what number would be paired with the bisector of ∠ *CDE*?

9.

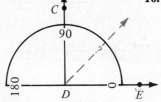

10.

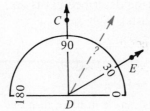

11.

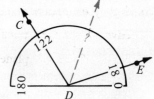

B 12. The diagram shows a cylindrical surface which we can imagine to extend indefinitely up and down. Is it possible to choose points *P* and *Q* in this surface so that $\overline{PQ}$ is in the surface? If *P* and *Q* are any two points in the surface, must $\overline{PQ}$ be in the surface?

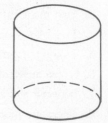

13. Copy and complete the following proof of the statement: If points *A* and *B* have coordinates *a* and *b,* then the midpoint of $\overline{AB}$ has coordinate $\dfrac{a+b}{2}$.

Given: Points *A* and *B* have coordinates *a* and *b*;
$\qquad$ $b > a$; midpoint *M* of $\overline{AB}$ has coordinate *x*.

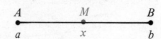

Prove: $x = \dfrac{a+b}{2}$

Proof:

Statements	Reasons
1. A, M, and B have coordinates a, x, and b; $b > a$	1. __?__
2. $AM = x - a$; $MB = b - x$	2. __?__
3. M is the midpoint of $\overline{AB}$.	3. __?__
4. $x - a = b - x$	4. __?__
5. $2x = $ __?__	5. By algebra
6. $x = \dfrac{a + b}{2}$	6. __?__

14. Given that M is the midpoint of $\overline{AB}$, copy and complete the table.

Coordinate of A	0	6	-12	-3	x	2	?
Coordinate of B	17	24	-28	13	y	?	4
Coordinate of M	?	?	?	?	?	20	-6

15. Supply the reasons.

Given: Line l and point P not in l.

Prove: There is a plane containing l and P.

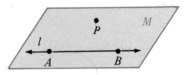

Proof:

Statements	Reasons
1. Line l contains at least two points, A and B.	1. __?__
2. A, B, and P are not collinear.	2. __?__
3. A, B, and P are contained in exactly one plane, M.	3. __?__
4. l is in M.	4. __?__
5. There is a plane containing l and P.	5. Plane M contains l (Step 4) and P (Step 3).

C 16. Write a proof in two-column form.
Given: Lines j and k intersect in point T.
Prove: There is a plane, M, containing j and k.

17. Suppose $\overrightarrow{OP}$ and $\overrightarrow{OQ}$ are paired with the real numbers p and q and that $p > q$. Let $\overrightarrow{OX}$, the bisector of $\angle POQ$, be paired with the real number x. Derive an expression for x in terms of p and q. (*Hint:* Refer to Postulate 7.)

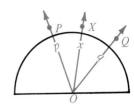

SELF-TEST

1. Given: $\overrightarrow{OB}$ bisects $\angle AOC$.

   Prove: $\angle 1 = \angle 3$

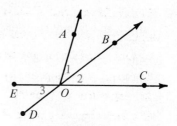

2. Two lines intersect. How many planes contain these two lines?
3. Three points are noncollinear. How many planes contain these three points?
4. Is there any limit to the number of planes that can contain three collinear points?
5. Is it possible for a line and a plane to have exactly one point in common?
6. Is it possible for two different lines to have two points in common?
7. Is it possible for two different planes to have exactly two points in common?

Chapter Summary

1. The concepts of point, line, and plane are basic to geometry. The terms "point," "line," and "plane" are undefined.
2. $\overleftrightarrow{AB}$ denotes a line; $\overline{AB}$, a segment; and $\overrightarrow{AB}$, a ray. AB denotes the length of $\overline{AB}$ and is a positive number.
3. Equal segments are segments with equal lengths.
4. Two rays with the same endpoint form an angle, which is measured in degrees from 0 through 180. Equal angles are angles with equal measures.
5. Angles are classified as acute, right, obtuse, or straight, according to their measures.
6. Postulates from algebra are used to work with the numbers that are measures of segments or angles.
7. Perpendicular lines form equal adjacent (90°) angles.
8. Two angles are equal if they are
   (a) complements of equal angles or of the same angle;
   (b) supplements of equal angles or of the same angle; or
   (c) vertical angles.
9. Statements that are accepted without proof are called axioms or postulates. Statements that are proved are called theorems.
10. Deductive reasoning is a process of logical reasoning from accepted statements to a conclusion.

Chapter Review

1-1 *Points, Lines, and Planes*

Sketch and label the figures described.

1. Line l and line k intersect in point A.

2. Line j contains points C and D.

3. Line h lies in plane M.

4. Line t intersects plane R in point P.

1-2 *Definitions*

5. How many endpoints does $\overline{CD}$ have?

6. Name a point which lies on $\overrightarrow{EC}$.

7. Find the length of $\overline{AD}$.

8. If E is the midpoint of $\overline{CD}$, what is the coordinate of E?

1-3 *Angles*

9. Name three angles which have D as the vertex.

10. Name an obtuse angle.

11. Name a straight angle.

12. If $\angle ABD = 115°$, then $\angle DBC = \underline{\ ?\ }°$.

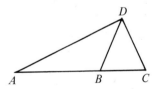

1-4 *Postulates from Algebra*

Justify each statement with a postulate from algebra.

13. If $AB = BC$ and $BC = CD$, then $AB = CD$.

14. If $\angle 8 = \angle 5$ and $\angle 7 = \angle 6$, then $\angle ABC = \angle BCD$.

15. If $\angle 1 + \angle 2 = 60°$ and $\angle 1 = \angle 4$, then $\angle 4 + \angle 2 = 60°$.

16. If $\angle BAD = \angle ADC$ and $\angle 1 = \angle 4$, then $\angle 2 = \angle 3$.

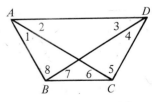

1-5 *Special Pairs of Angles*

17. Name two perpendicular lines.

18. Name two vertical angles which are acute.

19. Name a complement of $\angle 1$.

20. Name a supplement of $\angle 1$.

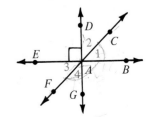

1-6 *Our First Theorems*

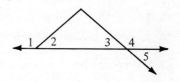

21. Statements accepted without proof are axioms or _?_ .

22. Statements which are proved are called _?_ .

23. If $\angle 2 = \angle 3$, write the theorem that allows you to conclude that $\angle 1 = \angle 4$.

24. Why is $\angle 3$ equal to $\angle 5$?

1-7 *Deductive Reasoning and Proof*

Write the definition, postulate, or theorem that supports each statement.

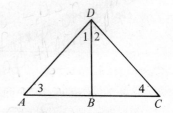

25. If $\overrightarrow{DB}$ bisects $\angle ADC$, then $\angle ADB = \angle BDC$.

26. If B is the midpoint of $\overline{AC}$, then $AB = BC$.

27. If $\angle 1$ is the complement of $\angle 3$, $\angle 2$ is the complement of $\angle 4$, and $\angle 3 = \angle 4$, then $\angle 1 = \angle 2$.

28. If $\angle 1 + \angle 3 = 90°$ and $\angle 3 = 46°$, then $\angle 1 = 44°$.

1-8 *Postulates of Points, Lines, and Planes*

Classify each statement as true or false.

29. Through any three points there is exactly one plane.

30. It is possible for the intersection of two planes to be a point.

31. There is only one plane which contains a line and a point which is not on the line.

32. Every line has a midpoint.

Chapter Test

1. What is another name for line *l*?

2. What is the intersection of $\overline{BE}$ and $\overline{AC}$?

3. If $AB = AE$, what can you say about point A?

4. What two rays form $\angle 1$?

5. Name a complement of $\angle 1$.

6. Name a supplement of $\angle 1$.

7. Name three collinear points.

8. Name two perpendicular lines.

9. Name two right angles.

10. Name two acute angles.

11. Name an obtuse angle adjacent to $\angle 1$.

12. If $\overrightarrow{AD}$ bisects $\angle CAE$, then $\angle 1 = $ _?_ °.

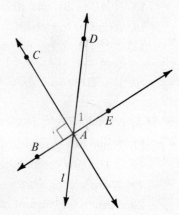

Exs. 1-12

13. If $\angle 1 = 56°$, then $\angle 3 = \underline{\ ?\ }°$.

14. If $\angle 8 = 124°$, then $\angle 5 = \underline{\ ?\ }°$.

15. If $\angle 1 = \angle 5$, name all supplements of $\angle 1$.

16. Name four pairs of vertical angles.

On a number line, G has coordinate -7, and H has coordinate 2.

17. What is the length of $\overline{GH}$?

18. What is the coordinate of the midpoint of $\overline{GH}$?

19. Write a proof in two-column form.

Given: $AD = BD$; $\overline{BD}$ bisects $\overline{AC}$.

Prove: $BD = DC$

Classify each statement as true or false.

20. If two planes intersect, then their intersection is a segment.

21. Through any three points there is exactly one plane.

Algebra Review

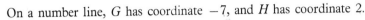

Solve each equation or inequality.

1. $a + 6 = 23$

2. $7 + b = 15$

3. $c - 12 = 8$

4. $19 = d - 2$

5. $3e = 12$

6. $8f = 4$

7. $2g + 5 = 15$

8. $3h - 2 = 10$

9. $6i = 5i + 11$

10. $12 - j = 2j$

11. $20 + k = 7k$

12. $6m + 12 = 4m + 18$

13. $2 - 3n = 4n - 12$

14. $30 + 2p = 180 - 3p$

15. $q + 1 > 5$

16. $4r > 12$

17. $2s - 3 > 7$

18. $5t + 9 < 19$

19. $1 + u < 2u$

20. $7v - 5 > 4v + 16$

21. $3(w - 5) = 12$

22. $4x = 2(x + 11)$

23. $3(3y + \frac{1}{3}) = 5y + 19$

24. $\frac{1}{2}(8z - 4) = 3(8 - 3z)$

25. $3(90 - x) = 180 - x$

26. $180 - (x + 10) = 70$

27. $2(90 - x) = 3(90 - 2x)$

28. $180 - 2(x + 20) = x - 10$

29. $(n - 2)180 = 165n$

30. $n(n + 5) = n^2 + 10$

2
Parallel Lines and Planes

When Lines and Planes Are Parallel

Objectives
1. Distinguish between parallel lines and skew lines.
2. State and apply the theorem about the intersection of two parallel planes by a third plane.
3. Identify the angles formed when two lines are cut by a transversal.
4. State and apply the postulates and theorems about parallel lines.
5. State and apply the theorems about a parallel and a perpendicular to a given line through a point outside the line.

2-1 *Definitions*

Two lines which do not intersect are either *parallel* or *skew*.

Parallel lines (∥ lines) do not intersect and are coplanar.
Skew lines do not intersect and are not coplanar.

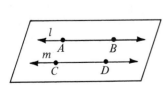

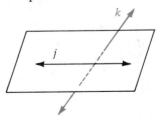

l and *m* are parallel lines.
l is parallel to *m* (*l*∥*m*).

j and *k* are skew lines.

Segments of parallel lines are called parallel. Above, $\overline{AB} \parallel \overline{CD}$.

Think of the top of the box pictured below as part of plane X, and the bottom as part of plane Y. Refer to the box to understand the following definitions.

Parallel planes (∥ planes) do not intersect.
Plane X is parallel to plane Y ($X \parallel Y$).

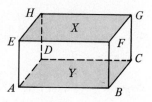

A line and a plane are parallel if they do not intersect.
For example, $\overleftrightarrow{EF} \parallel Y$ and $\overleftrightarrow{FG} \parallel Y$.
Also, $\overleftrightarrow{AB} \parallel X$ and $\overleftrightarrow{BC} \parallel X$.

Our first theorem about parallel lines and planes is given below. Notice the importance of definitions in the proof.

Theorem 2-1   If two parallel planes are cut by a third plane, then the lines of intersection are parallel.

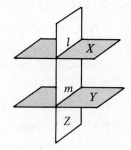

Given: Plane $X \parallel$ plane Y;
       plane Z intersects X in l;
       plane Z intersects Y in m.

Prove: $l \parallel m$

Proof:

Statements	Reasons
1. l is in X; m is in Y.	1. Given
2. $X \parallel Y$	2. Given
3. l and m do not intersect.	3. X and Y do not intersect. (Def. of ∥ planes)
4. l and m are coplanar.	4. Given (Both are in Z.)
5. $l \parallel m$	5. l and m are coplanar and do not intersect. (Def. of ∥ lines)

The following terms, which are needed for future theorems about parallel lines, apply only to coplanar lines.

A **transversal** is a line that intersects two or more coplanar lines in different points. In the next diagram, t is a transversal of h and k. The angles formed have special names.

Interior angles: angles 3, 4, 5, 6   *Exterior angles:* angles 1, 2, 7, 8

Alternate interior angles (alt. int. ∠) are two nonadjacent interior angles on opposite sides of the transversal.

∠3 and ∠6   ∠4 and ∠5

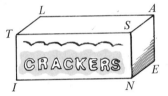

Same-side interior angles (s-s. int. ∠) are two interior angles on the same side of the transversal.

∠3 and ∠5   ∠4 and ∠6

Corresponding angles (corr. ∠) are two angles in corresponding positions relative to the two lines.

∠1 and ∠5   ∠2 and ∠6   ∠3 and ∠7   ∠4 and ∠8

Classroom Exercises

Exercises 1–7 refer to the box pictured at the right.

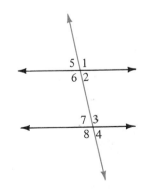

1. Name four pairs of segments which appear to be parallel.

2. Name four pairs of segments which appear to be parts of skew lines.

3. Name two segments which appear to be parallel to the plane containing points *S, A, L, T.*

Name three points that lie in a plane parallel to the plane described.

4. The plane containing *S, A, L, T*   5. The plane containing *S, A, N, E*

6. The plane containing *T, I, N, S*

7. Suppose the top and bottom of the box are parts of parallel planes. Explain how Theorem 2-1 can be used to prove $\overline{SA} \parallel \overline{NE}$.

Exercises 8–14 refer to the diagram.

8. Name four pairs of corresponding angles.

9. Name two pairs of alternate interior angles.

10. Name two pairs of same-side interior angles.

11. Although we have not defined *alternate exterior angles,* you can guess what they are. Name two pairs of them.

12. Name two pairs of angles we would call *same-side exterior angles.*

13. Suppose one pair of alternate interior angles are equal (say, ∠2 = ∠7). Explain why the other pair of alternate interior angles must also be equal.

14. Suppose a pair of same-side interior angles are supplementary (say, ∠2 + ∠3 = 180°). What must be true of any pair of corresponding angles?

Written Exercises

Name the two lines and the transversal which form each pair of angles.

A **1.** ∠1 and ∠2   **2.** ∠3 and ∠4   **3.** ∠5 and ∠6   **4.** ∠7 and ∠8

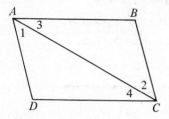 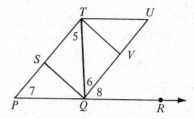

Classify each pair of angles as alternate interior angles, same-side interior angles, or corresponding angles.

5. ∠1 and ∠9

6. ∠6 and ∠9

7. ∠6 and ∠10

8. ∠10 and ∠11

9. ∠13 and ∠15

10. ∠7 and ∠12

11. ∠11 and ∠14

12. ∠8 and ∠16

13. ∠8 and ∠6

14. ∠8 and ∠12

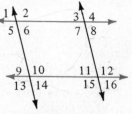

15. Make a drawing that shows two coplanar segments that do not intersect and yet are not parallel.

In Exercises 16–18, use two lines of notebook paper for parallel lines and draw any transversal. Use a protractor to measure.

16. Measure one pair of corresponding angles. Repeat the experiment with another transversal. What appears to be true?

17. Measure one pair of alternate interior angles. Repeat the experiment with another transversal. What appears to be true?

18. Measure one pair of same-side interior angles. Repeat the experiment with another transversal. What appears to be true?

Complete each statement with the word *always, sometimes,* or *never*.

B **19.** Two skew lines are ___?___ parallel.

20. Two parallel lines are ___?___ coplanar.

21. Two lines that are not coplanar ___?___ intersect.

22. A line in the plane of the ceiling and a line in the plane of the floor are ___?___ parallel.

23. Two lines in the plane of the floor are ___?___ skew.

24. If a line is parallel to a plane, a plane containing that line is ___?___ parallel to the given plane.

25. Two lines parallel to the same plane are ___?___ parallel to each other.

26. Two lines parallel to a third line are ___?___ parallel to each other.

27. Two lines skew to a third line are ___?___ skew to each other.

28. Two lines perpendicular to a third line are ___?___ perpendicular to each other.

29. Two planes parallel to the same line are ___?___ parallel to each other.

30. When there is a transversal of two lines, the three lines are ___?___ coplanar.

31. Two planes parallel to the same plane are ___?___ parallel to each other.

32. Write the reasons for the proof.

Given: $\angle 1 = \angle 5$

Prove: $\angle 1 = \angle 8$

Proof:

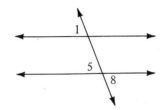

Statements	Reasons
1. $\angle 1 = \angle 5$	1. ___?___
2. $\angle 5 = \angle 8$	2. ___?___
3. $\angle 1 = \angle 8$	3. ___?___

33. Given: $\angle 3$ is supplementary to $\angle 5$.
Prove: $\angle 3 = \angle 6$

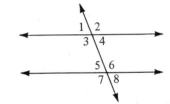

C 34. Given: $\angle 1 = \angle 3$;
$\angle 10$ is supplementary to $\angle 11$.
Prove: $\angle 1$ is supplementary to $\angle 6$.

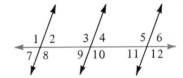

Draw the figures described.

35. Lines a and b are skew, lines b and c are skew, and $a \parallel c$.

36. Lines d and e are skew, lines e and f are skew, and $d \perp f$.

37. Line $l \parallel$ plane X, plane $X \parallel$ plane Y, and l is not parallel to Y.

38. Suppose that three lines are positioned so that each line is parallel to the other two lines and no plane contains all three lines. How many planes contain at least two of the lines?

39. Suppose that four lines are positioned so that every two are parallel and no three are coplanar. How many planes contain at least two of these lines?

2-2 *Properties of Parallel Lines*

By experimenting with parallel lines, transversals, and a protractor in the last set of exercises, you probably discovered that corresponding angles are equal. There is not enough information in our previous postulates and theorems to deduce this property as a theorem. We accept it as a postulate.

Postulate 8   If two parallel lines are cut by a transversal, then corresponding angles are equal.

From this postulate we can easily prove the following theorems.

Theorem 2-2   If two parallel lines are cut by a transversal, then alternate interior angles are equal.

Given: $k \parallel m$; transversal t cuts k and m.

Prove: $\angle 1 = \angle 2$

Proof:

Statements	Reasons
1. $k \parallel m$; t is a transversal.	1. Given
2. $\angle 1 = \angle 3$	2. Vert. $\angle$s are equal.
3. $\angle 3 = \angle 2$	3. If two parallel lines are cut by a transversal, then corr. $\angle$s are equal.
4. $\angle 1 = \angle 2$	4. Transitive Property

In the same way the other pair of alternate interior angles can be proved equal.

Theorem 2-3   If two parallel lines are cut by a transversal, then same-side interior angles are supplementary.

Given: $k \parallel m$; transversal t cuts k and m.

Prove: $\angle 1$ is supplementary to $\angle 4$.

The proof is left as Exercise 15.

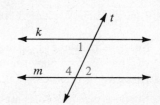

Theorem 2-3 can be proved for either pair of same-side interior angles.

Theorem 2-4 If a transversal is perpendicular to one of two parallel lines, then it is perpendicular to the other one also.

Given: Transversal t cuts l and n;
$\quad\quad t \perp l;\ l \| n$

Prove: $t \perp n$

Proof:

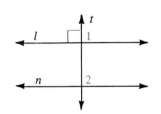

Statements	Reasons
1. $t \perp l$	1. Given
2. $\angle 1 = 90°$	2. Def. of $\perp$ lines
3. $l \| n$	3. Given
4. $\angle 2 = 90°$	4. If two parallel lines are cut by a transversal, then corr. $\angle$s are equal.
5. $t \perp n$	5. Def. of $\perp$ lines

At this point in your study of geometry, it is no longer necessary to use arrowheads when drawing a line to suggest that the line extends in both directions without ending. Instead, arrowheads will be used to indicate parallel lines as in the following examples.

Example 1: The double arrowheads show that $a \| b$. The single arrowheads show that $c \| d$. Find the values of x and y.

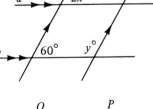

Solution: Since $a \| b$, $2x = 60$. (Why?)
$\quad\quad$ Thus, $x = 30$.
$\quad\quad$ Since $c \| d$, $60 + y = 180$. (Why?)
$\quad\quad$ Thus, $\quad y = 120$.

Example 2: Find the measure of $\angle PQR$.

Solution: The diagram shows that $\overleftrightarrow{QR} \perp \overleftrightarrow{RS}$ and $\overleftrightarrow{QP} \| \overleftrightarrow{RS}$.
$\quad\quad$ Thus $\overleftrightarrow{QR} \perp \overleftrightarrow{QP}$ and $\angle PQR = 90°$.

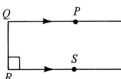

Diagrams in the text relate points, lines, and planes. From this diagram, you may conclude that the lines are coplanar, points *A*, *B*, and *C* are collinear, $\angle ABC$ is a straight angle, and $\angle 1$ is adjacent to $\angle 2$. No conclusions about equal segments, equal angles, perpendicular lines, or parallel lines may be reached, however, without the additional markings you have seen in examples.

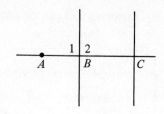

Classroom Exercises

1. What do the arrowheads in the diagram tell you?
2. How are lines *k* and *l* related?

State the postulate or theorem that justifies each statement.

3. $\angle 1 = \angle 5$ 4. $\angle 3 = \angle 6$

5. $\angle 1 = \angle 4$ 6. $\angle 4 = \angle 8$

7. $\angle 4 = \angle 5$ 8. $\angle 6 = \angle 7$

9. $k \perp m$ 10. $\angle 3$ is supplementary to $\angle 5$.

11. If $\angle 1 = 130°$, what are the measures of the other numbered angles?

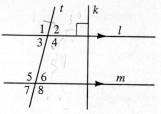

Exs. 1-11

Written Exercises

A 1. Name all angles equal to $\angle 1$.

2. Name all angles supplementary to $\angle 1$.

3. If $\angle 5 = 50°$, then $\angle 11 = \underline{\ ?\ }°$

4. If $\angle 4 = 48°$, then $\angle 10 = \underline{\ ?\ }°$

5. If $\angle 16 = 125°$, then $\angle 6 = \underline{\ ?\ }°$

6. If $\angle 5 = x°$, then $\angle 7 = \underline{\ ?\ }°$

7. If $\angle 10 = y°$, then $\angle 11 = \underline{\ ?\ }°$

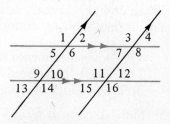

Find the values of *x* and *y*.

8.

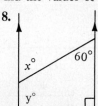

9.

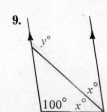

10.

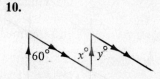

11.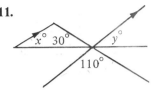

12.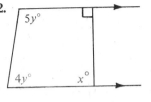

13.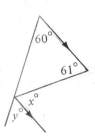

14. Write the reasons for the proof.

Given: $k \parallel m$

Prove: $\angle 6$ is supp. to $\angle 7$.

Proof:

Statements	Reasons
1. $k \parallel m$	1. ?
2. $\angle 6 = \angle 8$	2. ?
3. $\angle 8 + \angle 7 = 180°$	3. ?
4. $\angle 6 + \angle 7 = 180°$	4. ?
5. $\angle 6$ is supp. to $\angle 7$.	5. ?

B 15. Copy what is shown for Theorem 2-3 on page 46. Then write a proof in two-column form.

Use the diagram in Exercise 14. Write proofs in two-column form.

16. Given: $k \parallel m$
Prove: $\angle 2 = \angle 7$

17. Given: $k \parallel m$
Prove: $\angle 1$ is supplementary to $\angle 7$.

18. Given: $\overline{PQ} \parallel \overline{SR}$; $\overline{PS} \parallel \overline{QR}$
Prove: $\angle P = \angle R$

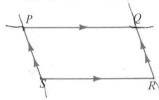

19. Given: $\overrightarrow{BA} \parallel \overline{ED}$; $\overline{BC} \parallel \overline{EF}$
Prove: $\angle ABC = \angle DEF$

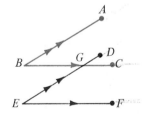

C 20. Given: $\overline{AE} \parallel \overline{BD}$; $\angle 3 = \angle 4$
Prove: $\overrightarrow{BD}$ bisects $\angle EBC$.

21. Given: $\overline{AE} \parallel \overline{BD}$;
$\overrightarrow{BD}$ bisects $\angle EBC$.
Prove: $\angle 5 = \angle 6 + \angle 2$

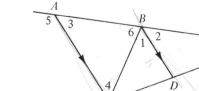

2-3 *Proving Lines Parallel*

Postulate 9, below, gives us a method for proving that two lines are parallel. The previous postulate is listed along with it so that you can see how closely related they are.

Postulate 8   If two parallel lines are cut by a transversal, then corresponding angles are equal.

Postulate 9   If two lines are cut by a transversal and corresponding angles are equal, then the lines are parallel.

The next three theorems can be deduced from Postulate 9.

Theorem 2-5   If two lines are cut by a transversal and alternate interior angles are equal, then the lines are parallel.

Given: Transversal t cuts lines k and m;
     $\angle 1 = \angle 2$

Prove: $k \parallel m$

Proof:

Statements	Reasons
1. Transversal t cuts k and m.	1. Given
2. $\angle 3 = \angle 1$	2. Vert. $\angle$ are equal.
3. $\angle 1 = \angle 2$	3. Given
4. $\angle 3 = \angle 2$	4. Transitive Property
5. $k \parallel m$	5. If two lines are cut by a transversal and corr. $\angle$ are equal, the lines are $\parallel$.

Theorem 2-6   If two lines are cut by a transversal and same-side interior angles are supplementary, then the lines are parallel.

Given: Transversal t cuts lines k and m;
     $\angle 1$ is supplementary to $\angle 2$.

Prove: $k \parallel m$

The proof is left as Exercise 21.

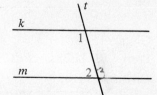

Theorem 2-7 If a transversal is perpendicular to two lines, then the two lines are parallel.

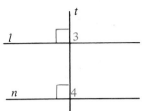

Given: Transversal $t \perp l$; $t \perp n$

Prove: $l \| n$

The proof is left as Exercise 22.

Example: Which lines are parallel?

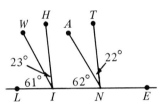

Solution: (1) $\overline{HI}$ and $\overline{TN}$ are parallel since corr. ∠s are equal.

$\angle HIL = 23° + 61° = 84°$
$\angle TNI = 22° + 62° = 84°$

(2) $\overline{WI}$ and $\overline{AN}$ are *not* parallel since $61° \neq 62°$.

The following theorems can be proved using previous postulates and theorems. We state the theorems without proof, however, for you to use in future work. Notice that if we substitute the word *perpendicular* for the word *parallel* in Theorem 2-8, we then have Theorem 2-9.

Theorem 2-8 Through a point outside a line, exactly one parallel can be drawn to the line.

Theorem 2-9 Through a point outside a line, exactly one perpendicular can be drawn to the line.

Given this: $\overset{\bullet}{P}$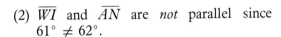

Then line *m* exists and is unique.

Given this: $P \bullet$

Then line *h* exists and is unique.

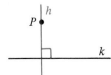

Classroom Exercises

Tell which lines are parallel. Then state the theorem or postulate that justifies your answer.

1.

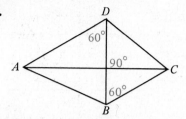

2.

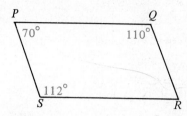

3.

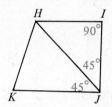

4.

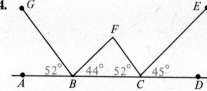

5. Which of the following statements can be used to prove $\overline{IT} \| \overline{NK}$?
   a. $\angle 4 = \angle 3$
   b. $\angle 5 = \angle 1$
   c. $\angle 6 = \angle 2$
   d. $\angle 4$ is supplementary to $\angle NHT$.
   e. $\angle 5 = \angle IHK = 90°$

6. Which statements can be used to prove $\overline{HN} \| \overline{TK}$?
   a. $\angle 2 = \angle 6$
   b. $\angle 1 = \angle 5$
   c. $\angle 3 = \angle 7$
   d. $\angle IHK$ is supplementary to $\angle 5$.
   e. $\angle NHT$ is supplementary to $\angle 7$.

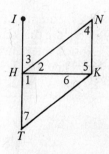

Exs. 5, 6

Exercises 7–11 refer to the diagram shown.

7. How many lines can be drawn through A that are perpendicular to $\overleftrightarrow{BC}$?

8. How many lines can be drawn through B that are parallel to $\overleftrightarrow{AC}$?

9. How many lines can be drawn through C that are parallel to $\overleftrightarrow{AB}$?

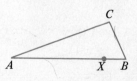

10. In the plane of the given figure, how many lines can be drawn through X that are perpendicular to $\overleftrightarrow{AB}$?

11. In space, how many lines can be drawn through X that are perpendicular to $\overleftrightarrow{AB}$?

Written Exercises

Use the information given to name the lines that must be parallel. If there are no such lines, write *none*.

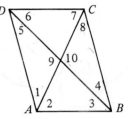

A 1. $\angle 1 = \angle 8$

3. $\angle 2 = \angle 6$

5. $\angle 6 = \angle 3$

7. $\angle 9 = \angle 10$

9. $\angle 9 = \angle 1 + \angle 8$

10. $\angle 5 = 20°$; $\angle 3 = 50°$; $\angle ABC = 70°$

2. $\angle 4 = \angle 5$

4. $\angle 3 = \angle 7$

6. $\angle 2 = \angle 7$

8. $\angle 4 = \angle 8$

Name one pair of parallel lines in each diagram.

11.

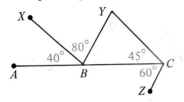

12.

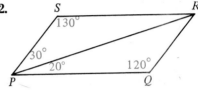

Find the value of x for which $l \parallel n$.

13.

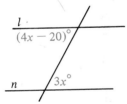

14.

15.

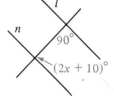

16. Write the reasons for the proof.

Given: Transversal t cuts lines l and n;
$\angle 1 = \angle 2$

Prove: $l \parallel n$

Proof:

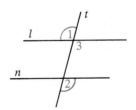

Statements	Reasons
1. Transversal t cuts l and n; $\angle 1 = \angle 2$	1. ___?___
2. $\angle 1 = \angle 3$	2. ___?___
3. $\angle 2 = \angle 3$	3. ___?___
4. $l \parallel n$	4. ___?___

B **17.** Given: $\angle 3 + \angle 6 = 180°$;
$\qquad\qquad \angle 2 + \angle 3 = 180°$
$\qquad$ Prove: $\overline{BC} \| \overline{EF}$

18. Given: $\overline{BC} \| \overline{EF}$; $\angle 6 = \angle 7$
$\qquad$ Prove: $\overline{AE} \| \overline{DC}$

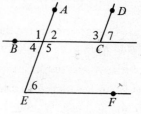

Exs. 17, 18

19. Given: $\angle 1 = \angle 2$; $\angle 4 = \angle 5$
$\qquad$ What can you prove about $\overrightarrow{PQ}$ and $\overrightarrow{RS}$?

20. Given: $\angle 3 = \angle 6$
$\qquad$ What can you prove about other angles?

21. Copy what is shown for Theorem 2-6 on page 50. Then write a proof in two-column form.

22. Copy what is shown for Theorem 2-7 on page 51. Then write a proof in two-column form.

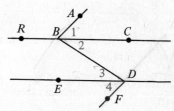

Exs. 19, 20

23. Given: $\angle 1 = \angle 4$; $\overline{BC} \| \overline{ED}$
$\qquad$ Prove: $\overline{AB} \| \overline{DF}$

24. Given: $\angle ABD = \angle FDB$; $\angle 1 = \angle 4$
$\qquad$ Prove: $\overline{BC} \| \overline{ED}$

C **25.** Given: $\angle 3$ is supplementary to $\angle ABR$;
$\qquad\qquad \angle ABR = \angle DBR$
$\qquad$ Prove: $\overline{BC} \| \overline{ED}$

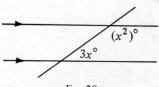

Exs. 23-25

26. Draw two parallel lines cut by a transversal. Then draw the bisectors of two corresponding angles. What appears to be true about the bisectors? Can you prove it?

27. Prove: If two lines are cut by a transversal so that the bisectors of a pair of corresponding angles are parallel, then the two lines are parallel.

28. Find the value of x.

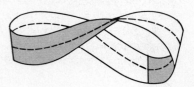

Ex. 28

Challenge _____

You can take a strip of paper and give it a half twist before taping the ends together. The result is called a *Möbius band.* Predict what would happen if you cut a Möbius band lengthwise. (*Hint:* think of cutting before taping.) Try it.

SELF-TEST

Exercises 1–7 refer to the diagram. Classify each statement as true or false.

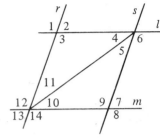

1. $\angle 4$ and $\angle 10$ are alternate interior angles.
2. $\angle 3$ and $\angle 6$ are corresponding angles.
3. If $r \| s$ and $\angle 8 = 120°$, then $\angle 13 = 120°$.
4. If $r \| s$ and $l \| m$ and $\angle 6 = 122°$, then $\angle 12 = 122°$.
5. If $\angle 6$ and $\angle 7$ are supplementary, then $l \| m$.
6. If $\angle 12 = \angle 13$, then $r \| s$.
7. If $\angle 5 = \angle 11$, then $l \| m$.

Complete each statement with the word *always*, *sometimes*, or *never*.

8. Two lines which do not intersect are __?__ parallel.
9. Two skew lines __?__ intersect.
10. Three lines are coplanar. If two of them are perpendicular to the third line, those two lines are __?__ parallel.
11. If a line is parallel to plane X and also to plane Y, then plane X and plane Y are __?__ parallel.
12. If two parallel lines are cut by a transversal, then the same-side interior angles are __?__ supplementary.

Career Notes

Glazier

Have you ever noticed the unusually shaped windows in some modern buildings? A glazier is the person who cuts, fits, and installs the glass. The great variety in shapes and sizes of windows means that glaziers need to use mathematics to fit the glass properly. Beautiful and intricate stained glass designs especially challenge the worker's knowledge of geometry.

A glazier may work independently, installing glass for homes and businesses, or with a construction company on a large project. Training requires a four-year apprenticeship, which includes on-the-job training, as well as classroom instruction in mathematics, blueprint reading, and general construction techniques.

Applying Parallel Lines to Polygons

Objectives

1. Classify triangles according to sides and to angles.
2. State and apply the theorem and the corollaries about the sum of the angles of a triangle.
3. State and apply the theorem about the measure of an exterior angle of a triangle.
4. Recognize and name convex polygons and regular polygons.
5. Know and use the formula for the interior angle sum of any convex polygon.
6. Use the fact that the exterior angle sum of any convex polygon is 360°.

2-4 *Angles of a Triangle*

A **triangle** (△) is the figure formed by three segments joining three noncollinear points. Each of the three points is a **vertex** of the triangle (plural: *vertices*). The segments are the **sides** of the triangle.

Triangle ABC ($\triangle ABC$) is shown.
Vertices of $\triangle ABC$: points A, B, C
Sides of $\triangle ABC$: $\overline{AB}$, $\overline{BC}$, $\overline{CA}$
Angles of $\triangle ABC$: $\angle A$, $\angle B$, $\angle C$

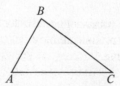

When two sides of a triangle have equal lengths, we say that these sides are equal. A triangle is sometimes classified by the number of equal sides it has.

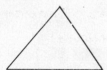

Scalene triangle
No sides equal

Isosceles triangle
At least two sides equal

Equilateral triangle
All sides equal

Triangles can also be classified by their angles.

Acute △
Three acute △

Obtuse △
An obtuse ∠

Right △
A right ∠

Equiangular △
All △ equal

The next theorem is one of the best-known theorems of geometry.

Theorem 2-10   The sum of the angles of a triangle is 180°.

Given: △ABC

Prove: $\angle 1 + \angle 2 + \angle 3 = 180°$

Proof:

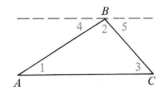

Statements	Reasons
1. Through *B* draw a line parallel to $\overleftrightarrow{AC}$.	1. Through a point outside a line, exactly one parallel can be drawn to the line.
2. $\angle 4 + \angle 2 + \angle 5 = 180°$	2. Def. of a straight angle
3. $\angle 4 = \angle 1$; $\angle 5 = \angle 3$	3. _____?_____
4. $\angle 1 + \angle 2 + \angle 3 = 180°$	4. Substitution property

---

At times, when a statement can be proved easily by applying a theorem, the statement is called a **corollary** of the theorem. Corollaries, like theorems, can be used as reasons in proofs. Each of the four statements below is a corollary of Theorem 2-10.

Corollary 1   If two angles of one triangle are equal to two angles of another triangle, then the third angles are equal.

Corollary 2   Each angle of an equiangular triangle has measure 60°.

Corollary 3   In a triangle, there can be at most one right angle or obtuse angle.

Corollary 4   The acute angles of a right triangle are complementary.

You will justify these corollaries in Exercises 15–18 on page 59.

Example: Does $\angle P$ equal $\angle V$?

Solution: $\angle R = \angle E$ (Given in diagram)
$\angle 1 = \angle 2$ (Vertical angles)
Thus, two angles of △PRO
equal two angles of △VEO,
and $\angle P = \angle V$ by Corollary 1.

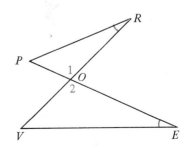

When one side of a triangle is extended, an **exterior angle** is formed. The diagrams and the theorem below show a special relationship between an exterior angle of a triangle and the two remote interior angles.

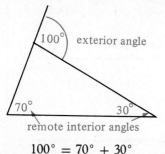

remote interior angles

$100° = 70° + 30°$

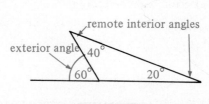

$60° = 40° + 20°$

Theorem 2-11 If one side of a triangle is extended, then the exterior angle formed equals the sum of the two remote interior angles.

Given: $\triangle ABC$ with exterior $\angle 4$

Prove: $\angle 4 = \angle 1 + \angle 2$

Proof:

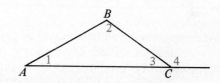

Statements	Reasons
1. $\angle 1 + \angle 2 + \angle 3 = 180°$	1. ?
2. $\angle 3 + \angle 4 = 180°$	2. Def. of a straight angle
3. $\angle 3 + \angle 4 = \angle 1 + \angle 2 + \angle 3$	3. ?
4. $\angle 4 = \angle 1 + \angle 2$	4. Subtraction Postulate

Classroom Exercises

In Exercises 1–15, classify each statement as true or false.

1. All equilateral triangles are isosceles triangles.
2. Some scalene triangles are isosceles triangles.
3. Some obtuse triangles are isosceles triangles.
4. Some triangles have two obtuse angles.
5. Every equilateral triangle is an acute triangle.
6. If two angles of a triangle are complementary, then the triangle is a right triangle.
7. If two angles of a triangle are equal, then the triangle is acute.

8. Some right triangles are isosceles triangles.

9. No right triangle has an obtuse angle.

10. No equiangular triangle is an obtuse triangle.

11. A corollary may be used as a reason in a proof.

12. ∠7 is an exterior angle of △PQR.

13. ∠1 and ∠3 are remote interior angles with respect to ∠5.

14. ∠5 = ∠1 + ∠3   **15.** ∠4 = ∠2 + ∠3

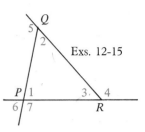

Exs. 12-15

Written Exercises

Draw a triangle that satisfies the conditions stated. If no triangle can satisfy the conditions, write *not possible*.

A **1.** An acute scalene triangle

2. An obtuse scalene triangle

3. An obtuse isosceles triangle

4. A scalene isosceles triangle

5. An obtuse equilateral triangle

6. An isosceles right triangle

7. A scalene right triangle

8. A triangle with three acute exterior angles

Find the values of *x* and *y*.

9.

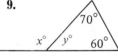

10.

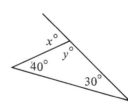

11.

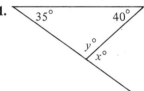

12.

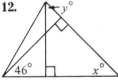

13.

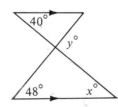

14.

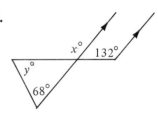

B **15.** Explain how Corollary 2 follows from Theorem 2-10.

16. Explain how Corollary 3 follows from Theorem 2-10.

17. Prove Corollary 1 of Theorem 2-10.   **18.** Prove Corollary 4 of Theorem 2-10.

19. In a given right triangle, one acute angle is four times as large as the other one. What is the measure of each acute angle?

20. One angle of a triangle is 30° more than another and 15° less than the third. What is the measure of each angle?

21. Given: $\overline{AB} \perp \overline{BC}$; $\overline{BD} \perp \overline{AC}$
  a. If ∠C = 22°, find ∠ABD.
  b. If ∠C = 23°, find ∠ABD.
  c. Explain why ∠ABD always equals ∠C.

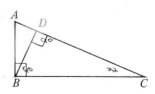

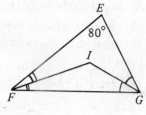

22. The bisectors of ∠F and ∠G meet at *I*.
   a. If ∠EFG = 40°, find ∠FIG.
   b. If ∠EFG = 50°, find ∠FIG.
   c. Explain why ∠FIG always equals 130°.

23. a. Draw two parallel lines and a transversal.
   b. Use a protractor to draw bisectors of two same-side interior angles.
   c. Measure the angle formed by the bisectors. What do you notice?
   d. Prove your answer to (c).

24. Given: ∠ABD = ∠AED
   Prove: ∠C = ∠F

25. Given: $\overrightarrow{GK}$ bisects ∠JGI; ∠H = ∠I
   Prove: $\overline{GK} \| \overline{HI}$

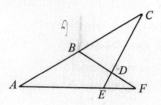

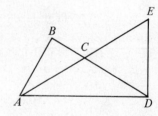

C 26. Given: $\overline{AB} \perp \overline{BD}$; $\overline{ED} \perp \overline{AD}$;
      $\overrightarrow{AE}$ bisects ∠BAD.
   Prove: ∠E = ∠ECD

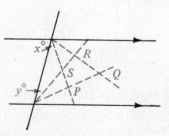

27. A pair of same-side interior angles are trisected (divided into three equal angles) by the dashed lines in the diagram. Find out what you can about the angles of *PQRS*.

Find the sum of the angles of each figure. (*Hint:* Divide each figure into triangles.)

28.

29.

30.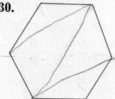

2-5 *Angles of a Polygon*

The word **polygon** means "many angles." Look at the figures below and note that each polygon is formed by coplanar segments (called *sides*) such that:

(1) Each segment intersects exactly two other segments, one at each endpoint.

(2) No two segments with a common endpoint are in the same line.

POLYGONS   NOT POLYGONS

Sides	Names
3	triangle
4	quadrilateral
5	pentagon
6	hexagon
8	octagon
10	decagon
n	n-gon

The table shows names given to polygons according to the number of sides or angles. A triangle is the simplest polygon. The terms which we applied to triangles (such as *vertex* and *exterior angle*) also apply to other polygons.

When referring to a polygon, we list its consecutive vertices in order. We can refer to the polygon shown as pentagon *ABCDE*.

A segment joining two nonconsecutive vertices is a **diagonal** of the polygon. The diagonals of the pentagon at the right are indicated by dashes.

Compare the two polygons below. A **convex polygon** is a polygon such that no line containing a side of the polygon contains a point in the interior of the polygon.

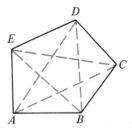

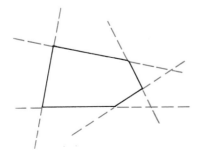

Convex

Not convex

If you draw all the diagonals from just one vertex of a convex polygon, you can figure out the sum of the angles of the polygon.

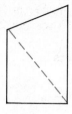

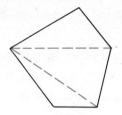

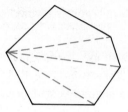

4 sides	5 sides	6 sides
2 triangles	3 triangles	4 triangles
Angle sum = 2(180°)	Angle sum = 3(180°)	Angle sum = 4(180°)

Note that the number of triangles formed in each polygon is two less than the number of sides. This result suggests the following theorem.

Theorem 2-12   The sum of the angles of a convex polygon with *n* sides is (*n* − 2)180°.

Theorem 2-12 deals with the interior angle sum of a polygon and states that the sum depends on the number of sides of the polygon. On the other hand, the exterior angle sum does *not* depend on the number of sides of the polygon. It is always 360°.

Theorem 2-13   The sum of the exterior angles of any convex polygon, one angle at each vertex, is 360°.

To prove Theorem 2-13, we reason as follows. At each vertex of a polygon, the interior and exterior angles are supplementary:

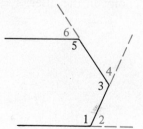

$\angle 1 + \angle 2 = 180°$, $\angle 3 + \angle 4 = 180°$, and so on.

If the polygon has *n* vertices, then there are *n* pairs of supplementary angles and the sum of all these angles is $n(180°)$.

Thus: $n(180°)$ = (interior angle sum) + (exterior angle sum)
$n(180°) = (n - 2)180°$ + (exterior angle sum)
$n(180°) = n(180°) - 360°$ + (exterior angle sum)
$360°$ = (exterior angle sum)

Example 1: A polygon has 22 sides. Find (a) the interior angle sum and (b) the exterior angle sum, one angle at each vertex.

Solution: (a) Interior angle sum = $(22 - 2)180° = 3600°$ (Theorem 2-12)
           (b) Exterior angle sum = $360°$ (Theorem 2-13)

Polygons can be equiangular or equilateral. If a polygon is both equiangular and equilateral, it is called a **regular polygon.**

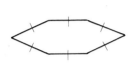

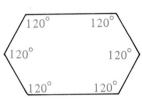

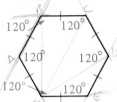

Equilateral hexagon   Equiangular hexagon   Regular hexagon

Example 2: A regular polygon has 12 sides. Find the measure of each angle.

Solution 1: Angle sum = $(12 - 2)180° = 1800°$
Each of the 12 equal angles has measure $1800° \div 12$, or $150°$.

Solution 2: Exterior angle sum = $360°$
Each of the 12 equal exterior angles has measure $360° \div 12$, or $30°$.
Since each interior angle is a supplement of an exterior angle, each interior angle has measure $180° - 30°$, or $150°$.

Classroom Exercises

Is the figure a polygon? If so, is it convex?

1.

2.

3.

4.

5.

6.

7. Imagine stretching a rubber band around the figures in Exercises 1–6. What is the relationship between the rubber band and the figure when the figure is a convex polygon?

Classify each statement as true or false.

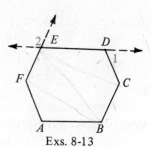

Exs. 8-13

8. Polygon *ABCDEF* is a pentagon.

9. It is possible to draw three diagonals from vertex *E*.

10. If polygon *ABCDEF* is a regular polygon, then $\angle A = \angle B$.

11. If polygon *ABCDEF* is equiangular, then $\angle A = 60°$.

12. $\angle EDC + \angle 1 = 180°$

13. $\angle 2$ is an exterior angle of polygon *ABCDEF*.

Written Exercises

Each polygon is convex. What is its interior angle sum?

A **1.** Triangle

2. Octagon

3. Decagon

4. Pentagon

5. Hexagon

6. 7-gon

7. What is the exterior angle sum of each polygon in Exercises 1–6?

8. a. Draw a quadrilateral which is equiangular but not equilateral.
b. Draw a quadrilateral which is equilateral but not equilangular.
c. Is it possible to draw a triangle which is equilateral but not equiangular?

9. Complete the table for regular polygons.

Number of sides	9	15	30	?	?	?	?	?
Each exterior angle	?	?	?	20°	10°	8°	?	?
Each interior angle	?	?	?	?	?	?	165°	178°

10. Find the measure of each interior angle of a regular decagon.

B **11.** Find the measure of each interior angle of a regular polygon with *n* sides. Answer in terms of *n*.

12. The face of a honeycomb consists of interlocking regular hexagons. What is the measure of each angle of these hexagons?

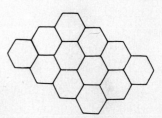

13. a. What is the measure of each angle of a STOP sign?
b. Can you interlock STOP signs as you can hexagons in Exercise 12?

14. a. What is the measure of each interior angle of a regular pentagon?
b. Can you tile a floor with tiles shaped like regular pentagons?

15. **a.** Is it possible to tile a floor with tiles shaped like equilateral triangles? (Ignore the difficulty in tiling along the edges of the room.)
    b. Make a sketch showing how such tiles could be placed together to form the shape of a regular hexagon.

16. Make a sketch showing how to tile a floor using both squares and regular octagons.

17. The cover of a soccer ball consists of interlocking regular pentagons and regular hexagons, as shown at the right. The second diagram shows that pentagons and hexagons cannot be interlocked in the same pattern to tile a floor. Why not?

 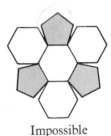

Possible   Impossible

18. The sum of the angles in a polygon is known to be between 2500° and 2600°. How many sides does the polygon have?

C 19. Given: *ABCDEF* is a regular hexagon.
    Prove: $\overleftrightarrow{AB} \| \overleftrightarrow{DE}$

20. Given: *ABCDEFGH* is a regular octagon.
    Prove: $\overleftrightarrow{AB} \perp \overleftrightarrow{DC}$

21. The measure of each exterior angle of a convex polygon with *n* sides is *x*°. The measure of each exterior angle of a convex polygon with *m* sides is *y*°.

    Show that $\dfrac{n}{m} = \dfrac{y}{x}$.

SELF-TEST

1. Give the sum of the measures of the acute angles in a right triangle.
2. If a triangle has three equal sides, it is called ___?___.
3. If a triangle has no equal sides, it is called ___?___.
4. In the diagram, find the values of *x* and *y*.

5. In a decagon, the sum of the interior angles is ___?___° and the sum of the exterior angles is ___?___°.
6. What is the definition of a regular polygon?
7. Find the measure of each angle in a regular decagon.
8. Each angle of a polygon is 160°. How many sides does the polygon have?

If-Then Statements

Objectives

1. State conditionals in if-then form.
2. Recognize the hypothesis and the conclusion of a conditional.
3. State the converse of an if-then statement.
4. Understand the meaning of *if and only if.*
5. Use circle diagrams to represent if-then statements.
6. State the contrapositive and the inverse of an if-then statement.
7. Understand the relationship between logically equivalent statements.
8. Draw correct conclusions from given statements.

2-6 *If-Then Statements and Their Converses*

Each of the following statements is an **if-then statement.**

> *If* Jane is on our team, *then* we are sure to win.
> *If* two lines intersect, *then* exactly one plane contains them.

If-then statements are also called **conditionals.** Conditionals are sometimes expressed without the words "if" and "then" but may be restated in if-then form. Here are three examples:

Conditional: All panthers are cats.
If-then form: If an animal is a panther, then it is a cat.

Conditional: The diagonals of a square are equal.
If-then form: If a figure is a square, then its diagonals are equal.

Conditional: The sum of two odd integers is even.
If-then form: If two numbers are odd integers, then their sum is even.

If-then statements have this basic form:

$$\text{If } p \text{, then } q.$$

$$\uparrow \qquad \uparrow$$

statement p: **hypothesis**	statement q: **conclusion**

When the hypothesis and conclusion of an if-then statement are interchanged, the new statement formed is called the **converse** of the original statement.

Statement: If p, then q.
Converse: If q, then p.

Some true statements have false converses.

True statement: If a figure is a triangle, then it is a polygon.
False converse: If a figure is a polygon, then it is a triangle.

An if-then statement is false if an example can be found for which the hypothesis is true and the conclusion is false. Such an example is called a *counterexample*. It takes only one counterexample to disprove a statement. To show that the converse above is false, we can give the example of a quadrilateral, which is a polygon but is not a triangle.

Some true statements have true converses.

True statement: If a triangle is equilateral, then it is equiangular.
True converse: If a triangle is equiangular, then it is equilateral.

When a statement and its converse are both true, as in the preceding example, we can combine them into one statement by using the words "if and only if" as follows:

A triangle is equilateral *if and only if* it is equiangular.

The converse of a definition is *always* a true statement. All the definitions you have learned could have been stated as if-and-only-if statements. Consider, for example, the following definition: *Perpendicular lines are two lines that form equal adjacent angles.* This definition tells us two things:

(1) If two lines are perpendicular, then the lines form equal adjacent angles.

(2) If two lines form equal adjacent angles, then the lines are perpendicular.

Thus, the definition could have been stated in this form:
Two lines are perpendicular *if and only if* the lines form equal adjacent angles.

Remember, you cannot assume that a true conditional necessarily has a true converse unless the statement is a definition.

Classroom Exercises ⎯⎯⎯⎯⎯⎯⎯⎯⎯⎯⎯⎯⎯⎯⎯⎯⎯⎯⎯⎯⎯

State the hypothesis and conclusion of each statement.

1. If $2x + 5 = 13$, then $x = 4$.

2. If it rains, it pours.

3. If good food is served, then the party is a success.

4. I'll go if you go. (*Hint:* The hypothesis is not always mentioned first.)

5. You can if you try.

6. If the perimeter of a square is 8 cm, each side is 2 cm long.

7. Give the converse of each statement in Exercises 1–6.

Express each statement in if-then form.

8. I go whenever it isn't raining.

9. $x = 7$ if $3x - 9 = 12$.

10. All quarterbacks are athletes. (Begin, "If a person is . . .")

11. All students love vacations.

12. All equilateral triangles are isosceles. (Begin, "If a figure is . . .")

13. The diagonals of a rectangle are equal.

14. Each angle of a square is 90°.

Classify each statement as true or false. Then state the converse and classify it as true or false.

15. If two lines are skew, then they don't intersect.

16. If a polygon is regular, then it is equilateral.

17. If $x = -7$, then $x^2 = 49$.

18. If $\triangle ABC$ is equilateral, then it is equiangular.

19. If quadrilateral $ABCD$ is equilateral, then it is equiangular.

Rewrite each statement as two if-then statements which are converses of each other.

20. An angle is a right angle if and only if its measure is 90°.

21. Two angles are supplementary if and only if their sum is 180°.

Written Exercises

For each statement, state (a) the hypothesis, (b) the conclusion, and (c) the converse.

A   **1.** If $x = 3$, then $5x - 9 = 6$.   **2.** I will, if you won't.

    3. If $\overline{AB} \perp \overline{BC}$, then $\angle ABC = 90°$.   **4.** Ann shouldn't go if Connie doesn't go.

    5. If a line contains two points of a plane, then the line is in the plane.

Express each statement in if-then form.

6. All rectangles are quadrilaterals. (Begin, "If a figure is . . .")

7. A car with poor brakes is a menace on the highway.

8. Every positive number has two square roots.

9. $x = 3$ or $x = -3$ whenever $x^2 = 9$.

10. The product of two odd numbers is odd.

11. All integers divisible by 2 are even.

Rewrite each statement as two if-then statements which are converses of each other.

12. Two angles are complementary if and only if their sum is 90°.

13. $ab > 0$ if and only if a and b are both positive or both negative.

14. $(x - 5)(x + 7) = 0$ if and only if $x = 5$ or $x = -7$.

15. $\angle 1$ and $\angle 2$ are called equal angles if and only if their measures are equal.

16. Postulate 8 may also be worded in this way: "If two lines are parallel, then corresponding angles formed by the two lines and a transversal are equal." Write the converse of this statement. Is the converse true?

17. Theorem 2-5 may also be worded in this way: "If alternate interior angles formed by two lines and a transversal are equal, then the lines are parallel." Write the converse of this statement. Is the converse true?

B 18. a. Write the converse of the Subtraction Postulate: If $a = b$ and $c = d$, then $a - c = b - d$.
b. Choose values of a, b, c, and d to show that the converse is false.

19. a. Write the converse of Theorem 1-3: Vertical angles are equal.
b. Make a sketch which shows that the converse of Theorem 1-3 is false.

Tell whether you think each statement is true or false.
If false, draw a diagram that shows a counterexample.
If true, draw a diagram with the "Given" and "Prove." Do *not* write a proof.

Example 1: All isosceles triangles are acute.

Solution: False (The diagram shows an isosceles triangle which is not acute.)

Example 2: The diagonals of an equiangular quadrilateral are equal.

Solution: True

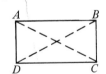

Given: Quad. *ABCD*;
$\angle A = \angle B = \angle C = \angle D$

Prove: $AC = BD$

20. If a triangle has two equal sides, then the angles opposite those sides are equal.

21. If a triangle has two equal angles, then the sides opposite those angles are equal.

22. If two triangles have equal perimeters, then they have equal sides.

23. All diagonals of a regular pentagon are equal.

24. If the diagonals of a quadrilateral are equal and also perpendicular, then it is a regular quadrilateral.

25. If the opposite sides of a quadrilateral are parallel, then they are equal.

26. The diagonals of an equilateral quadrilateral are equal.

27. The diagonals of an equilateral quadrilateral are perpendicular.

Classify each statement as true or false.

C 28. Two lines are parallel if and only if they are coplanar.

29. Two lines are skew if and only if they are not coplanar.

30. $a > 0$ if and only if $a^2 > 0$.

31. A number is divisible by 4 if and only if its square is divisible by 4.

2-7 *Converse, Contrapositive, Inverse*

To show the relationship between an if-then statement and its converse, it is helpful to use circle diagrams (also called Venn diagrams or Euler diagrams).

To represent a statement p, we draw a circle named p. If p is true, we think of a point inside circle p. If p is not true, we think of a point outside circle p.

In the diagram at the left below, a point that lies inside circle p must also lie inside circle q. In other words: *If p, then q.* Check to see that the middle diagram represents the converse: *If q, then p.* Check the diagram at the right also.

If p, then q.

If q, then p.

p if and only if q.

Compare the following if-then statements.

Statement:	If p, then q.
Contrapositive:	If not q, then not p.

You already know that the diagram at the right represents "If p, then q." The diagram also represents "If not q, then not p," because a point that isn't inside circle q can't be inside circle p either. Since the statement and its contrapositive are both true or else both false, they are called **logically equivalent.** The following statements are logically equivalent.

True statement:	If a figure is a triangle, then it is a polygon.
True contrapositive:	If a figure is not a polygon, then it is not a triangle.

Since a statement and its contrapositive are logically equivalent, we may prove a statement by proving its contrapositive. Sometimes that is easier.

There is one more conditional related to "If p, then q" that we shall consider. Compare the following statements.

Statement:	If p, then q.
Inverse:	If not p, then not q.

The next two sentences show that a statement and its inverse are not logically equivalent.

True statement:	If a figure is a triangle, then it is a polygon.
False inverse:	If a figure is not a triangle, then it is not a polygon.

Summary of Related If-Then Statements

Given statement: If *p*, then *q*.
Contrapositive: If not *q*, then not *p*.
Converse: If *q*, then *p*.
Inverse: If not *p*, then not *q*.

A statement and its contrapositive are logically equivalent.
A statement is *not* logically equivalent to its converse or to its inverse.

The relationships summarized above permit us to base conclusions on the contrapositive of a true if-then statement but *not* on the converse or inverse. For example, suppose we accept this statement as true:

If it is an ostrich, then it is a fast runner.   (If *p*, then *q*.)

Let's see what we can conclude when we pair this if-then statement with each of the four given statements below, which we also accept as true.

(1) Given *p*:   Ossie is an ostrich.
   Conclude *q*:   Ossie is a fast runner.

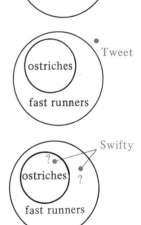

(2) Given *not q*:   Tweet is not a fast runner.
   Conclude *not p*:   Tweet is not an ostrich.

(3) Given *q*:   Swifty is a fast runner.
   No conclusion follows.

(4) Given *not p*:   Bruno is not an ostrich.
   No conclusion follows.

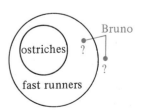

The circle diagrams show that we cannot conclude in (3) that Swifty is an ostrich, or in (4) that Bruno is not a fast runner.

Classroom Exercises

State the converse, contrapositive, and inverse of each statement.

1. If $x = 5$, then $3x - 7 = 8$.

2. If it is snowing, then it is cold.

3. If a figure is a convex polygon, then the sum of its exterior angles is $360°$.

4. All mice like cheese. (Begin, "If an animal is . . .")

Classify each statement as true or false. Then classify the converse, contrapositive, and inverse as true or false.

5. If $x = 4$, then $x^2 = 16$.

6. If M is the midpoint of $\overline{AB}$, then $AM = MB$.

7. If $\angle 1 = \angle 2$, then $l \parallel m$. (See diagram.)

8. If each angle of a hexagon is $120°$, then the hexagon is regular.

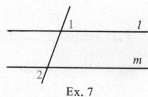

Ex. 7

Given statement: All whales are mammals.

9. Reword the statement in if-then form.

10. Explain how the diagram shown represents both the statement and its contrapositive.

11. Pair the given statement with each statement below. Accept the statements as true. What can you conclude? If no conclusion is possible, say so.
  a. Moby is a whale. **b.** Mabel is not a whale.
  c. Manfred is a mammal. **d.** Myrtle is not a mammal.

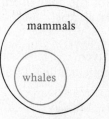

Exs. 10-12

12. Copy the diagram on the chalkboard. Locate points to represent statements (a)–(d) in Exercise 11. Use the points to check your conclusions in Exercise 11.

Accept the first two statements as true. Does the third statement follow as a conclusion?

13. All dogs love Biscos. Wags is not a dog. Therefore, Wags doesn't love Biscos.

14. All dogs love Biscos. Wags does not love Biscos. Therefore, Wags isn't a dog.

Written Exercises

State the converse, contrapositive, and inverse of each statement.

A **1.** If $a = 12$, then $5a - 17 = 43$.

2. If $\overline{AB} \parallel \overline{CD}$, then $AB = CD$.

3. If the diagonals of a quadrilateral bisect each other, then it is a parallelogram.

4. If Juanita is not here, she is not well.

Classify each statement as true or false. Then classify the converse, contrapositive, and inverse as true or false.

5. If the value of b is 3 or -3, then $2b^2 = 18$.

6. If $a^2 > b^2$, then $a > b$ for positive numbers a and b.

7. If all sides of a pentagon are equal, then all angles of the pentagon are equal.

Make a circle diagram to illustrate each statement.

8. If an animal is a wombat, then it is a marsupial.

9. All poets are philosophers.

10. If it is a sunny Sunday, then it is a good day for swimming.

11. I am a cousin if and only if I have a cousin.

Given statement: All senators are at least 30 years old.

B **12.** Reword the statement in if-then form.

13. Make a circle diagram to illustrate the statement.

14. Pair the given statement with each statement below. Accept the statements as true. What can you conclude? If no conclusion is possible, say so.
    a. Lee Hong is 48 years old.   **b.** Joan Baker is a senator.
    c. Pam Rivera is not a senator.   **d.** Hank Foster is 29 years old.

In Exercises 15–17, pair the given statement with each of statements (a)–(d). Accept the statements as true and tell what you can conclude. If no conclusion is possible, say so.

15. Given: If I am guilty, then I was at the scene of the crime.
    a. I was at the scene of the crime.   **b.** I was not at the scene of the crime.
    c. I am guilty.   **d.** I am not guilty.

16. Given: All my students love geometry.
    a. Juan is my student.   **b.** Joanna is not my student.
    c. Wilma loves geometry.   **d.** Walter does not love geometry.

C **17.** Given: Anyone who is eligible is at least 16 years old.
    a. Ed is eligible.   **b.** Ellen is over 16 years old.
    c. Jose is not yet 16 years old.   **d.** Eli is ineligible.

Complete with the word *converse, contrapositive,* or *inverse.*

18. The converse of the inverse of a statement is the ___?___ .

19. The converse of the contrapositive of a statement is the ___?___ .

20. The converse of the ___?___ of a statement is the original statement.

21. **a.** Draw a circle diagram for the converse of the statement "If p, then q."
    b. Does the diagram you have drawn also represent the statement "If not p, then not q"? Explain.
    c. Is the converse of a statement logically equivalent to the inverse of the statement?

Prove each of the following statements by proving its contrapositive.

22. If *l* is not parallel to *m*,
then $\angle 1 + \angle 2 + \angle 3 \neq 180°$.

23. If a^2 is odd, then *a* is odd.

SELF-TEST

Given the statement: If Ned can't go, then Valerie can go.

1. What is the hypothesis of the statement?
2. What is the conclusion of the statement?
3. What is the converse of the statement?
4. What is the inverse of the statement?
5. What is the contrapositive of the statement?
6. Pair the given statement with each statement below. Accept the statements as true. What can you conclude? If no conclusion is possible, say so.
   a. Ned can't go.   **b.** Ned can go.
   c. Valerie can go.   **d.** Valerie can't go.

7. Given the statement: All squares are rectangles.
   a. Rewrite the statement in if-then form.
   b. Make a circle diagram to illustrate the statement.

8. Rewrite the following statement as two if-then statements: Two angles are complementary if and only if their sum is 90°.

Extra Finite Geometry _____

Recall that this text treats *point, line,* and *plane* as undefined terms. Since the terms aren't defined, we are free to interpret them in any way that we wish. Our interpretation has been the ordinary one, the one in which a line is filled with points. Let's try to find some other interpretations.

Geometry A

We use a diagram to indicate the way to interpret the terms *point, line,* and *plane.* You should read the diagram in such a way that you identify the following:

Exactly four points: *A, B, C, D*
Exactly six lines: *(AB), (AC), (AD), (BC), (BD), (CD)*
Exactly four planes: *(ABC), (ABD), (ACD), (BCD)*

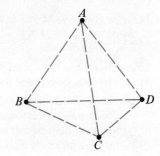

Note that each line consists of exactly two points and each plane consists of exactly three points. "Intersects" means what you would expect. For example, plane (*ABC*) intersects plane (*BCD*) in line (*BC*).

Check the diagram against Postulates 1–5, page 31. Since all five postulates are satisfied, the diagram shows that Postulates 1–5 can be interpreted in such a way that a *finite geometry* is possible. All of space could consist of just four points! (But the postulates don't limit us to that kind of space.)

In *Geometry A,* the terms *point, line,* and *plane* had interpretations not entirely unrelated to the usual ones. The next geometry will be far different. This time we won't use a diagram at all, but we'll explain in words how to interpret the basic undefined terms.

Geometry B

Term:	point	line	plane
Interpretation:	person	committee	club

Let's make some minor adjustments in vocabulary. For example, Postulate 2 states that there is exactly one line through any two points. We can interpret this to mean that there is exactly one committee containing any two persons. Also, "two intersecting planes" can be interpreted to mean "two clubs having at least one person in common." Any statement already established, under some other interpretation, is automatically established under our new interpretation. For instance, take four people: Ann, Bea, Carlos, and Dick. Refer to the finite geometry shown in *Geometry A.* You can, in a mechanical way, list six committees and four clubs so that Postulates 1–5 are satisfied. Furthermore, theorems based on Postulates 1–5 are automatically available. For example, you know: If two committees have a person in common, then they have exactly one person in common (Theorem 1–4, page 33).

This book has more than five postulates, of course. Consider Postulate 6, page 32. That postulate calls for infinitely many points, so you see that the finite geometry shown in *Geometry A* does not satisfy all the postulates of this book. How about *Geometry B?* Since the number of living persons is finite, you know that we cannot use the interpretation discussed in *Geometry B* when all the postulates of the book are to be satisfied.

Exercises

1. Refer to *Geometry B* and restate Postulates 1–5 in terms of *person, committee,* and *club.* Also restate Theorems 1-4, 1-5, and 1-6.

2. a. Check the five-point geometry shown to decide whether Postulates 1–5 are satisfied.

   b. Apply the *Geometry B* interpretation to the diagram. For the fifth point *X*, use a fifth person, Xavera. List the members of a three-person committee and the members of two four-member clubs.

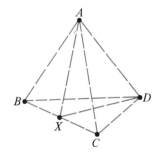

3. Limiting yourself to Postulates 1–5, write the key steps of a proof of the theorem: If there are five points not all in one plane, then there are at least eight lines and five planes. (*Hint:* Let A, B, C, and D be four points not all in one plane. Consider three cases for a fifth point X.

(1) X lies on one of the lines—use (BC)—determined by A, B, C, D.

(2) X lies on one of the planes, but not on any line, determined by A, B, C, D.

(3) X does not lie on any plane determined by A, B, C, D.

4. Refer to Exercise 3. State the theorem in terms of persons. Note that no further proof is necessary.

5. For this exercise, we define two lines to be parallel when they do not have any point in common. Decide whether each of the finite geometries pictured below satisfies the following statement: Through a point outside a line, there is exactly one parallel to the line.

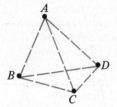
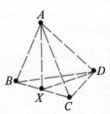

Chapter Summary

1. Lines which do not intersect are either parallel or skew.

2. When two parallel lines are cut by a transversal, the angles formed have these properties:

   a. Corresponding angles are equal.

   b. Alternate interior angles are equal.

   c. Same-side interior angles are supplementary.

3. The properties listed in (2) can be used to prove lines parallel.

4. Through a point outside a line, exactly one parallel and one perpendicular can be drawn to the given line.

5. The sum of the angles of a convex polygon with n sides is $(n-2)180°$. For triangles the angle sum is $180°$.

6. An exterior angle of a triangle equals the sum of the two remote interior angles.

7. The sum of the exterior angles of any convex polygon, one angle at each vertex, is $360°$.

8. The relationships between an if-then statement and its converse, its contrapositive, and its inverse can be represented by circle diagrams. An if-then statement and its contrapositive are logically equivalent.

Chapter Review

2-1 *Definitions*

Line *l* is a transversal cutting lines *m* and *n*.

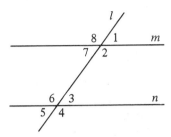

1. Name two pairs of alternate interior angles.
2. Name four pairs of corresponding angles.
3. Name two pairs of same-side interior angles.
4. Coplanar lines *m* and *n* do not intersect. Must they be parallel?

2-2 *Properties of Parallel Lines*

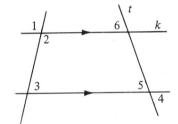

5. If $\angle 6 = 78°$, then $\angle 5 = \underline{\ ?\ }°$.
6. If $\angle 2 = 115°$, then $\angle 3 = \underline{\ ?\ }°$.
7. If $\angle 1 = 110°$, then $\angle 3 = \underline{\ ?\ }°$.
8. If $t \perp k$, then $\angle 4 = \underline{\ ?\ }°$.

2-3 *Proving Lines Parallel*

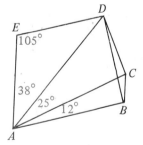

9. Name a pair of parallel lines in the diagram.
10. $\angle ADE = \underline{\ ?\ }°$.
11. If $\overline{BC} \| \overline{AE}$, then $\angle ABC = \underline{\ ?\ }°$.
12. If $\overline{DB} \perp \overline{AB}$, then $\angle ADB = \underline{\ ?\ }°$.

2-4 *Angles of a Triangle*

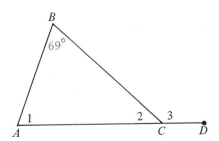

13. If $\angle 2 = 40°$, then $\angle 1 = \underline{\ ?\ }°$.
14. If $\angle 1 = 69°$, then $\angle 3 = \underline{\ ?\ }°$.
15. If $\angle 3 = 141°$, then $\angle 1 = \underline{\ ?\ }°$.
16. If $\angle 2 = (x + 23)°$ and $\angle 1 = (2x + 7)°$, what is the value of *x*?

2-5 *Angles of a Polygon*

17. If a regular polygon has 10 sides, find the measure of each interior angle.

18. Find the measure of each exterior angle in Exercise 17.

19. Draw a pentagon which is not convex.

20. If *ABCDE* is a regular pentagon, what is the measure of ∠*F* in the diagram?

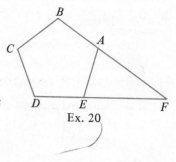
Ex. 20

2-6 *If-Then Statements and Their Converses*

21. Write the converse of the statement: If a polygon is a square, then it is a rectangle.

22. Tell whether the statement is true or false: If a triangle is isosceles, then it is equilateral.

23. Write the hypothesis of the statement: If a number is greater than 20, then it is greater than 10.

24. Rewrite the following statement as two if-then statements which are converses of each other: A whole number is divisible by 5 if and only if its last digit is 0 or 5.

2-7 *Converse, Contrapositive, Inverse*

In Exercises 25–28, take the following statement to be true: *If I am hungry, then I am grumpy.* In each exercise, pair the statement above with the additional given statement. What can you conclude? If no conclusion is possible, say so.

25. I am grumpy.

26. I am not grumpy.

27. I am hungry.

28. I am not hungry.

Chapter Test _____

Complete each statement with the word *always, sometimes,* or *never.*

1. If a transversal of two parallel lines is perpendicular to one of the lines, then it is ___?___ perpendicular to the other one also.

2. Through a point outside a line, you can ___?___ draw two different perpendiculars to the line.

3. An isosceles triangle is ___?___ an equilateral triangle.

4. A scalene triangle is ___?___ an obtuse triangle.

5. An obtuse triangle is ___?___ a right triangle.

6. An interior angle of a convex quadrilateral is ___?___ less than 90°.

7. An interior angle of a regular polygon is ___?___ less than 90°.

8. The sum of the exterior angles of a convex polygon is ___?___ equal to 180°.

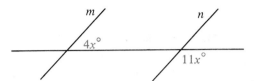

9. Find the value of *x* so that *m*‖*n* in the diagram.

Find the measures of the numbered angles.

10.

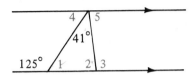

11.

12.

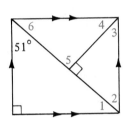

13.

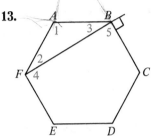

ABCDEF is a regular hexagon.

14. For the statement "If $x = 4$, then $x^2 = 16$," write the following:
    a. hypothesis   **b.** conclusion   **c.** converse   **d.** contrapositive

15. Throughout this exercise, take the following statement to be true: *If a car is new, then it is shiny.* In each part, pair the statement above with the additional given statement. What can you conclude? If no conclusion is possible, say so.
    a. Jill's car is new.   **b.** Joe's car is shiny.
    c. Mike's car is not new.   **d.** Marcia's car is not shiny.

16. Given: $\overleftrightarrow{AB} \parallel \overleftrightarrow{CD}$; $\overrightarrow{BF}$ bisects $\angle ABC$;
            $\overrightarrow{CE}$ bisects $\angle BCD$.
    Prove: $\overleftrightarrow{BF} \parallel \overleftrightarrow{CE}$

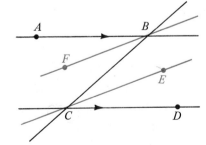

3

Congruent Triangles

Corresponding Parts in a Congruence

Objectives

1. Identify the corresponding parts of congruent figures.
2. Use the SSS Postulate, the SAS Postulate, and the ASA Postulate to prove two triangles congruent.
3. Deduce information about segments or angles by first proving that two triangles are congruent.

3-1 *Congruent Figures*

Have you ever watched a duplicate key being made? The original key and a blank are held firmly in a movable clamp. The clamp is moved so that a guide traces the edge of the key, while a cutting disc duplicates the pattern along the edge of the blank.

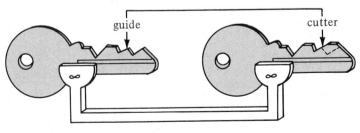

guide   cutter

Original key   Duplicate key

Whenever two figures have the same size and shape, they are called **congruent figures.** You can always match up the points of congruent figures as you can for the congruent keys above.

Triangles *ABC* and *DEF* are congruent. Let's see what this means.

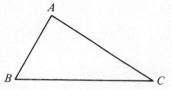 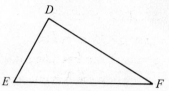

If you mentally slide △*ABC* to the right, you can fit it exactly over △*DEF* by matching up the vertices like this: $A \longleftrightarrow D$, $B \longleftrightarrow E$, $C \longleftrightarrow F$. The sides and angles will then match up like this:

Corresponding angles	Corresponding sides
$\angle A \longleftrightarrow \angle D$	$\overline{AB} \longleftrightarrow \overline{DE}$
$\angle B \longleftrightarrow \angle E$	$\overline{BC} \longleftrightarrow \overline{EF}$
$\angle C \longleftrightarrow \angle F$	$\overline{AC} \longleftrightarrow \overline{DF}$

Do you see that
  (1) since congruent triangles have the same shape, their corresponding angles are equal, and
  (2) since congruent triangles have the same size, their corresponding sides are equal?

We have the following definition:

Two triangles are **congruent** if their vertices can be matched up so that the corresponding parts of the triangles are equal.

The equal parts of the following triangles are marked alike. Imagine sliding △*SUN* to the right and then flipping it over so that its vertices are matched with the vertices of △*RAY* like this:

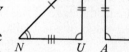

$$S \longleftrightarrow R \qquad U \longleftrightarrow A \qquad N \longleftrightarrow Y$$

The corresponding parts are then equal and we see that the triangles are congruent.

When referring to congruent triangles, we name their corresponding vertices in the same order. The symbol ≅ means "is congruent to." For the triangles above,

△*SUN* is congruent to △*RAY*.
△*SUN* ≅ △*RAY*

The following statements about these triangles are also correct, since corresponding vertices of the triangles are named in the same order.

$$\triangle NUS \cong \triangle YAR \qquad\qquad \triangle SNU \cong \triangle RYA$$

What has been said about congruent triangles applies to all congruent polygons.

Classroom Exercises

1. Read this aloud: $\triangle RKS \cong \triangle LNJ$

Suppose you know that $\triangle FAL \cong \triangle TIC.$

2. Name the three pairs of corresponding vertices.

3. Name the three pairs of corresponding angles.

4. Name the three pairs of corresponding sides.

5. Is it correct to say $\triangle LAF \cong \triangle CIT$?

6. Is it correct to say $\triangle ALF \cong \triangle TCI$?

Complete for the given diagram.

7. $\triangle HOE \cong \triangle \,\underline{?}$ **8.** $\triangle EHO \cong \triangle \,\underline{?}$

9. $\triangle OSU \cong \triangle \,\underline{?}$

The pentagons shown are congruent. Complete.

10. B corresponds to $\underline{?}$.

11. $BLORS \cong \underline{?}$

12. $WK = \underline{?}$

The triangles shown at the left below are congruent. Complete.

13. $\angle Q = \angle \,\underline{?}$ **14.** $\angle R = \angle \,\underline{?}$ **15.** $\angle P = \angle \,\underline{?}$

16. $MN = \underline{?}$ cm   **17.** $RQ = \underline{?}$ cm   **18.** $PR = \underline{?}$ cm

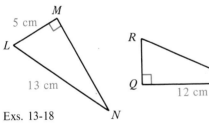

Exs. 13-18

Ex. 19

19. The five leaves shown at the right above are all congruent, but one differs from the others. Which one is different and how?

Written Exercises

Suppose $\triangle FAT \cong \triangle BUG.$ Complete each statement.

A **1.** $\angle F = \underline{?}$ **2.** $\angle G = \underline{?}$

3. $FA = \underline{?}$ **4.** $UG = \underline{?}$

5. $\overline{BU}$ and $\underline{?}$ are corresponding segments.

6. $\angle T$ and $\underline{?}$ are corresponding angles.

7. $\angle F = \angle B$ because $\underline{?}$.   **8.** $\triangle AFT \cong \triangle \,\underline{?}$ **9.** $\triangle BGU \cong \triangle \,\underline{?}$

10–15. Suppose $\triangle SLO \cong \triangle DWN$. Write six equations that follow from this statement and the definition of congruent triangles.

16. Do you think the two stars are congruent? If so, tell how the tips of the stars correspond.

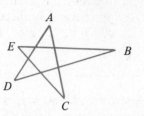
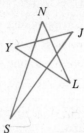

Plot the given points on graph paper. Draw $\triangle ABC$ and $\triangle DEF$. Decide from appearances whether $\triangle ABC \cong \triangle DEF$.

17. $A(1, 0)$ $B(3, 4)$ $C(0, 4)$
     $D(6, 1)$ $E(8, 5)$ $F(6, 5)$

18. $A(0, 1)$ $B(2, 1)$ $C(2, 5)$
     $D(5, 1)$ $E(3, 1)$ $F(3, 5)$

B **19.** In the diagram, $\overline{PQ} \parallel \overline{RS}$.

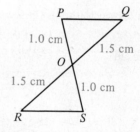

   a. How do you know $\angle P = \angle S$?
   b. What angle equals $\angle Q$?
   c. How do you know $\angle POQ = \angle SOR$?
   d. If $PQ = RS$, is $\triangle POQ \cong \triangle SOR$? Explain.
   e. If $PQ = RS$, is $\triangle POQ \cong \triangle ROS$? Explain.

20. a. Use a protractor to draw a triangle whose angles equal $50°$, $60°$, and $70°$.
   b. Can you draw another triangle whose angles equal $50°$, $60°$, and $70°$ but which is not congruent to the first triangle?

Accurately draw the pairs of triangles described. Measure all the corresponding parts of each pair. Are the two triangles congruent?

21. In $\triangle ABC$, $AB = 4$ cm, $\angle B = 45°$, and $BC = 6$ cm.
   In $\triangle DEF$, $DE = 4$ cm, $\angle E = 45°$, and $EF = 6$ cm.

22. In $\triangle RST$, $\angle R = 30°$, $RS = 5$ cm, and $\angle S = 100°$.
   In $\triangle XYZ$, $\angle X = 30°$, $XY = 5$ cm, and $\angle Y = 100°$.

C **23.** Given: $\triangle ABC \cong \triangle DEF$; $\triangle DEF \cong \triangle GHI$
   Do you think that $\triangle ABC \cong \triangle GHI$? Explain.

24. Suppose twelve toothpicks are arranged as shown to form a regular hexagon.
   a. Copy the figure. Sketch six more toothpicks of the same size inside the hexagon so that it is divided into three congruent regions.
   b. Now "move" only four toothpicks so that the hexagon is divided into two congruent regions.

3-2 *Some Ways to Prove Triangles Congruent*

If you know that two triangles are congruent, you can conclude that the six parts of one triangle are equal to the six parts of the other triangle. If you are not sure whether two triangles are congruent, however, it is not necessary to compare all six pairs of parts. You need to check only three pairs of parts. The following postulates will tell you which three pairs of corresponding parts to compare.

Postulate 10 If three sides of one triangle are equal to three sides of another triangle, then the triangles are congruent. (SSS Postulate)

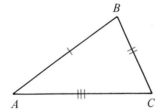 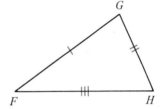

The equal parts of the triangles above are marked alike: $AB = FG$, $BC = GH$, and $AC = FH$. Thus, $\triangle ABC \cong \triangle FGH$ by the SSS Postulate.

Only two pairs of sides are marked as equal in $\triangle NOP$ and $\triangle MOP$. Since $OP = OP$ by the Reflexive Property, we can conclude that $\triangle NOP \cong \triangle MOP$ by the SSS Postulate.

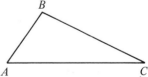

Sometimes it is helpful to describe the parts of a triangle in terms of their relative positions.

$\overline{AB}$ is *opposite* $\angle C$.
$\overline{AB}$ is *included* between $\angle A$ and $\angle B$.
$\angle A$ is *opposite* $\overline{BC}$.
$\angle A$ is *included* between $\overline{AB}$ and $\overline{AC}$.

Postulate 11 If two sides and the included angle of one triangle are equal to two sides and the included angle of another triangle, then the triangles are congruent. (SAS Postulate)

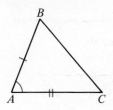 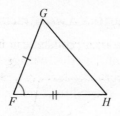 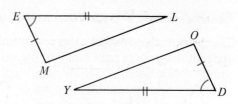

By the SAS Postulate, $\triangle ABC \cong \triangle FGH$ and $\triangle MEL \cong \triangle ODY$.

Postulate 12 If two angles and the included side of one triangle are equal to two angles and the included side of another triangle, then the triangles are congruent. (ASA Postulate)

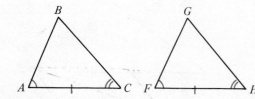

By the ASA Postulate, $\triangle ABC \cong \triangle FGH$ and $\triangle POE \cong \triangle TRY$.

Classroom Exercises

1. In $\triangle ABC$, what angle is included between $\overline{AB}$ and $\overline{BC}$?
2. In $\triangle ABC$, what angle is included between $\overline{AC}$ and $\overline{BC}$?
3. In $\triangle ABC$, what side is included between $\angle B$ and $\angle C$?
4. In $\triangle ABC$, what side is included between $\angle C$ and $\angle A$?

Can the two triangles be proved congruent? If so, what postulate can be used?

5.   6. 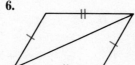   7.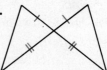

8.   9. 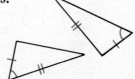   10.

Written Exercises

Can the two triangles be proved congruent? If so, what postulate can be used?

A **1.**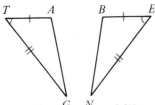

2.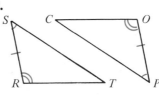

3.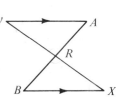

4.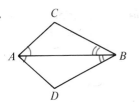

5.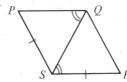

6.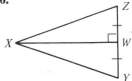

Copy everything shown. Complete each proof.

7. Given: $\overline{AB} \parallel \overline{DC}$; $AB = DC$

Prove: $\triangle ABC \cong \triangle CDA$

Proof:

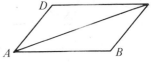

Statements	Reasons
1. $AB = DC$	1. ___?___
2. $AC = AC$	2. ___?___
3. $\overline{AB} \parallel \overline{DC}$	3. ___?___
4. $\angle BAC = \angle DCA$	4. ___?___
5. $\triangle ABC \cong \triangle CDA$	5. ___?___

8. Given: $\overline{PT} \parallel \overline{SR}$; $PQ = RQ$

Prove: $\triangle PQT \cong \triangle RQS$

Proof:

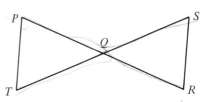

Statements	Reasons
1. $\overline{PT} \parallel \overline{SR}$	1. ___?___
2. $\angle P = \angle R$	2. ___?___
3. $PQ = RQ$	3. ___?___
4. $\angle PQT = \angle RQS$	4. ___?___
5. $\triangle PQT \cong \triangle RQS$	5. ___?___

Write proofs in two-column form.

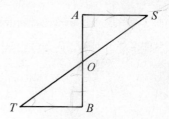

B 9. Given: $\overline{SA} \parallel \overline{TB}$; $TB = SA$
   Prove: $\triangle TOB \cong \triangle SOA$

10. Given: $\overline{SA} \perp \overline{AB}$; $\overline{TB} \perp \overline{AB}$;
    O is the midpoint of $\overline{AB}$.
    Prove: $\triangle TOB \cong \triangle SOA$

11. Given: Plane M bisects $\overline{AB}$; $PA = PB$
    Prove: $\triangle POA \cong \triangle POB$

12. Given: Plane M bisects $\overline{AB}$; $\overline{PO} \perp \overline{AB}$
    Prove: $\triangle POA \cong \triangle POB$

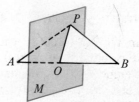

Plot the given points on graph paper. Draw $\triangle ABC$ and $\overline{DE}$. Locate point F so that $\triangle ABC \cong \triangle DEF$.

13. $A(-1, 0)$ $B(-3, 3)$ $C(-4, 1)$ $D(1, 0)$ $E(3, 3)$
14. $A(1, 2)$ $B(4, 2)$ $C(2, 4)$ $D(6, 4)$ $E(9, 4)$

The following diagrams represent three-dimensional figures. Copy each figure and with colored pencils outline the pair of triangles listed. Tell which postulate guarantees that these triangles are congruent.

C 15.

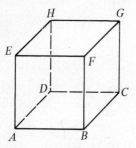

Given: Cube
Show: $\triangle ABF$, $\triangle BCG$

16.

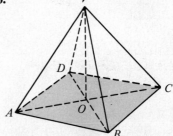

Given: Pyramid with square base;
$VA = VB = VC = VD$
Show: $\triangle VAB$, $\triangle VBC$

17. You can use your hat to find the width of a river. Stand on one bank and face directly across the river. Adjust your hat until the edge of its brim is in line with your eye and the opposite bank. Do an about-face and note the spot on the ground now in line with your eye and the hat brim. Pace off the distance to the spot and you will know the width of the river. What postulate is this method based on? Draw a diagram to explain.

Challenge ———————————————

Using six toothpicks, form four congruent equilateral triangles.

3-3 *Using Congruent Triangles*

Our goal in the last section was to learn when to conclude that two triangles are congruent. Our goal in this section is to deduce information about two segments or angles once we have shown that they are corresponding parts of congruent triangles. The following example will illustrate this technique.

Example:

Given: $\overline{AB}$ and $\overline{CD}$ bisect each other.

Prove: $\overline{AD} \| \overline{BC}$

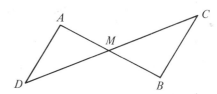

Analysis: We can prove $\overline{AD} \| \overline{BC}$ if we can show that alternate interior angles $\angle A$ and $\angle B$ are equal. To show that they are equal, we can show that they are corresponding parts of congruent triangles. Thus our first goal is to prove $\triangle AMD \cong \triangle BMC$.

Proof:

Statements	Reasons
1. $\overline{AB}$ and $\overline{CD}$ bisect each other.	1. Given
2. $AM = MB;\ CM = MD$	2. Def. of bisect
3. $\angle AMD = \angle BMC$	3. Vert. $\angle$s are equal.
4. $\triangle AMD \cong \triangle BMC$	4. SAS Postulate
5. $\angle A = \angle B$	5. Corr. parts of $\cong$ $\triangle$s are =.
6. $\overline{AD} \| \overline{BC}$	6. If two lines are cut by a transversal and alt. int. $\angle$s are equal, then the lines are $\|$.

Some proofs require the idea of a line perpendicular to a plane. **A line and a plane are perpendicular** if they intersect and the line is perpendicular to all lines in the plane that pass through the point of intersection. Suppose you are given $\overleftrightarrow{PQ} \perp$ plane X. Then you know that $\overleftrightarrow{PQ} \perp \overleftrightarrow{QA}$, $\overleftrightarrow{PQ} \perp \overleftrightarrow{QB}$, $\overleftrightarrow{PQ} \perp \overleftrightarrow{QC}$, and so on.

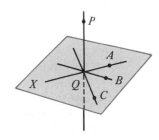

A Way to Prove Two Segments or Two Angles Equal

1. Identify two triangles in which the two segments or angles are corresponding parts.
2. Prove that the triangles are congruent.
3. State that the two parts are equal, supporting the statement with the reason: Corr. parts of ≅ △ are =.

Classroom Exercises

1. Supply the reasons.

Given: $\overleftrightarrow{RP}$ bisects $\angle QRS$ and $\angle QPS$.

Prove: $RQ = RS$

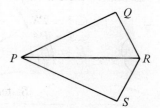

Statements	Reasons
1. $\overleftrightarrow{RP}$ bisects $\angle QRS$ and $\angle QPS$.	1. _____?_____
2. $\angle QRP = \angle SRP$; $\angle QPR = \angle SPR$	2. _____?_____
3. $RP = RP$	3. _____?_____
4. $\triangle QRP \cong \triangle SRP$	4. _____?_____
5. $RQ = RS$	5. _____?_____

In each of Exercises 2–4, you are given more information than you need. State one of the pieces of given information which you do *not* need for each proof. Then present a proof that does not use that piece of information.

2. Given: $PQ = PS$;
$\qquad\quad$ $QR = SR$;
$\qquad\quad$ $\angle Q = \angle S$

$\quad$ Prove: $\angle QPR = \angle SPR$

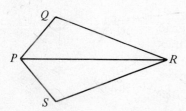

3. Given: $LM = LN$;
$\qquad\quad$ $KM = KN$;
$\qquad\quad$ $\overrightarrow{KO}$ bisects $\angle MKN$.

$\quad$ Prove: $\overrightarrow{LO}$ bisects $\angle MLN$.

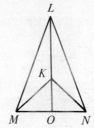

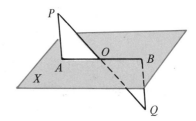

4. Given: $\overline{PA}$ and $\overline{QB}$ are perpendicular to plane X;
       O is the midpoint of $\overline{AB}$;
       O is the midpoint of $\overline{PQ}$.
  Prove: $PA = QB$

Written Exercises

A   **1.** Supply the reasons.

    Given: $LA = PS$;
         $LS = PA$

    Prove: $\overline{LA} \parallel \overline{PS}$

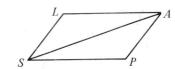

Statements	Reasons
1. $LA = PS$; $LS = PA$	1. ___?___
2. $AS = AS$	2. ___?___
3. $\triangle LAS \cong \triangle PSA$	3. ___?___
4. $\angle LAS = \angle PSA$	4. ___?___
5. $\overline{LA} \parallel \overline{PS}$	5. ___?___

Write proofs in two-column form.

2. Given: $\angle 1 = \angle 2$;
       $\angle 3 = \angle 4$
  Prove: $AP = AQ$

3. Given: $PM = MQ$;
       $AP = AQ$
  Prove: $\angle P = \angle Q$

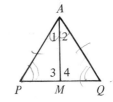

4. Given: $AO = CO$; $DO = BO$
  Prove: $AD = CB$

5. Given: $AO = CO$; $\angle A = \angle C$
  Prove: $\angle D = \angle B$

6. Given: $\angle 1 = \angle 2$; $\angle 3 = \angle 4$
  Prove: $KE = KI$

7. Given: $\overrightarrow{KT}$ bisects $\angle EKI$; $KE = KI$
  Prove: $\overrightarrow{TK}$ bisects $\angle ETI$.

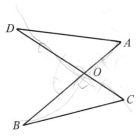

Exs. 4, 5

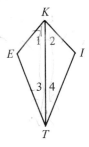

Exs. 6, 7

B 8. Given: $\overline{CA} \perp \overline{AB}$; $\overline{DB} \perp \overline{AB}$;
   M is the midpoint of $\overline{AB}$.
   Prove: $CA = DB$

9. Given: $\overline{AC} \| \overline{BD}$;
   M is the midpoint of $\overline{AB}$.
   Prove: M is the midpoint of $\overline{CD}$.

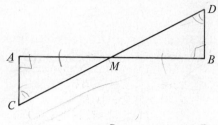

10. Given: $\overline{AD} \| \overline{BE}$; $\overline{BD} \| \overline{CE}$;
    B is the midpoint of $\overline{AC}$.
    Prove: $\angle D = \angle E$

11. Given: $\angle A = \angle EBC$; $\angle D = \angle E$;
    B is the midpoint of $\overline{AC}$.
    Prove: $\overline{BD} \| \overline{CE}$

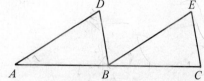

12. Given: $\overline{WX} \perp \overline{UV}$; $\overline{WX} \perp \overline{YZ}$;
    $WU = WV$
    Prove whatever you can about angles 1, 2, 3, and 4.

13. Given: $\overline{WX} \perp \overline{YZ}$; $\angle 1 = \angle 2$;
    $UX = VX$
    Prove: $\overline{WX} \perp \overline{UV}$

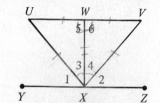

14. A young tree on level ground is supported at P by three
    wires of equal length. The wires are staked to the ground at
    points A, B, and C, which are equally distant from the base
    of the tree, T. Explain in a paragraph how you can prove
    that the angles the wires make with the ground are all equal.

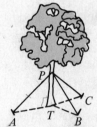

In each of Exercises 15–18, first draw a figure and state what is given and what is to be
proved. Then write a proof in two-column form.

15. In $\triangle ABC$, D is the midpoint of $\overline{BC}$, and $\overline{AD} \perp \overline{BC}$. Prove the triangle is
    isosceles.

16. $\overline{PQ}$ and $\overline{PR}$ are the equal sides of isosceles triangle PQR. The bisector of $\angle P$
    meets $\overline{QR}$ at K. Prove $\overline{PK} \perp \overline{QR}$.

17. Prove: If both pairs of opposite sides of a quadrilateral are equal, then they are
    parallel. (*Hint:* When you draw the quadrilateral, also draw one
    of its diagonals.)

18. State and prove the converse of the statement in Exercise 17.

C 19. Copy the cube shown. Then draw $\overline{BE}$, $\overline{BG}$, and $\overline{EG}$. What
    kind of triangle is $\triangle BEG$? Write a paragraph explaining
    your reasoning. (You may use the fact that the faces of a cube
    are congruent squares.)

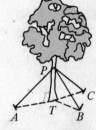

20. Given: $\overline{AB} \perp$ plane M;
        O is the midpoint of $\overline{AB}$.
  Prove: (1) $AD = BD$
         (2) $AC = BC$
         (3) $\angle CAD = \angle CBD$

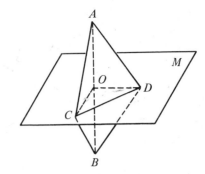

SELF-TEST

In Exercises 1–3, name two congruent triangles, listing corresponding vertices in the same order. Also tell whether you use the SSS, SAS, or ASA postulate.

1.

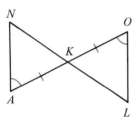

2.

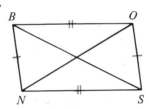

3.

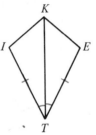

Write proofs in two-column form.

4. Given: $QS = RS$; $\overline{PS} \perp \overline{QR}$
    Prove: $\overrightarrow{PS}$ bisects $\angle QPR$.

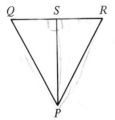

5. Given: $\overline{BL} \parallel \overline{SH}$; $LA = SA$
    Prove: A is the midpoint of $\overline{BH}$.

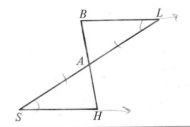

Challenge

Draw three lines, each one intersecting two sides of a triangle, that divide the triangle into four parts; five parts; six parts; seven parts.

   Suppose you can draw four lines. Then what is the greatest number of parts you can form?

Some Theorems Based on Congruent Triangles

Objectives

1. Apply the theorems and corollaries about isosceles triangles.
2. Use the AAS Theorem to prove two triangles congruent.
3. Use the HL Theorem to prove two right triangles congruent.
4. State and apply the theorem about a point on the bisector of an angle, and the converse.

3-4 *The Isosceles Triangle Theorems*

Recall that an isosceles triangle has two equal sides. These equal sides are called **legs** and the third side is called the **base.** The angles at the base are called *base angles* and the angle opposite the base is called the *vertex angle* of the isosceles triangle.

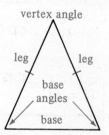

We can prove that the base angles of an isosceles triangle are equal.

Theorem 3-1　If two sides of a triangle are equal, then the angles opposite those sides are equal.

Given: $AB = AC$

Prove: $\angle B = \angle C$

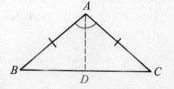

Analysis: For this proof, draw $\overrightarrow{AD}$, the bisector of $\angle A$. A line, ray, or segment added to a diagram to help in a proof is called an **auxiliary line.** Can you now prove $\triangle BAD \cong \triangle CAD$? Why does $\angle B = \angle C$?

The proof is left as Exercise 11.

Corollary 1　An equilateral triangle is also equiangular.

Corollary 2　An equilateral triangle has three 60° angles.

Corollary 3　The bisector of the vertex angle of an isosceles triangle is perpendicular to the base at its midpoint.

Theorem 3-2   If two angles of a triangle are equal, then the sides opposite those angles are equal.

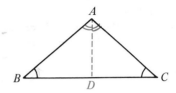

Given: $\angle B = \angle C$

Prove: $AB = AC$

Proof:

Statements	Reasons
1. Draw the bisector of $\angle A$.	1. An angle has exactly one bisector.
2. $\angle BAD = \angle CAD$	2. Def. of angle bisector
3. $\angle B = \angle C$	3. Given
4. $\angle BDA = \angle CDA$	4. If two ∠s of one △ equal two ∠s of another △, then the third ∠s are =.
5. $AD = AD$	5. Reflexive Property
6. $\triangle BAD \cong \triangle CAD$	6. ASA Postulate
7. $AB = AC$	7. Corresponding parts of ≅ △s are =.

Corollary   An equiangular triangle is also equilateral.

Classroom Exercises

1. Given $NA = NL$, you can conclude that $\angle\,\underline{?} = \angle\,\underline{?}$. Give a reason to support your conclusion.

2. Given $\angle A = \angle L$, you can conclude that $\underline{?} = \underline{?}$. Give a reason to support your conclusion.

3. If $\triangle NAL$ is an isosceles triangle and base $\angle A = 40°$, then vertex $\angle N = \underline{?}°$.

4. If $\triangle NAL$ is an isosceles right triangle, then base $\angle A = \underline{?}°$.

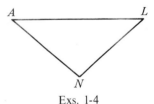

Exs. 1-4

What is wrong with each of the following instructions?

5. Draw the bisector of $\angle J$ to the midpoint of $\overline{PE}$.

6. Draw the line from P perpendicular to $\overline{JE}$ at its midpoint.

7. Draw the line through P and X parallel to $\overleftrightarrow{JE}$.

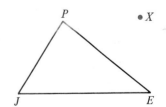

8. Explain how Corollary 1 follows from Theorem 3-1.

9. Explain how Corollary 2 follows from Corollary 1.

10. What is the converse of Corollary 1? Is this converse true?

11. Use the diagram for Theorem 3-1 on page 94 to explain why Corollary 3 is true.

Written Exercises

Find the value of *x* for each diagram.

A **1.**

2.

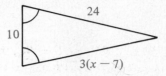

3.

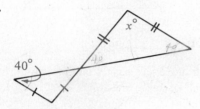

4.

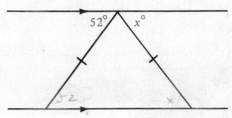

Write proofs in two-column form.

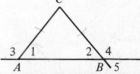

5. Given: $AC = BC$
  Prove: $\angle 1 = \angle 5$

6. Given: $\angle 3 = \angle 4$
  Prove: $AC = BC$

7. Given: $EJ = EK$; $JM = KN$
  Prove: $\angle M = \angle N$

8. Given: $EJ = EK$; $\overline{JK} \parallel \overline{MN}$
  Prove: $\angle M = \angle N$

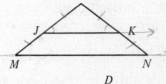

B **9.** Given: $\overline{AD} \parallel \overline{BC}$; $AB = AC$
  Prove: $\angle 1 = \angle 2$

10. Given: $\overline{AD} \parallel \overline{BC}$; $\angle 1 = \angle 2$
  Prove: $AB = AC$

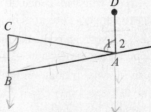

11. Write a complete proof of Theorem 3-1.
  Use the diagram on page 94.

12. Given: $PQ = PR$; $\angle 3 = \angle 4$
  Prove: $QS = RS$

13. Given: $\angle 1 = \angle 2$; $\angle 3 = \angle 4$
  Prove: $\overline{PS}$ bisects $\angle QPR$.

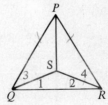

14. Given: $\angle 1 = \angle 2$; $\angle 3 = \angle 4$
  Prove: $\angle 5 = \angle 6$

15. Given: $AO = BO$; $DO = CO$
  Prove: $\overline{AB} \parallel \overline{DC}$
  (*Hint:* Do not base your proof on congruent triangles.)

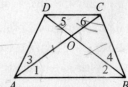

Complete.

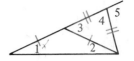

16. If $\angle 1 = x°$, then $\angle 3 = \underline{?}°$.

    (*Hint:* Use the "Exterior Angle Theorem," Theorem 2-11.)

17. If $\angle 1 = x°$, then $\angle 5 = \underline{?}°$.

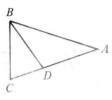

18. If $AB = AC$ and $AD = DB = BC$, then $\angle A = \underline{?}°$.

    (*Hint:* Let $\angle A = x°$ and express all other angles in terms of x.)

In Exercises 19 and 20, a regular pentagon and two diagonals are shown.

19. $\angle 1 = \underline{?}°$

20. $\angle 2 = \underline{?}°$

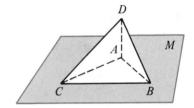

21. Given: $\triangle ABC$ in plane M; D not in plane M;

        $\angle ACB = \angle ABC$; $\angle DCB = \angle DBC$

    Name a pair of congruent triangles.

    Prove your answer correct.

C 22. Prove that the triangle formed by joining the midpoints of the three sides of an isosceles triangle is also isosceles.

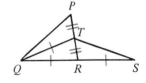

23. Given: $QT = QR = RS$;

        $RT = TP$

    Prove: $QP = TS$

24. Given: Equilateral $\triangle ABC$;

        $\angle BAE = \angle CBF = \angle ACD$.

    Prove something interesting about $\triangle DEF.$

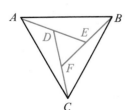

25. Given: $AC = AB$; $DE = DB$;

        $\overline{DE} \parallel \overline{AC}$

    Find the measure of $\angle EBC$.

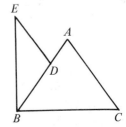

3-5 *Other Methods of Proving Triangles Congruent*

The SSS, SAS, and ASA Postulates give us three methods of proving triangles congruent. In this section we shall discuss two other methods.

Theorem 3-3   If two angles and a non-included side of one triangle are equal to the corresponding parts of another triangle, then the triangles are congruent. (AAS Theorem)

Given: $\triangle ABC$ and $\triangle DEF$; $\angle B = \angle E$;
$\angle C = \angle F$; $AC = DF$

Prove: $\triangle ABC \cong \triangle DEF$

The proof is left as Exercise 11.

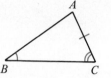 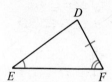

Now that we have an AAS method for proving two triangles congruent, it is natural to ask whether there is an SSA method (for two sides and a non-included angle). The answer is No. For example, both triangles below have a side 3 cm long and a side 2 cm long and a non-included angle of 30°, but the triangles are not congruent.

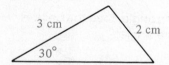

Our final method of proving triangles congruent applies only to right triangles. In a right triangle the side opposite the right angle is called the **hypotenuse** (hyp.). The other two sides are called **legs**.

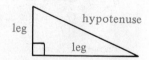

Theorem 3-4   If the hypotenuse and a leg of one right triangle are equal to the corresponding parts of another right triangle, then the triangles are congruent. (HL Theorem)

Given: $\triangle ABC$ and $\triangle DEF$;
$\angle C = 90°$; $\angle F = 90°$;
$AB = DE$; $BC = EF$

Prove: $\triangle ABC \cong \triangle DEF$

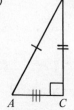 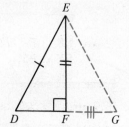

Proof:

Statements	Reasons
1. Let *G* be the point on the ray opposite $\overrightarrow{FD}$ such that $FG = CA$.	1. The Ruler Postulate guarantees exactly one such point.
2. Draw $\overline{EG}$.	2. Through any two points there is exactly one line.
3. $\angle EFG = 90°$	3. $\angle EFG$ is the supplement of a rt. $\angle$.
4. $BC = EF$	4. Given
5. $\triangle ABC \cong \triangle GEF$	5. SAS Postulate
6. $GE = AB$	6. Corr. parts of $\cong$ $\triangle$ are $=$.
7. $AB = DE$	7. Given
8. $GE = DE$	8. Substitution Property
9. $\angle G = \angle D$	9. If a $\triangle$ has two equal sides, then the $\angle$ opposite those sides are $=$.
10. $\angle G = \angle A$	10. Corr. parts of $\cong$ $\triangle$ are $=$.
11. $\angle D = \angle A$	11. Substitution Property
12. $\triangle ABC \cong \triangle DEF$	12. AAS Theorem

The **distance from a point to a line** (or plane) is defined to be the length of the perpendicular segment from the point to the line (or plane). You will need this definition in proving the following theorems.

Theorem 3-5 If a point lies on the bisector of an angle, then the point is equidistant from the sides of the angle.

Given: $\overrightarrow{BZ}$ bisects $\angle ABC$; P lies on $\overrightarrow{BZ}$; $\overline{PX} \perp \overrightarrow{BA}$; $\overline{PY} \perp \overrightarrow{BC}$

Prove: $PX = PY$

The proof is left as Exercise 12.

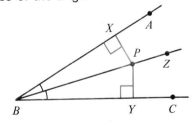

Theorem 3-6 If a point is equidistant from the sides of an angle, then the point lies on the bisector of the angle.

Given: $\overline{PX} \perp \overrightarrow{BA}$; $\overline{PY} \perp \overrightarrow{BC}$; $PX = PY$

Prove: $\overrightarrow{BP}$ bisects $\angle ABC$.

The proof is left as Exercise 13.

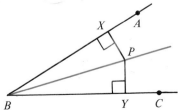

Classroom Exercises

Can the triangles be proved congruent? If so, by which method, ASA or AAS?

1. 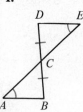 **2.** 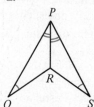 **3.** 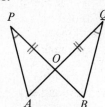 **4.**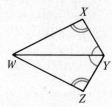

Can the right triangles be proved congruent? If so, by which method, HL, AAS, ASA, or SAS?

5.

6.

7.

8.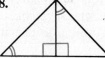

9.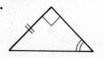

10.

Refer to the diagram to complete Exercises 11–13.

11. If *P* is on the bisector of ∠*B*, then *P* is equidistant from _?_ and _?_.

12. If *P* is on the bisector of ∠*C*, then *P* is equidistant from _?_ and _?_.

13. If *P* is equidistant from $\overrightarrow{AB}$ and $\overrightarrow{AC}$, then *P* is on the _?_.

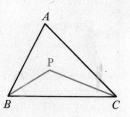

14. Combine the statements of Theorems 3-5 and 3-6 into a single "if and only if" statement.

15. Given: ∠1 = ∠2; ∠5 = ∠6
Prove as much as you can about the diagram.

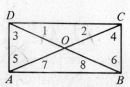

Written Exercises

Write proofs in two-column form.

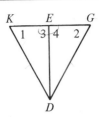

A **1.** Given: $\angle 1 = \angle 2$; $\angle 3 = \angle 4$
  Prove: $\triangle KED \cong \triangle GED$

2. Given: $KD = GD$;
    $\angle 3$ and $\angle 4$ are right angles.
  Prove: $\triangle KED \cong \triangle GED$

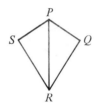

3. Given: $\angle S$ and $\angle Q$ are right angles;
    $SR = QR$
  Prove: $SP = QP$

4. Given: $\angle S$ and $\angle Q$ are right angles;
    $\overline{PR}$ bisects $\angle SPQ$.
  Prove: $\overline{PR}$ bisects $\angle SRQ$.

5. Given: B, O, and C lie in plane M;
    $\overline{AO} \perp$ plane M;
    $AB = AC$
  Prove: $OB = OC$

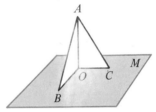

6. Given: B, O, and C lie in plane M;
    $\overline{AO} \perp$ plane M;
    $\angle B = \angle C$
  Prove: $AB = AC$

Exs. 5-7

B **7.** Given: B, O, and C lie in plane M;
    $AB = AC$; $OB = OC$
  What method can be used to prove $\triangle AOB \cong \triangle AOC$?

8. Copy the figure for Exercises 5–7 and draw $\overline{BC}$. Suppose $\overline{AO} \perp \overline{OB}$ and
  $\overline{AO} \perp \overline{OC}$. Classify the following as true or false.
  (a) If $AB = AC$, then $\angle OBC = \angle OCB$.
  (b) The converse of the statement in (a).

9. Given: $\angle A$ and $\angle B$ are right angles;
    $AP = BQ$; $\angle PSQ = \angle PQS$
  Prove: $\angle APS = \angle BQP$

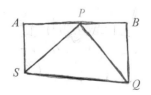

10. Suppose in Exercise 9 that $\angle APS = x°$. Give the measures
  of these angles. (Some answers will be in terms of x.)
  a. $\angle BQP$ **b.** $\angle BPQ$ **c.** $\angle SPQ$ **d.** $\angle PSQ$

11. Prove Theorem 3-3. Use the diagram on page 98. (*Hint:* You can use
  Corollary 1 of Theorem 2-10 to prove that $\angle A = \angle D$.)

12. Prove Theorem 3-5. Use the diagram on page 99.

13. Prove Theorem 3-6. Use the diagram on page 99.

C **14.** Draw an isosceles triangle. From the midpoint of each leg draw a perpendicular segment to the base. Prove that these segments are equal. (First label your figure and state what is given and what is to be proved.)

15. Given: $\overrightarrow{DP}$ bisects $\angle ADE$;

$\overrightarrow{EP}$ bisects $\angle DEC$.

Prove: $\overrightarrow{BP}$ bisects $\angle ABC$.

(*Hint:* There are two theorems which will make your proof short.)

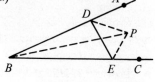

16. Given: Planes E and F intersect in $\overleftrightarrow{AB}$;

C is in F; D is in E; $CB = DA$;

$\overline{CA} \perp \overline{AB}$; $\overline{DB} \perp \overline{AB}$

Prove: $CA = DB$

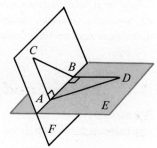

SELF-TEST

Find the value of x for each diagram.

1.

2.

Exercises 3–7 refer to the diagram.

3. If $AD = DC$, what two triangles can be proved congruent by the HL Theorem?

4. If $\overrightarrow{BD}$ bisects $\angle ABC$, then $DX = \underline{\ ?\ }$.

5. If $DX = DY$, then $\underline{\ ?\ }$ bisects $\angle \underline{\ ?\ }$.

6. If $AD = DC$, then $\angle DAC = \angle \underline{\ ?\ }$.

7. If $\angle ABD = \angle BAD$, then $BD = \underline{\ ?\ }$.

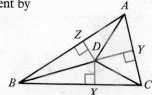

8. If you know that $\angle Q = \angle R$, can you conclude that $PQ = PR$? Explain.

9. Given: $\angle Q = \angle R$; $\overline{PS} \perp$ plane X

Prove: $\triangle PQS \cong \triangle PRS$

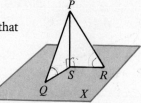

More about Proof in Geometry

Objectives

1. Define the median and the altitude of a triangle.
2. Prove congruence in the case of two overlapping triangles.
3. State and apply the theorem about a point on the perpendicular bisector of a segment, and the converse.
4. Prove two triangles congruent by first proving two other triangles congruent.
5. Understand and use inductive reasoning.

3-6 *Medians and Altitudes; Overlapping Triangles*

A **median** of a triangle is a segment from a vertex to the midpoint of the opposite side. The three medians of △*ABC* are shown below in color.

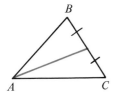 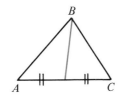 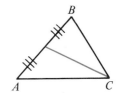

An **altitude** of a triangle is a segment from a vertex perpendicular to the line that contains the opposite side. The three altitudes of acute △*ABC* are shown below.

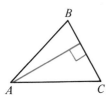 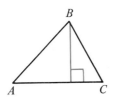 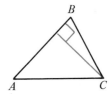

Two of the altitudes of a right triangle are legs of the triangle. (See Exercise 3.)

Two of the altitudes of an obtuse triangle are outside of the triangle. For obtuse △*KLN*, $\overline{LH}$ is the altitude from *L*, and $\overline{NI}$ is the altitude from *N*.

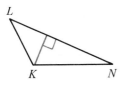 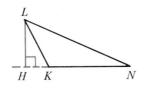 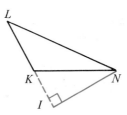

Medians (or altitudes) having the same length are called *equal* medians (or altitudes). Do you think the medians drawn to the legs of an isosceles triangle are equal? The answer is given in the next example.

Example: Given: $\overline{BN}$ and $\overline{CM}$ are medians; $AB = AC$
Prove: $BN = CM$

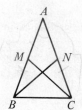

Key steps of proof:

$AB = AC$	(Why?)
$\angle A = \angle A$	(Why?)
$AN = AM$	(Why?)
$\triangle BAN \cong \triangle CAM$	(Why?)
$BN = CM$	(Why?)

Theorem 3-7 If a point lies on the perpendicular bisector of a segment, then the point is equidistant from the endpoints of the segment.

Given: A is on l, the perpendicular bisector of $\overline{BC}$.

Prove: $AB = AC$

The proof is left as Exercise 6.

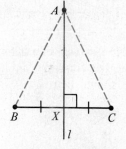

Theorem 3-8 If a point is equidistant from the endpoints of a segment, then the point lies on the perpendicular bisector of the segment.

Given: $AB = AC$

Prove: A is on the perpendicular bisector of $\overline{BC}$.

Analysis: We can draw median $\overline{AX}$ and use corresponding parts of congruent triangles to show that $\overline{AX} \perp \overline{BC}$.

The proof is left as Exercise 7.

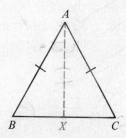

We have seen that the two medians to the legs of an isosceles triangle are equal. Now let us look at the median to the base of an isosceles triangle. In isosceles △*ABC*, it is easy to prove the following:

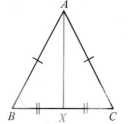

(1) Median $\overline{AX}$ is the bisector of ∠*BAC*.
(2) Median $\overline{AX}$ is the altitude to $\overline{BC}$.
(3) Median $\overline{AX}$ is the perpendicular bisector of $\overline{BC}$.

You can also prove that if statement (1), (2), or (3) is true, then △*ABC* must be isosceles.

Classroom Exercises

Refer to the diagram to complete Exercises 1–4.

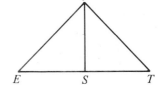

1. If *S* is the midpoint of $\overline{ET}$, then $\overline{JS}$ is called a(n) __?__ .

2. If $\overline{JS} \perp \overline{ET}$, then $\overline{JS}$ is called a(n) __?__ .

3. If *J* is on the perpendicular bisector of $\overline{ET}$, then *J* is equidistant from __?__ and __?__ .

4. If *J* is equidistant from *E* and *T*, then *J* is on the __?__ .

5. Draw an obtuse triangle and its three altitudes.

6. Draw a right triangle and its three altitudes.

7. Is it possible for a triangle to have three equal medians?

8. Study the diagram at the top of this page. Explain how to prove that median $\overline{AX}$ has the three properties listed.

9. Combine the statements of Theorems 3-7 and 3-8 into a single "if and only if" statement.

For each diagram, name a pair of overlapping congruent triangles. Tell whether the triangles are congruent by the SSS, SAS, ASA, AAS, or HL method.

10.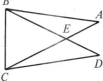

Given: *AB* = *DC*;
       AC = *DB*

11.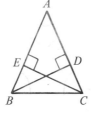

Given: ∠*ABC* = ∠*ACB*

12.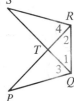

Given: ∠1 = ∠2;
       ∠3 = ∠4

Written Exercises

A **1. a.** Draw a large scalene triangle ABC. Carefully draw the bisector of $\angle A$, the altitude from A, and the median from A. These three should all be different.

    b. Draw an isosceles triangle ABC with vertex angle A. Carefully draw the bisector of $\angle A$, the altitude from A, and the median from A. Are these three different?

2. Draw an obtuse triangle. Then draw its three altitudes in color.

3. Draw a right triangle. Then draw its three altitudes in color.

Write proofs in two-column form.

4. Given: $WX = ZY$; $WY = ZX$
   Prove: $\triangle WXY \cong \triangle ZYX$

5. Given: $AD = BC$; $\angle DAB = \angle CBA$
   Prove: $\triangle ABD \cong \triangle BAC$

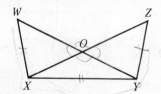 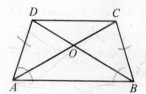

6. Prove Theorem 3-7. Use the diagram on page 104.

B **7.** Prove Theorem 3-8. Use the diagram on page 104.

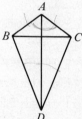

8. Given: A is equidistant from B and C;
         D is equidistant from B and C.
   Prove: $\overleftrightarrow{AD}$ is the perpendicular bisector of $\overline{BC}$.
      (*Hint:* Use Theorem 3-8 twice.)

9. Given: $BX = XY = YC$; $AB = AC$;
         M and N are midpoints.
   Prove: $XN = YM$

10. Given: $AB = AC$;
          $\overline{BD}$ and $\overline{CE}$ are altitudes.
   Prove: $BD = CE$

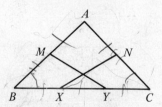 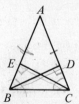

11. Given: $PQRST$ is a regular pentagon.
   a. Prove: $PS = TQ$
   b. You have proved two diagonals are equal. Are they all equal?
   c. Are all the diagonals of a regular hexagon equal?

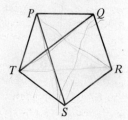

12. Given: $\angle 1 = \angle 2 = \angle 3$;
  $ES = DT$
Prove: $\angle 4 = \angle 5$

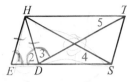

13. Given: $\overleftrightarrow{SR}$ is the perpendicular bisector of $\overline{QT}$;
  $\overleftrightarrow{QR}$ is the perpendicular bisector of $\overline{SP}$.
Prove: $PQ = ST$

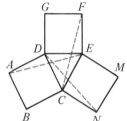

Ex. 13

14. a. Draw an isosceles triangle RST with base $\overline{ST}$. Also draw the medians $\overline{SM}$ and $\overline{TN}$ and label their common point O. Now draw $\overline{MN}$.
  b. Name 4 *pairs* of congruent triangles.

C 15. Given: $\overline{QM}$ and $\overline{RN}$ are altitudes to the legs of isosceles $\triangle PQR$;
  $\overline{QM}$ and $\overline{RN}$ intersect at O.
Prove: $\triangle MNO$ is isosceles.

16. The diagram shows three squares and an equilateral triangle.
Prove: $AE = FC = ND$

17. Use the results of Exercise 16 to prove $\triangle FAN$ is equilateral.

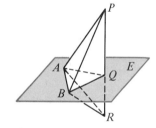

18. Given: A, B, and Q are in plane E;
  $\overline{AQ}$ is an altitude of $\triangle PAR$:
  $\overline{BQ}$ is an altitude of $\triangle PBR$:
  $\angle PAB = \angle PBA$
Prove: $\angle PAR = \angle PBR$

Biographical Note Maria Gaetana Agnesi

Maria Gaetana Agnesi (1718–1799) was born in Milan, Italy. A child prodigy, she had mastered seven languages by the age of thirteen. Between the ages of twenty and thirty she compiled the works of the mathematicians of her time into two volumes on calculus, called *Analytical Institutions*. This was an enormous task, since the mathematicians had originally published their results in different languages and had used a variety of methods of approach.

Her volumes were praised as clear, methodical, and comprehensive. They were translated into English and French and were widely used as textbooks. Agnesi was elected to the Bologna Academy of Sciences and was appointed honorary lecturer in mathematics at the University of Bologna.

3-7 *Using More than One Pair of Congruent Triangles*

Sometimes you have to prove one pair of triangles congruent in order to prove a second pair congruent. For example:

Given: $\angle 1 = \angle 2$; $\angle 5 = \angle 6$

Prove: $\overline{AC} \perp \overline{BD}$

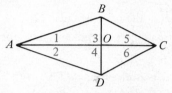

Analysis: We can show $\overline{AC} \perp \overline{BD}$ if we can show $\angle 3 = \angle 4$. $\angle 3$ and $\angle 4$ are corresponding parts of $\triangle ABO$ and $\triangle ADO$. We can prove $\triangle ABO \cong \triangle ADO$ by SAS if we first prove $AB = AD$. $\overline{AB}$ and $\overline{AD}$ are corresponding parts of $\triangle ABC$ and $\triangle ADC$. We can prove $\triangle ABC \cong \triangle ADC$ by ASA. We shall do this first.

Proof:

Statements	Reasons
1. $\angle 1 = \angle 2$; $\angle 5 = \angle 6$	1. Given
2. $AC = AC$	2. Reflexive Property
3. $\triangle ABC \cong \triangle ADC$	3. ASA
4. $AB = AD$	4. Corr. parts of $\cong$ ⚠ are =.
5. $AO = AO$	5. Reflexive Property
6. $\triangle ABO \cong \triangle ADO$	6. SAS
7. $\angle 3 = \angle 4$	7. Corr. parts of $\cong$ ⚠ are =.
8. $\overline{AC} \perp \overline{BD}$	8. Def. of $\perp$ lines

---

Summary of Ways to Prove Two Triangles Congruent

All triangles: SSS   SAS   ASA   AAS
Right triangles: HL

Classroom Exercises

In Exercises 1–3 you are given a diagram and the key steps of a proof. Give the reason for each key step.

1. Prove: $AS = DT$
Key steps of proof:
a. $\triangle ABC \cong \triangle DEF$
b. $\angle C = \angle F$
c. $\triangle ACS \cong \triangle DFT$
d. $AS = DT$

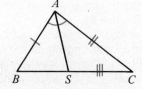
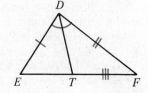

2. Prove: $AX = AY$
Key steps of proof:
  a. $\triangle PAL \cong \triangle KAN$
  b. $\angle L = \angle N$
  c. $\triangle LAX \cong \triangle NAY$
  d. $AX = AY$

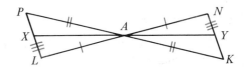

3. Prove: $\angle 3 = \angle 4$
Key steps of proof:
  a. $\triangle LOB \cong \triangle JOB$
  b. $\angle 1 = \angle 2$
  c. $\triangle LBA \cong \triangle JBA$
  d. $\angle 3 = \angle 4$

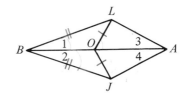

Written Exercises

In Exercises 1–4, you are given a diagram and the key steps of a proof. Give the reason for each key step.

A **1.** Prove: $NE = OS$
Key steps of proof:
  a. $\triangle RNX \cong \triangle LOY$
  b. $\angle X = \angle Y$
  c. $\triangle NEX \cong \triangle OSY$
  d. $NE = OS$

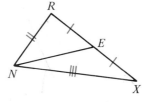
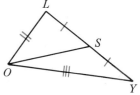

2. Prove: $BE = DF$
Key steps of proof:
  a. $\triangle ABC \cong \triangle CDA$
  b. $\angle 1 = \angle 2$
  c. $\triangle ABE \cong \triangle CDF$
  d. $BE = DF$

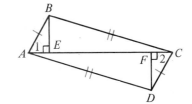

3. Prove: $\angle G = \angle T$
Key steps of proof:
  a. $\triangle RAJ \cong \triangle NAK$
  b. $RJ = NK$
  c. $\triangle GRJ \cong \triangle TNK$
  d. $\angle G = \angle T$

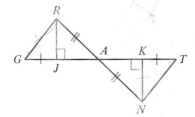

4. Prove: $DX = EX$
Key steps of proof:
  a. $\triangle POD \cong \triangle POE$
  b. $PD = PE$
  c. $\triangle PDX \cong \triangle PEX$
  d. $DX = EX$

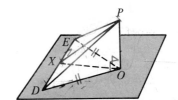

5. Refer to the diagram for Exercise 2. Without using congruent triangles, explain why $\overline{BE} \parallel \overline{DF}$.

6. Refer to Exercise 3. **a.** Explain why $\overline{RJ} \parallel \overline{NK}$. **b.** Explain why $\overline{GR} \parallel \overline{TN}$.

Write proofs in two-column form.

B   7. Given: $\angle 1 = \angle 2$; $\angle 3 = \angle 4$
Prove: $QT = ST$

8. Given: $PQ = PS$; $QR = SR$
Prove: $QT = ST$

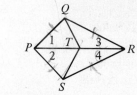

9. Given: $\overline{BD}$ and $\overline{KJ}$ bisect each other.
Prove: $AE = CE$

10. Given: $AB = CD$; $\overline{AB} \parallel \overline{CD}$
Prove: $KE = JE$

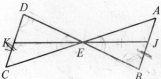

11. Given: $\angle 1 = \angle 2 = \angle 3 = \angle 4$;
   $RS = TS$
Prove: $SA = SB$

12. Given: $SR = ST$; $AU = BU$; $RA = TB$
Prove whatever you can about $\angle 1$, $\angle 2$, $\angle 3$, and $\angle 4$.

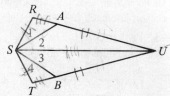

13. The figure is like the one which Euclid used to prove that the base angles of an isosceles triangle are equal (our Theorem 3-1). Write a proof following his key steps shown below.

Given: $AB = AC$;
   $\overline{AB}$ and $\overline{AC}$ are extended so $BD = CE$.
Prove: $\angle ABC = \angle ACB$

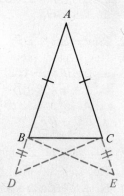

Euclid's Key Steps
1. Prove $\triangle DAC \cong \triangle EAB$.
2. Prove $\triangle DBC \cong \triangle ECB$.
3. Prove $\angle DBC = \angle ECB$ and then
   $\angle ABC = \angle ACB$.

14. Given: $\angle PQR = \angle PSR = 90°$;
   $PQ = PS$
Prove: O is the midpoint of $\overline{QS}$.

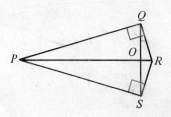

15. Given: $AM = MB$; $AD = BC$;
   $\angle MDC = \angle MCD$
Prove: $AC = BD$

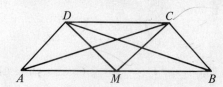

16. Given: ∠1 = ∠2;
  ∠3 = ∠4;
  ∠5 = ∠6
  Prove: $BC = ED$

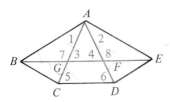

17. Given: The diagram as marked
  Prove: ∠I = ∠M
  (*Hint:* Draw two auxiliary lines.)

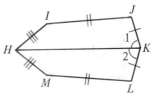

18. Given: $\overline{PQ}$ ⊥ plane M;
  plane M bisects $\overline{PQ}$ at O.
  a. List as many pairs of congruent triangles as you can.
  b. Prove each pair congruent.

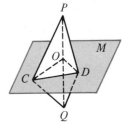

C 19. Given: $\overline{AE} \parallel \overline{BD}$; $\overline{BC} \parallel \overline{AD}$;
  $AE = BC$; $AD = BD$
  Prove: (a) $AC = BE$
  (b) $\overline{EC} \parallel \overline{AB}$

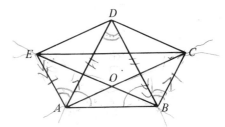

20. Given: $\overline{AM}$ is both a median and an altitude of △ABC;
  $\overline{AE} \perp \overline{BD}$; $\overline{AF} \perp \overline{DF}$;
  ∠1 = ∠2
  Prove: $BE = CF$

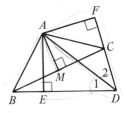

21. Given: The three-dimensional figure shown;
  X is the midpoint of $\overline{AB}$;
  $\overline{RX} \perp \overline{AB}$; $\overline{TX} \perp \overline{AB}$
  Prove: $\overline{SX} \perp \overline{AB}$

22. Given: $\overline{BE}$ and $\overline{CD}$ are medians of scalene △ABC;
  P is on $\overrightarrow{BE}$ such that $BE = EP$;
  Q is on $\overrightarrow{CD}$ such that $CD = DQ$.
  Prove: (a) $AQ = AP$
  (b) $\overline{AQ}$ and $\overline{AP}$ are both parallel to $\overline{BC}$.
  (c) P, A, and Q are collinear.

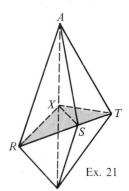

Ex. 21

3-8 *Inductive Reasoning*

Throughout these first three chapters, we have been using deductive reasoning. Now we'll consider **inductive reasoning,** a kind of reasoning that is widely used in science and in everyday life.

Example 1: After picking marigolds for the first time, Connie began to sneeze. She also began sneezing the next four times she was near marigolds. Based on this past experience, Connie reasons inductively that she is allergic to marigolds.

Example 2: Every time Pitch has thrown a high curve ball to Slugger, Slugger has gotten a hit. Pitch concludes from this experience that it is not a good idea to pitch high curve balls to Slugger.

In coming to this conclusion, Pitch has used inductive reasoning. It may be that Slugger just happened to be lucky those times, but Pitch is too bright to feed another high curve to Slugger.

From these examples you can see how inductive reasoning differs from deductive reasoning.

Deductive Reasoning	Inductive Reasoning
Conclusion based on accepted statements (postulates, previous theorems, and given information)	Conclusion based on several past observations
Conclusion *must* be true if hypotheses are true.	Conclusion is *probably* true, but not necessarily true.

Often in mathematics we reason inductively by observing a pattern. For example, can you see the pattern and predict the next number in each sequence below?

a. 3, 6, 12, 24, _?_   b. 11, 15, 19, 23, _?_   c. 5, 6, 8, 11, 15, _?_

In (a), we see that each number is twice the preceding number. We expect that the next number will be 24 × 2, or 48.

In (b), we see that each number is 4 more than the preceding number. We expect that the next number will be 23 + 4, or 27.

To see a pattern in (c), we look at the differences between the numbers. Can you now predict the next number?

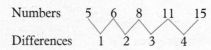

Numbers   5   6   8   11   15

Differences   1   2   3   4

Classroom Exercises

Tell whether the reasoning process is deductive or inductive.

1. Ramon noticed that spaghetti had been on the school menu for the past five Wednesdays. Ramon decides that the school always serves spaghetti on Wednesday.

2. Suppose you believe that when a person wishes on a shooting star, the wish will come true. You know that Becky wished on a shooting star. You conclude that Becky's wish will come true.

3. A manufacturer tested a new toothpaste on 5000 children for a year. None of the children in the test had tooth decay that year. The manufacturer concludes that the new toothpaste is effective in preventing tooth decay.

4. Ricardo did his assignment, adding the lengths of the sides of triangles to find the perimeters. Noticing the results for equilateral triangles, he guesses that the perimeter of every equilateral triangle is three times the length of a side.

5. By using the definitions of equilateral triangle (a triangle with three equal sides) and of perimeter (the sum of the lengths of the sides of a figure), Lillian concludes that the perimeter of every equilateral triangle is three times the length of a side.

6. Look at the discussion leading up to the statement of Theorem 2-12 on page 62. Is the thinking inductive or deductive?

Written Exercises

A 1. Give two examples of inductive reasoning. Do not use any from this book.

Guess the next two numbers of each sequence.

2. 27, 24, 21, 18, ...
3. 1, 3, 9, 27, ...
4. $1, \frac{1}{2}, \frac{1}{4}, \frac{1}{8}, \ldots$
5. 1, 4, 9, 16, ...
6. 8, 10, 14, 20, 28, ...
7. 17, 25, 32, 38, 43, ...
8. $\frac{2}{3}, \frac{4}{3}, 2, \ldots$
9. 0, 2, 6, 14, 30, ...

10. Study the sequence of diagrams. Draw the fourth diagram in the sequence. Then complete the table.

1 line

2 regions

2 lines

4 regions

3 lines

_?_ regions

lines	1	2	3	4	n
regions	2	4	?	?	?

B 11. a. Study the diagrams below. Then guess the number of regions for the fourth diagram. Check your answer by counting.

2 points	3 points	4 points	5 points
2 regions	4 regions	8 regions	? regions

b. Using 6 points on a circle as shown, guess the number of regions within the circle. Carefully check your answer by counting.

Important note: This exercise shows that a pattern predicted on the basis of a few cases may be incorrect. To be sure of a conclusion, use a deductive proof.

For each exercise, write the equation that you think should come next. Then check that the left side of your equation really does equal the right side.

12. $1 \times 9 + 2 = 11$
$12 \times 9 + 3 = 111$
$123 \times 9 + 4 = 1111$

13. $9 \times 9 + 7 = 88$
$98 \times 9 + 6 = 888$
$987 \times 9 + 5 = 8888$

14. $9^2 = 81$
$99^2 = 9801$
$999^2 = 998001$

15. A piece of string is laid out to form an "S." As you make more and more vertical cuts through the string, more and more pieces are formed. Guess how many pieces will be formed when 100 cuts are made.

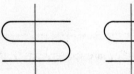

1 cut	2 cuts	3 cuts	4 cuts	100 cuts
4 pieces	7 pieces	10 pieces	? pieces	? pieces

16. Draw several quadrilaterals whose opposite sides are parallel. With ruler and protractor measure the opposite sides and the opposite angles of each figure. On the basis of the diagrams and measurements, what do you guess is true for all such quadrilaterals?

C 17. a. Substitute each of the integers from 1 to 9 for n in the expression $n^2 + n + 11$.

b. Using inductive reasoning, guess what kind of number you will get when you substitute any positive integer for n in the expression $n^2 + n + 11$.

c. Test your guess by substituting 10 and 11 for n.

18. Complete the table for convex polygons.

Number of sides	3	4	5	6	7	8	n
Number of diagonals	0	2	?	?	?	?	?

SELF-TEST

Complete.

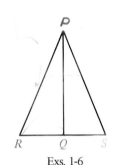

1. If $\overline{PQ}$ is a median, then by definition, ___?___.
2. If $\overline{PQ}$ is an altitude, then by definition, ___?___.
3. If $\overleftrightarrow{PQ}$ is the perpendicular bisector of $\overline{RS}$, then by definition, ___?___ and ___?___.
4. If P is on the perpendicular bisector of $\overline{RS}$, a theorem allows you to conclude that ___?___.
5. If $\overrightarrow{PQ}$ bisects $\angle P$, can you conclude that $RQ = QS$?
6. If $RQ = QS$, can you conclude that $\overrightarrow{PQ}$ bisects $\angle P$?

Exs. 1–6

7. Given: $\angle D = \angle C$; $\angle 1 = \angle 2$;
$\quad\quad AD = BC$
  a. Prove: $\triangle ADX \cong \triangle BCY$
  b. Prove: $\triangle AXY \cong \triangle BYX$

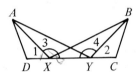

Use inductive reasoning to guess the next number in each sequence.
8. 1, 5, 25, 125, . . .
9. 1, 2, 4, 5, 7, 8, 10, 11, . . .

Chapter Summary

1. Congruent figures have the same size and shape. Two triangles are congruent if their corresponding sides and angles are equal.

2. Ways to prove two triangles congruent:

           SSS   SAS   ASA   AAS   HL

3. A way to prove that two angles or two segments are equal is to show that they are corresponding parts of congruent triangles.

4. If two sides of a triangle are equal, then the angles opposite those sides are equal. An equilateral triangle has three 60° angles.

5. If two angles of a triangle are equal, then the sides opposite those angles are equal.

6. If a point lies on the bisector of an angle, then the point is equidistant from the sides of the angle. The converse is also true.

7. If a point lies on the perpendicular bisector of a segment, then the point is equidistant from the endpoints of the segment. The converse is also true.

8. Inductive reasoning is the process of observing several individual cases and then reaching a general conclusion suggested by them. The conclusion is probably, but not necessarily, true.

Chapter Review

3-1 Congruent Figures

Suppose $\triangle ABC \cong \triangle DEF$. Complete each statement.

1. $\angle A = \angle$ _?_
2. $DE =$ _?_
3. $\triangle FDE \cong \triangle$ _?_
4. If $\angle C = 30°$ and $\angle B = 80°$, then $\angle D =$ _?_ $°$, $\angle E =$ _?_ $°$, and $\angle F =$ _?_ $°$.

3-2 Some Ways to Prove Triangles Congruent

5. In $\triangle JKL$, what angle is included between $\overline{JK}$ and $\overline{KL}$?

Complete, and tell which postulate you used, SSS, SAS, or ASA.

6. $\triangle ABC \cong \triangle$ _?_
7. $\triangle XCB \cong \triangle$ _?_
8. $\triangle ABX \cong \triangle$ _?_

3-3 Using Congruent Triangles

9. Identify two pairs of triangles in which $\angle 1$ and $\angle 2$ appear to be corresponding parts.
10. Given: $QR = QS$; $PR = PS$
    Prove: $\angle 1 = \angle 2$
11. Given: $QR = QS$; $\angle 1 = \angle 2$
    Prove: $\overline{QN} \perp \overline{RS}$

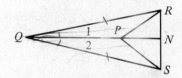

3-4 The Isosceles Triangle Theorems

12. If $AB = AC$, then $\angle$ _?_ $= \angle$ _?_.
13. If $\angle 1 = \angle 3$, find the value of x.
14. If $\triangle ABC$ is equilateral, then $\angle 4 =$ _?_.

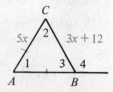

3-5 Other Methods of Proving Triangles Congruent

15. The hypotenuse of $\triangle MNP$ is _?_.
16. The legs of $\triangle ONP$ are _?_ and _?_.
17. If $MN = ON$, then $\triangle MNP \cong \triangle ONP$ by the _?_ method.

3-6 Medians and Altitudes; Overlapping Triangles

D is the midpoint of $\overline{AC}$, and $\overline{AE} \perp \overline{BC}$. Complete.

18. $\overline{BD}$ is a median of $\triangle$ _?_.
19. $\overline{AE}$ is an altitude of $\triangle$ _?_, $\triangle$ _?_, and $\triangle$ _?_.

3-7 *Using More than One Pair of Congruent Triangles*

20. You are given a diagram and the key steps of a proof that $PX = RY$. Give a reason for each step.

1. $\triangle PQR \cong \triangle RSP$
2. $\angle Q = \angle S$
3. $\triangle PQX \cong \triangle RSY$
4. $PX = RY$

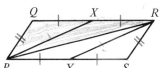

3-8 *Inductive Reasoning*

Use inductive reasoning to guess the next number in each sequence.

21. 10, 11, 13, 16, ...

22. 1, 4, 16, 64, ...

Chapter Test ────────────

1. If $\triangle ABC \cong \triangle DEF$, then $\angle BCA = \angle \underline{?}$ and $AC = \underline{?}$.

2. In $\triangle ABC$, $\overline{AC}$ is opposite $\angle \underline{?}$.

3. In $\triangle ABC$, $\overline{AB}$ is included between $\angle \underline{?}$ and $\angle \underline{?}$.

4. In $\triangle GET$, $\overline{ET}$ and $\overline{TG}$ include $\angle \underline{?}$.

5. The equal sides of an isosceles triangle are called the ___?___.

6. The angles opposite equal sides of an isosceles triangle are called the ___?___ angles, and the third angle is called the ___?___ angle.

7. A segment from a vertex of a triangle to the midpoint of the opposite side is a(n) ___?___ of the triangle.

8. If a point is equidistant from the sides of an angle, then the point lies on the ___?___.

9. Use inductive reasoning to complete: 10, 13, 18, 25, ___?___, ___?___.

Can the triangles be proved congruent? If so, by which method, SSS, SAS, ASA, AAS, or HL?

10.
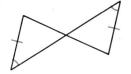

11.
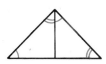

12.

13.

14.

15.

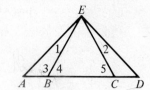

16. If $BE = CE$, then $\angle\,\underline{?} = \angle\,\underline{?}$.
17. If $\angle A = \angle D$, then $\underline{?} = \underline{?}$.
18. If $\triangle BCE$ is equilateral and $\angle A = 43°$, then $\angle 1 = \underline{?}°$.

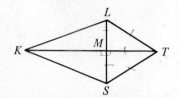

19. $\overline{KT}$ is the perpendicular bisector of $\overline{LS}$.
    a. K is equidistant from $\underline{?}$ and $\underline{?}$.
    b. T is equidistant from $\underline{?}$ and $\underline{?}$.
    c. Name two isosceles triangles.
    d. $\triangle LMT \cong \triangle\,\underline{?}$
    e. $\triangle MLK \cong \triangle\,\underline{?}$

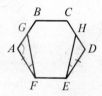

20. Supply the reasons in the proof.

    Given: $ABCDEF$ is a regular hexagon;
    $\quad\quad\ \angle AFG = \angle DEH$

    Prove: $AG = DH$

    Proof:

	Statements		Reasons
1.	$ABCDEF$ is a reg. hexagon.	1.	?
2.	$\angle A = \angle D$; $AF = DE$	2.	?
3.	$\angle AFG = \angle DEH$	3.	?
4.	$\triangle AFG \cong \triangle DEH$	4.	?
5.	$AG = DH$	5.	?

Write proofs in two-column form.

21. Given: $\angle 1 = \angle 2$; $\angle 3 = \angle 4$
    Prove: $PR = QR$

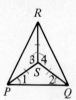

22. Given: $\angle B$ and $\angle D$ are rt. $\angle\!s$;
    $\quad\quad\ AB = CD$
    Prove: $AD = CB$

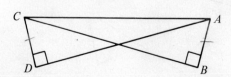

Algebra Review _____

Solve each equation.

1. $\dfrac{a}{3} = 3$

2. $\dfrac{b}{2} = 45$

3. $\dfrac{4c}{3} = 8$

4. $\dfrac{2d}{5} - 4 = 0$

5. $\dfrac{3e}{4} + 4 = 7$

6. $\dfrac{f}{2} - 5 = 6$

7. $g = \dfrac{g}{6} + 5$

8. $\dfrac{3h}{2} - \dfrac{1}{2} = \dfrac{h}{2}$

9. $\dfrac{3}{5}(j + 1) = \dfrac{2}{5}(j + 2)$

10. $\dfrac{k}{2} = \dfrac{1}{4}$

11. $\dfrac{2m}{15} = \dfrac{1}{5}$

12. $\dfrac{5}{2} = \dfrac{n}{8}$

13. $\dfrac{p}{8} - \dfrac{1}{4} = 3$

14. $\dfrac{q}{2} + \dfrac{1}{4} = 4$

15. $\dfrac{r}{5} - \dfrac{1}{10} = \dfrac{1}{15}$

16. $\dfrac{s}{6} - \dfrac{s}{15} = \dfrac{1}{2}$

17. $\dfrac{7t}{2} - \dfrac{5t}{3} = 11$

18. $\dfrac{5u}{12} = \dfrac{3u}{8} + 1$

19. $\dfrac{v + 2}{6} = \dfrac{v + 1}{5}$

20. $\dfrac{2w - 3}{3} = \dfrac{w + 4}{2}$

21. $\dfrac{3}{2w - 3} = \dfrac{2}{w + 4}$

22. $\dfrac{x - 1}{3} + \dfrac{x - 2}{2} = 2$

23. $\dfrac{y - 2}{2} = \dfrac{3}{4} + \dfrac{7 - 3y}{8}$

24. $\dfrac{3z - 1}{2} - \dfrac{2 + z}{5} = z + \dfrac{1}{10}$

Cumulative Review: Chapters 1-3 _____

In Exercises 1-18, complete each sentence with the most appropriate word, phrase, or value.

A 1. Statements which are accepted without proof are called ___?___.

2. If a statement is easily proved by applying a theorem, the statement is called a ___?___ of the theorem.

3. A triangle with no sides equal is a ___?___ triangle.

4. Non-intersecting lines which are not coplanar are called ___?___ lines.

5. Points which are all in one line are ___?___ points.

6. If two intersecting lines form equal adjacent angles, the lines are ___?___.

7. When two lines intersect, two non-adjacent angles formed are called ___?___ angles.

8. The acute angles of a right triangle are ___?___ angles.

9. The statement "If $x = 4$, then $x - 1 = 3$" is the ___?___ of the statement "If $x - 1 = 3$, then $x = 4$."

10. If a triangle has two equal angles, then the triangle is ___?___.

11. If two parallel planes are cut by a third plane, then the lines of intersection ___?___.

12. If each angle of a regular polygon measures 140°, then the polygon has ___?___ sides.

13. If a point is equidistant from the endpoints of a segment, then the point lies on the ___?___ of the segment.

14. If $\angle PQR$ is a right angle, then two of the altitudes of $\triangle PQR$ are ___?___ and ___?___.

15. Two angles are called adjacent angles if they have a common vertex and a common side, but no common ___?___

16. An angle with measure 180° is called a ___?___.

17. If $\angle A = 18°$ and $\angle B$ is the complement of $\angle A$, then $\angle B =$ ___?___°.

18. If $\angle C = 67°$ and $\angle D$ is the supplement of $\angle C$, then $\angle D =$ ___?___°.

19. On a number line, points J and K have coordinates -7 and 12. What is the coordinate of the midpoint of $\overline{JK}$?

20. In the diagram, $\overleftrightarrow{WX} \| \overleftrightarrow{YZ}$ and $\angle 1 = 48°$.

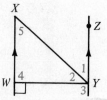

   a. $\angle 2 =$ ___?___°   **b.** $\angle 3 =$ ___?___°

   c. $\angle 4 =$ ___?___°   **d.** $\angle 5 =$ ___?___°

21. The key steps of a proof are given. Write the reason for each key step.

Given: $\angle 1 = \angle 2$; $\overline{AB} \| \overline{DC}$; P is the midpoint of $\overline{AB}$.

Prove: $\angle A = \angle B$

Key steps of proof:

   a. $DP = CP$ **d.** $AP = BP$

   b. $\angle 1 = \angle 3$; $\angle 2 = \angle 4$ **e.** $\triangle APD \cong \triangle BPC$

   c. $\angle 3 = \angle 4$ **f.** $\angle A = \angle B$

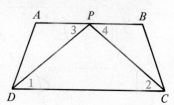

B **22.** A certain angle has measure $(2x - 16)°$. The complement of the angle has measure $(x - 2)°$. Find the value of x.

23. In the diagram at the left below, $\angle 1 = (2x + 19)°$, $\angle 2 = (4x - 1)°$, and $\angle 3 = 42°$. Find the value of x.

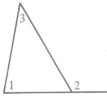

Ex. 23

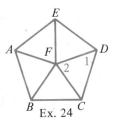

B Ex. 24 C

24. In the diagram at the right above, $ABCDE$ is a regular pentagon. The five segments drawn from F are equal. Find the measures of $\angle 1$ and $\angle 2$.

25. Complete the proof by supplying the reasons.

Given: $\overline{AB} \| \overline{CD}$; $\overline{CF} \| \overline{EB}$; $AE = FD$

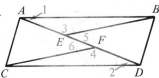

Prove: $AB = CD$

Proof:

Statements		Reasons
1. $\overline{AB} \| \overline{CD}$	1.	?
2. $\angle 1 = \angle 2$	2.	?
3. $\overline{CF} \| \overline{EB}$	3.	?
4. $\angle 6 = \angle 5$	4.	?
5. $\angle 3 + \angle 5 = 180°$; $\angle 4 + \angle 6 = 180°$	5.	?
6. $\angle 3 = \angle 4$	6.	?
7. $AE = FD$	7.	?
8. $\triangle AEB \cong \triangle DFC$	8.	?
9. $AB = CD$	9.	?

C **26.** Given: $\overleftrightarrow{PR} \| \overleftrightarrow{US}$; $\overleftrightarrow{QU} \| \overleftrightarrow{RT}$; $QP = TS$;
$\angle UPQ = \angle RST$

Prove: $UP = RS$

27. Given: $AD = EC = AC$;
$\angle DAC = \angle ECA$

Prove: $\overline{DE} \| \overline{AC}$

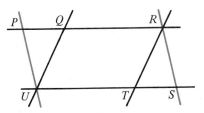

Ex. 26

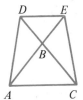

Ex. 27

4

Using Congruent Triangles

Parallelograms

Objectives

1. Define a parallelogram.
2. State and apply the theorems about properties of a parallelogram.
3. State and apply the theorems that can be used to prove that a quadrilateral is a parallelogram.
4. Identify the special properties of a rectangle, a rhombus, and a square.

4-1 *Properties of Parallelograms*

A **parallelogram** (□) is a quadrilateral with both pairs of opposite sides parallel.

The following theorems state some properties common to all parallelograms. The proofs are based on what you have learned about parallel lines and congruent triangles.

For Theorem 4-1, all the steps in the proof are supplied. You will be asked to prove Theorems 4-2 and 4-3 as Exercises 15 and 16. Use the analysis of Theorem 4-3 to help you start your proof in Exercise 16.

Theorem 4-1   Opposite sides of a parallelogram are equal.

Given: $\square ABCD$

Prove: $AB = CD$; $DA = BC$

Proof:

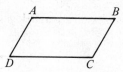

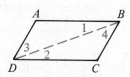

Statements	Reasons
1. Draw $\overline{DB}$.	1. Through any two points there is exactly one line.
2. $\overline{AB} \parallel \overline{DC}$; $\overline{AD} \parallel \overline{BC}$	2. By definition, opp. sides of a $\square$ are $\parallel$.
3. $\angle 1 = \angle 2$; $\angle 3 = \angle 4$	3. If two parallel lines are cut by a transversal, then ___?___.
4. $DB = DB$	4. Reflexive Property
5. $\triangle ABD \cong \triangle CDB$	5. ___?___
6. $AB = CD$; $DA = BC$	6. ___?___

Corollary   If two lines are parallel, then all points on each line are equidistant from the other line.

We can show this easily.

In the diagram, $j \parallel k$. P and Q are any two points on j. The distances from P to k and from Q to k are the lengths of the perpendiculars $\overline{PA}$ and $\overline{QB}$. In $\square PQBA$, $PA = QB$ by Theorem 4-1.

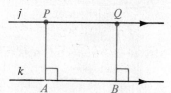

Theorem 4-2   Opposite angles of a parallelogram are equal.

Theorem 4-3   The diagonals of a parallelogram bisect each other.

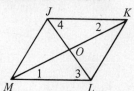

Given: $\square JKLM$ with diagonals $\overline{JL}$ and $\overline{KM}$

Prove: $JO = LO$; $KO = MO$

Analysis: We can prove $JO = LO$ and $KO = MO$ by showing that they are corresponding parts of congruent triangles. Since $\overline{JK} \parallel \overline{LM}$, $\angle 1 = \angle 2$ and $\angle 3 = \angle 4$. $JK = LM$ by Theorem 4-1. We can now prove that $\triangle JKO \cong \triangle LMO$.

Classroom Exercises

1. Quad. *GRAM* is a parallelogram.
   a. Why is ∠*G* supplementary to ∠*M*?
   b. Why is ∠*M* supplementary to ∠*A*?
   c. Complete: Consecutive angles of a parallelogram are __?__ .

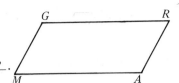

2. In ▱*WXYZ*, ∠*W* = 70°. Find the measures of ∠*X*, ∠*Y*, ∠1, and ∠2.
3. Repeat Exercise 2, using ∠*W* = *c*°.

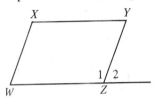

Exs. 2, 3

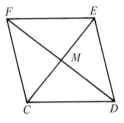

Ex. 4

4. Name all pairs of equal segments in ▱*CDEF*, above.
5. Draw the figure described.
   a. A quadrilateral with two 70° angles and two 110° angles which is not a parallelogram.
   b. A parallelogram with two 70° angles and two 110° angles.
6. In ▱*RECT*, ∠*R* is a right angle. Explain why each statement is true.
   a. ∠*C* is a right angle.
   b. ∠*E* is a right angle.
   c. ∠*T* is a right angle.

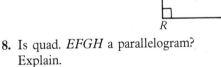

7. Is quad. *ABCD* a parallelogram? Explain.

8. Is quad. *EFGH* a parallelogram? Explain.

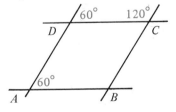

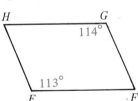

Exercises 9–13 refer to ▱*RSTU*. Justify each statement with a definition or theorem.

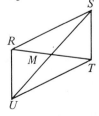

9. $\overline{RS} \parallel \overline{UT}$
10. *RS* = *UT*
11. *MR* = *MT*
12. *UM* = ½*US*
13. ∠*URS* = ∠*STU*

Written Exercises

Exercises 1–14 refer to $\square GHJK$. Find the lengths or angle measures.

A 1. If $KH = 14$, $KM = \underline{}$.

    2. If $MJ = 9$, $GJ = \underline{}$.

    3. If $\angle KGH = 81°$, $\angle KJH = \underline{}°$.

    4. If $\angle GKJ = 104°$, $\angle HGK = \underline{}°$.

    5. If $\angle 1 = 36°$ and $\angle 2 = 32°$, $\angle KGH = \underline{}°$.

    6. If $\angle 3 = 40°$ and $\angle 4 = 30°$, $\angle 2 = \underline{}°$.

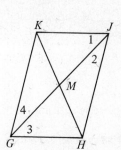

Find the value of x or y in each.

    7. If $KJ = 2x - 3$ and $GH = 11$, $x = \underline{}$.

    8. If $KM = 8x$ and $HM = 5x + 15$, $x = \underline{}$.

    9. If $GJ = 10x$ and $GM = 3x + 4$, $x = \underline{}$.

  10. If $GK = 3x - 7$ and $HJ = 2x + 4$, $x = \underline{}$.

  11. If $\angle 1 = 2y°$ and $\angle 3 = (y + 20)°$, $y = \underline{}$.

  12. If $\angle 2 = (y + 5)°$ and $\angle 4 = (2y - 20)°$, $y = \underline{}$.

  13. If $\angle GKJ = 144°$ and $\angle GHJ = 6y°$, $y = \underline{}$.

  14. If $\angle KGH = (4y + 3)°$ and $\angle GHJ = (7y + 12)°$, $y = \underline{}$.

Exs. 1–14

B 15. Prove Theorem 4-2.

  16. Prove Theorem 4-3. Use the diagram on page 124.

In Exercises 17–19, use the given information to prove something about each diagram.

  17. Given: $\square WXYZ$; $\angle 1 = \angle 2$

      Prove: $\underline{}$

  18. Given: $\square WXYZ$; $\overline{ZP} \perp \overline{WY}$; $\overline{XQ} \perp \overline{WY}$

      Prove: $\underline{}$

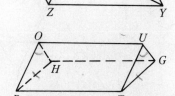

  19. Given: The solid shown;

         $\square TROU$; $\square OUGH$;

         $\angle ROH = \angle TUG$

      Prove: $\underline{}$

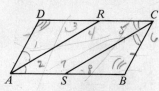

C 20. Given: $\square ABCD$; $\overrightarrow{AR}$ bisects $\angle DAB$;

         $\overrightarrow{CS}$ bisects $\angle DCB$.

      Prove: $\overleftrightarrow{AR} \parallel \overleftrightarrow{CS}$

  21. In Exercise 20, suppose that all four sides of $\square ABCD$ are equal. Is it still true that the bisectors of $\angle DAB$ and $\angle DCB$ are parallel? Explain.

22. Prove that if the bisectors of two consecutive angles of a quadrilateral are *not* perpendicular, then the quadrilateral is *not* a parallelogram. (*Hint:* What is the contrapositive?)

23. The opposite faces of the box shown are in parallel planes. Points *T*, *W*, *O*, and *I* are coplanar. State the theorem and the definition which can be used to justify the statement that quad. *TWOI* is a parallelogram.

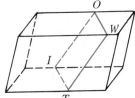

4-2 *Ways to Prove That Quadrilaterals Are Parallelograms*

If you know that the opposite sides of a quadrilateral are parallel, then you know that the figure is a parallelogram by definition. There are other methods for proving that a given quadrilateral is a parallelogram.

Theorem 4-4   If one pair of opposite sides of a quadrilateral are equal and parallel, then the quadrilateral is a parallelogram.

Given: $PL = AN$; $\overline{PL} \parallel \overline{AN}$

Prove: Quad. *PLAN* is a $\square$.

Proof:

Statements	Reasons
1. Draw $\overline{PA}$.	1. Through any two points there is ___?___ .
2. $PL = AN$; $\overline{PL} \parallel \overline{AN}$	2. Given
3. $\angle 1 = \angle 2$	3. ___?___
4. $PA = PA$	4. Reflexive Property
5. $\triangle PLA \cong \triangle ANP$	5. ___?___
6. $\angle 3 = \angle 4$	6. ___?___
7. $\overline{LA} \parallel \overline{NP}$	7. If two lines are cut by a trans. and alt. int. $\angle$ are =, then ___?___ .
8. Quad. *PLAN* is a $\square$.	8. Def. of a $\square$

Theorem 4-5   If both pairs of opposite sides of a quadrilateral are equal, then the quadrilateral is a parallelogram.

The proof is left as Exercise 16.

Theorem 4-6 If both pairs of opposite angles of a quadrilateral are equal, then the quadrilateral is a parallelogram.

The proof is left as a Classroom Exercise.

Theorem 4-7 If the diagonals of a quadrilateral bisect each other, then the quadrilateral is a parallelogram.

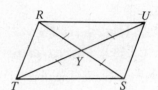

Given: Quad. $RUST$; $TY = UY$; $RY = SY$

Prove: Quad. $RUST$ is a $\square$.

The proof is left as Exercise 17.

Five Ways to Prove That a Quadrilateral Is a Parallelogram

1. Show both pairs of opposite sides parallel.
2. Show one pair of opposite sides parallel and equal.
3. Show both pairs of opposite sides equal.
4. Show both pairs of opposite angles equal.
5. Show that the diagonals bisect each other.

The proof of the following theorem is based on what you have learned about parallelograms. The proof is left as Exercise 6.

Theorem 4-8 If three parallel lines cut off equal segments on one transversal, then they cut off equal segments on every transversal.

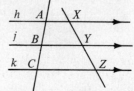

Given: $h \parallel j \parallel k$; trans. $\overleftrightarrow{AC}$ and $\overleftrightarrow{XZ}$;
$AB = BC$

Prove: $XY = YZ$

Classroom Exercises

In each of Exercises 1–5, state the definition or theorem that allows you to conclude that quad. *ABCD* is a parallelogram.

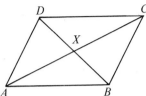

1. $DC = AB$; $\overline{DC} \| \overline{AB}$
2. $DX = XB$; $AX = XC$
3. $\overline{DC} \| \overline{AB}$; $\overline{AD} \| \overline{BC}$
4. $DC = AB$; $AD = BC$
5. $\angle DAB = \angle BCD$; $\angle ADC = \angle CBA$

6. **a.** Theorem 4-1 can be worded: "If a quadrilateral is a parallelogram, then both pairs of opposite sides are equal." Which theorem in this section is a converse of Theorem 4-1?
   b. Which other theorems in this section are converses of earlier theorems?

7. Given: Quad. *ABCD*;
   $\angle A = \angle C = x°$;
   $\angle B = \angle D = y°$

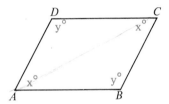

   a. Tell why $2x + 2y = 360$.
   b. $x + y = \underline{\ ?\ }$
   c. Tell why $\overline{AB} \| \overline{DC}$ and why $\overline{AD} \| \overline{BC}$.
   d. Tell why quad. *ABCD* must be a parallelogram.
   e. What theorem have you proved?

8. The cardboard strips shown below are joined so that $AB = DC$ and $AD = BC$. Although the joints are such that $\overline{AB}$ can be moved with respect to $\overline{DC}$, $\overline{AB}$ is always parallel to $\overline{DC}$. State the theorem that tells you this is so.

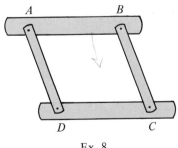

Ex. 8

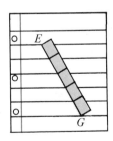

Ex. 9

9. An easy way to divide a segment into a number of equal parts without a ruler is shown in the diagram above. A cardboard strip with edge $\overline{EG}$ is placed on a sheet of lined paper so that $\overline{EG}$ is separated into five parts. State the theorem that tells you the parts are equal.

Written Exercises

In each of Exercises 1–5, state the definition or theorem that allows you to conclude that quadrilateral $WXYZ$ is a parallelogram.

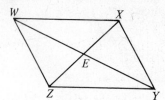

A 1. $\overline{WZ} \parallel \overline{XY}$; $\overline{WX} \parallel \overline{ZY}$

2. $\overline{WX} \parallel \overline{ZY}$; $WX = ZY$

3. $ZE = XE$; $WE = YE$

4. $WX = ZY$; $WZ = XY$

5. $\angle ZWX = \angle ZYX$; $\angle WZY = \angle WXY$

6. Copy everything shown. Complete the proof of Theorem 4-8.

Given: $h \parallel j \parallel k$; trans. $\overleftrightarrow{AC}$ and $\overleftrightarrow{XZ}$;
$\quad\quad AB = BC$

Prove: $XY = YZ$

Proof:

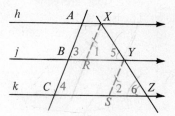

Statements	Reasons
1. Through X and Y draw lines parallel to $\overleftrightarrow{AC}$.	1. Through a point outside a line, __?__.
2. $h \parallel j \parallel k$	2. __?__
3. Quad. $AXRB$ is a $\square$; Quad. $BYSC$ is a $\square$.	3. Def. of a $\square$
4. $XR = AB$; $YS = BC$	4. __?__
5. $AB = BC$	5. __?__
6. $XR = YS$	6. __?__
7. $\angle 1 = \angle 3$	7. __?__
8. $\angle 3 = \angle 4$	8. __?__
9. $\angle 4 = \angle 2$	9. __?__
10. $\angle 1 = \angle 2$	10. __?__
11. $\angle 5 = \angle 6$	11. __?__
12. $\triangle XYR \cong \triangle YZS$	12. __?__ (See Steps 6, 10, 11)
13. $XY = YZ$	13. __?__

Lines AR, BS, CT, and DV are parallel; $AB = BC = CD$. Complete.

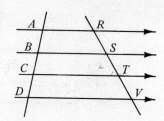

7. If $RS = 5$, $RT = $ __?__ .

8. If $RT = 12$, $ST = $ __?__ .

9. If $ST = 7x + 2$ and $TV = 4x + 20$, $x = $ __?__ .

10. If $RS = 3x$ and $RT = 5x + 11$, $x = $ __?__ .

Tell which of the five methods listed on page 128 you would use to prove each of
Exercises 11–14. (No proofs are required.)

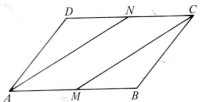

B **11.** Given: $\square ABCD$; M and N are the midpoints of $\overline{AB}$ and $\overline{DC}$.

Prove: $AMCN$ is a $\square$.

12. Given: $\square ABCD$; $\overline{AN}$ and $\overline{CM}$ bisect $\angle A$ and $\angle C$.

Prove: $AMCN$ is a $\square$.

13. Given: $\square ABCD$; W, X, Y, Z are midpoints of $\overline{AO}, \overline{BO}, \overline{CO},$ and $\overline{DO}$.

Prove: $WXYZ$ is a $\square$.

14. Given: $\square ABCD$; $DE = BF$

Prove: $AFCE$ is a $\square$.

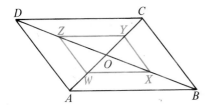

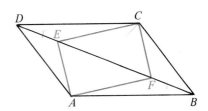

15. The legs of the ironing board are built so that $BR = DR = OR = AR$. For any position of the legs, the top of the board is parallel to the floor. Explain why.

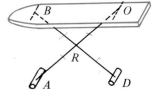

16. List what is given and what is to be proved in Theorem 4-5. (Refer to the diagram at the right.) Then give the key steps of a proof of the theorem.

17. Write a complete proof of Theorem 4-7. Use the diagram on page 128.

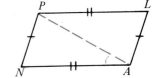

C **18.** Given: $\overline{BC}\|\overline{EF}$; $\angle 1 = \angle 2$;
quad. $ADEB$ is a $\square$;
quad. $ADFC$ is a $\square$.

Prove: Quad. $BEFC$ is a $\square$.

19. Given: $h\|j\|k$; $m\|n$;
transversal l; $AB = BC$

Prove: $CE = BF$

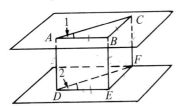

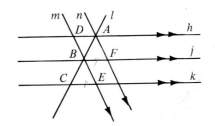

4-3 *Special Parallelograms*

A quadrilateral with four right angles is a **rectangle.** Since a rectangle has both pairs of opposite angles equal, a rectangle is a parallelogram.

A quadrilateral with four equal sides is a **rhombus.** Since a rhombus has both pairs of opposite sides equal, a rhombus is a parallelogram.

A quadrilateral with four equal sides and four right angles is a **square.** Notice that a square is a special kind of rectangle. It is also a special kind of rhombus.

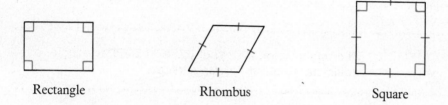

Rectangle Rhombus Square

Since rectangles, rhombuses, and squares are parallelograms, you may use the properties of parallelograms as reasons in proofs involving these figures. The figures also have some special properties stated in the following theorems.

Theorem 4-9 The diagonals of a rectangle are equal.

The proof is left as Exercise 14.

Theorem 4-10 The diagonals of a rhombus are perpendicular.

The proof is left as a Classroom Exercise.

Theorem 4-11 Each diagonal of a rhombus bisects a pair of opposite angles.

The proof is left as Exercise 15.

Theorem 4-10 tells us that in rhombus $QRST$, $\overline{QS} \perp \overline{RT}$.

Theorem 4-11 tells us that $\angle 1 = \angle 2$, $\angle 3 = \angle 4$, $\angle 5 = \angle 6$, and $\angle 7 = \angle 8$.

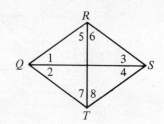

The properties of rectangles lead to an interesting conclusion about any right triangle *RST*.

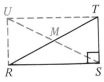

1. Building onto rt. $\triangle RST$, draw rectangle *RSTU*. (How?)
2. Draw $\overline{SU}$. Then $RT = SU$. (Why?)
3. $\overline{RT}$ and $\overline{SU}$ bisect each other. (Why?)
4. $MR = MS = MT = MU$.

The fact that $MR = MS = MT$ can be stated as follows.

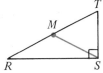

Theorem 4-12   The midpoint of the hypotenuse of a right triangle is equidistant from the three vertices.

Classroom Exercises

1. Quadrilateral *SEAL* is a parallelogram known to have a right angle at *S*. Explain why the quadrilateral must be a rectangle.

2. Quadrilateral *WXYZ* is known to be a parallelogram with two consecutive sides equal. Explain why it must be a rhombus.

3. Draw a quadrilateral which is clearly *not* a rectangle but which does have equal diagonals.

4. Draw a quadrilateral which is *not* a rhombus but which has a diagonal that bisects two angles of the quadrilateral.

5. Is it correct to refer to the figure shown as a polygon? a quadrilateral? a parallelogram? a rhombus? a rectangle? a square? Which term describes the figure best?

$\overline{DE}$ is the hypotenuse of rt. $\triangle DEF$. *M* is the midpoint of $\overline{DE}$. Find the lengths.

6. $MD = \underline{\ ?\ }$

7. $MF = \underline{\ ?\ }$

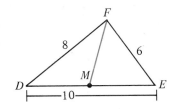

8. Given: Rhombus $QRST$ with diagonals $\overline{QS}$ and $\overline{RT}$

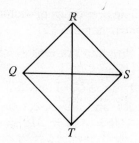

a. Since R is equidistant from Q and S, R is on the _?_ of $\overline{QS}$.

b. Since T is equidistant from Q and S, T is on the _?_ of $\overline{QS}$.

c. From (a) and (b) you can conclude that the perpendicular bisector of $\overline{QS}$ is _?_ .

d. State the theorem of this section which you have proved by (a)–(c).

Written Exercises

Copy and complete the chart by placing check marks in the appropriate spaces.

	Property	Parallelogram	Rectangle	Rhombus	Square
A 1.	Opposite sides are parallel.				
2.	Opposite sides are equal.				
3.	Opposite ∠ are equal.				
4.	Diagonal forms two ≅ △.				
5.	Diagonals bisect each other.				
6.	Diagonals are =.				
7.	Diagonals are ⊥.				
8.	Diagonals bisect opposite angles.				
9.	All ∠ are rt. ∠.				
10.	All sides are =.				

11. Explain why every equiangular quadrilateral must be a rectangle.

12. Explain why every regular quadrilateral must be a square.

13. Draw any $\triangle ABC$. Draw altitude $\overline{AT}$. Let M be the midpoint of $\overline{AC}$ and draw $\overline{MT}$. What kind of triangle is $\triangle MTC$? Why?

14. Copy everything shown and complete the proof of Theorem 4-9.

Given: Rect. *ABCD* with diagonals $\overline{AC}$ and $\overline{BD}$

Prove: $AC = BD$

Proof:

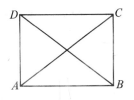

Statements	Reasons
1. *ABCD* is a rectangle with diags. $\overline{AC}$ and $\overline{BD}$.	1. ___?___
2. $\angle ABC = \angle BAD = 90°$	2. Def. of a rectangle
3. $AD = \underline{?}$	3. Opp. sides of a $\square$ are $=$.
4. $AB = AB$	4. ___?___
5. $\triangle\,\underline{?}\, \cong \triangle\,\underline{?}$	5. SAS Postulate
6. $AC = BD$	6. ___?___

B 15. Prove Theorem 4-11. The beginning is shown below.

Given: Rhombus *QRST* with diagonal $\overline{QS}$

Prove: $\angle 1 = \angle 2;\ \angle 3 = \angle 4$

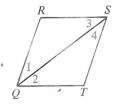

16. Given: $\square ABCD;\ \angle 1 = \angle 2$

Prove: *ABCD* is a rhombus.

17. Given: $\square TENG;\ TN = EG$

Prove: *TENG* is a rectangle.

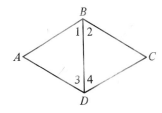

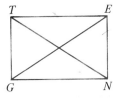

18. Exercises 16 and 17 deal with statements that could be classified as theorems. Write the two statements.

M is the midpoint of hypotenuse $\overline{AB}$ of rt. $\triangle ABC$.

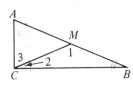

19. If $\angle A = 60°,\ \angle 1 = \underline{\ ?\ }°$.

20. If $\angle B = 28°,\ \angle 3 = \underline{\ ?\ }°$.

21. Given: $\triangle DEF$; altitudes $\overline{DG}$ and $\overline{EH}$; *M* is the midpoint of $\overline{DE}$.

Prove: $\triangle MGH$ is isosceles.

C 22. Draw a regular pentagon *ABCDE*. Draw diagonals $\overline{BD}$ and $\overline{CE}$ intersecting at *X*. What kind of special quadrilateral is *ABXE*? Write the key steps of a proof.

23. Given: $\angle ACB = 90°$;

$\overline{CD}$ is an altitude of $\triangle ACB$;

$\overline{CE}$ is the bisector of $\angle ACB$;

$\overline{CF}$ is a median of $\triangle ACB$.

Prove: $\angle 2 = \angle 3$

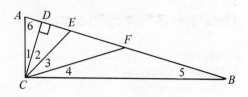

24. Given: Quad. *WXYZ* is a square;

$XA = YE = ZI = WO = a$;

$AY = EZ = IW = OX = b$

Prove: Quad. *AEIO* is a square.

25. Given: $\square ABCD$; $\square BEDF$

Prove: *AECF* is a $\square$.

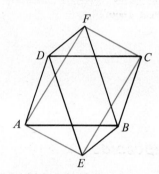

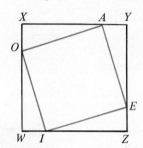

SELF-TEST

The diagonals of $\square JUMP$ intersect at X. State whether each statement is *always*, *sometimes*, or *never* true.

1. $\overline{JU} \parallel \overline{PM}$

2. $UM = MP$

3. $JX = MX$

4. $\angle JUM = \angle MPJ$

5. If $\angle JUM = 70°$, then $\angle UMP = 100°$.

6. If a perpendicular from M intersects $\overleftrightarrow{JU}$ in N, and a perpendicular from P intersects $\overleftrightarrow{JU}$ in Q, then $MN = PQ$.

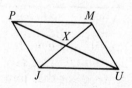

Can you conclude from the information given that quadrilateral *RSTQ* is a parallelogram?

7. $RS = TQ$; $\overline{RS} \parallel \overline{TQ}$

8. $\overline{RS} \parallel \overline{TQ}$; $QR = TS$

9. Diagonal $\overline{RT}$ bisects diagonal $\overline{SQ}$.

10. $\overline{RS} \parallel \overline{TQ}$; $\angle QRS = 100°$; $\angle RST = 80°$

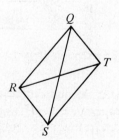

Write a proof in two-column form.

11. Given: Rectangle *ABCD* with diagonals $\overline{AC}$ and $\overline{BD}$.

Prove: $\angle 1 = \angle 2$

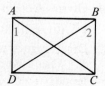

Other Polygons

Objectives
1. Define a trapezoid.
2. State and apply the theorem about the median of a trapezoid.
3. State and apply the theorem about the segment that joins the midpoints of two sides of a triangle.
4. State and apply the concurrency theorems.
5. Write indirect proofs.

4-4 *Trapezoids*

Each quadrilateral below has just one pair of parallel sides.

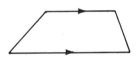 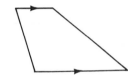

A quadrilateral with exactly one pair of parallel sides is a **trapezoid.**
The parallel sides are **bases,** and the other two sides are **legs.**
The segment that joins the midpoints of the legs is the **median.**

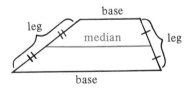 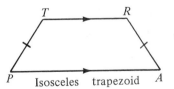

An **isosceles trapezoid** is a trapezoid with equal legs. In isosceles trapezoid *TRAP* above, you may guess that $\angle P = \angle A$ and $\angle T = \angle R$. In Exercises 15 and 16 you will prove this property:

Theorem 4-13   Base angles of an isosceles trapezoid are equal.

The next theorem could be proved now but a shorter, more appealing proof will come in Chapter 11.

Theorem 4-14　The median of a trapezoid has two properties:
(1) It is parallel to the bases.
(2) Its length equals half the sum of the lengths of the bases.

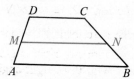

Given: Trapezoid $ABCD$; median $\overline{MN}$

Prove: (1) $\overline{MN} \| \overline{AB}$; $\overline{MN} \| \overline{DC}$
(2) $MN = \frac{1}{2}(AB + DC)$

In the diagrams below, $\overline{MN}$ joins the midpoints of two sides. Study the figures from left to right and you notice that the upper base of trapezoid $ABCD$ becomes shorter and shorter. Finally, in figure (d), $\overline{DC}$ has shrunk to a single point and trapezoid $ABCD$ has become $\triangle ABC$.

(a)

(b)

(c)

(d)

If we think of figure (d) as a trapezoid with $DC = 0$, then Theorem 4-14 suggests that $\triangle ABC$ has these properties:

(1) $\overline{MN} \| \overline{AB}$
(2) $MN = \frac{1}{2}(AB + 0) = \frac{1}{2}AB$

These properties, stated in Theorem 4-15, will be proved in Chapter 11.

Theorem 4-15　The segment that joins the midpoints of two sides of a triangle has two properties:
(1) It is parallel to the third side.
(2) Its length equals half the length of the third side.

Classroom Exercises

1. In the isosceles trapezoid, $\angle A = 70°$. Find the measures of the other angles.

2. In the isosceles trapezoid, $\angle A = 5k°$. Find the measures of the other angles in terms of k.

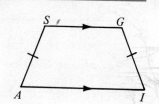

$\overline{MN}$ is the median of trapezoid *GHJK*. Find the lengths indicated.

3. If $HM = 6$ and $JN = 7$, $MG = \underline{\ ?\ }$ and $JK = \underline{\ ?\ }$.

4. If $HJ = 5$ and $GK = 11$, $MN = \underline{\ ?\ }$.

5. If $HJ = 16 - y$ and $GK = 16 + y$, $MN = \underline{\ ?\ }$.

6. If $HJ = 4z$ and $GK = 7z$, $MN = \underline{\ ?\ }$.

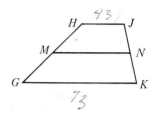

Draw the figure described. If it cannot be drawn, state why not.

7. A trapezoid with two right angles

8. A trapezoid with three right angles

9. A non-isosceles trapezoid with two equal sides

10. A trapezoid with three equal sides

X, *Y*, and *Z* are the midpoints of the sides of $\triangle RST$. Find the lengths indicated.

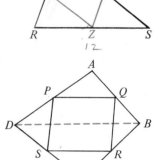

11. $RS = 12$; $XY = \underline{\ ?\ }$ **12.** $ST = 10$; $YZ = \underline{\ ?\ }$

13. $RT = 8$; $XZ = \underline{\ ?\ }$ **14.** $YX = j$, $RS = \underline{\ ?\ }$

15. Suppose $RS = 12$, $ST = 10$, and $RT = 8$. What are the lengths of the sides of $\triangle XYZ$? of $\triangle XYT$? of $\triangle RZY$? of $\triangle ZSX$? How are the four triangles related to each other?

16. State a theorem suggested by Exercise 15.

17. *P*, *Q*, *R*, and *S* are the midpoints of the sides of quad. *ABCD*.

   a. Explain how diagonal $\overline{BD}$ helps you prove that $PQ = SR$.

   b. How could you prove that $PS = QR$?

   c. State the theorem that tells you quad. *PQRS* is a parallelogram.

Written Exercises

$\overline{MN}$ is the median of trapezoid *ABCD*. Complete the table.

	AB	*CD*	*MN*
A 1.	14	8	?
2.	12.1	6.5	?
3.	?	13	18
4.	27.3	?	20.1
5.	?	$7\frac{1}{5}$	$9\frac{3}{10}$

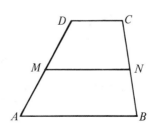

M and N are the midpoints of $\overline{FG}$ and $\overline{FE}$. Find the measures.

6. If $\angle G = 40°$ and $\angle E = 60°$, $\angle FNM = \underline{\;?\;}°$.

7. If $GE = 12.2$, $MN = \underline{\;?\;}$.

8. If $MN = 4\frac{2}{3}$, $GE = \underline{\;?\;}$.

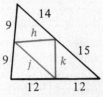

For each figure, one of the values h, j, or k is easy to find. Find it.

9.

10.

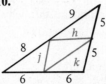

11.

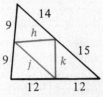

In Exercises 12–14, a trapezoid and its median are shown. Find the value of x.

12.

13.

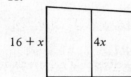

14.

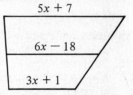

B 15. Copy everything shown and complete the proof of Theorem 4-13.

Given: Trapezoid $TRAP$; $PT = AR$

Prove: $\angle P = \angle A$; $\angle PTR = \angle ART$

Proof:

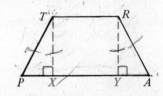

Statements	Reasons
1. Draw $\overline{TX} \perp \overline{PA}$ and $\overline{RY} \perp \overline{PA}$.	1. Through a point outside a line, $\underline{\;?\;}$.
2. $TRAP$ is a trapezoid.	2. $\underline{\;?\;}$
3. $\overline{TR} \parallel \overline{PA}$	3. $\underline{\;?\;}$
4. $TX = RY$	4. If two lines are $\parallel$, then all points on each line are equidistant from the other line.
5. $PT = AR$	5. $\underline{\;?\;}$
6. $\triangle PTX \cong \triangle ARY$	6. $\underline{\;?\;}$
7. $\angle P = \angle A$	7. $\underline{\;?\;}$
8. $\angle PTR$ is supp. to $\angle P$; $\angle ART$ is supp. to $\angle A$.	8. $\underline{\;?\;}$
9. $\angle PTR = \angle ART$	9. $\underline{\;?\;}$

16. Write a proof of Theorem 4-13 which is different from the one in Exercise 15. Use a method suggested by the diagram below.

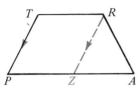
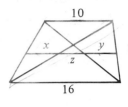

Ex. 16   Exs. 17–18   Ex. 19

J and *K* are the midpoints of $\overline{RT}$ and $\overline{ST}$.

17. If $JK = 3x$ and $RS = x^2$, then $x = \underline{\ ?\ }$.

18. If $JK = x - 1$ and $RS = x^2 - 3x - 2$, then $x = \underline{\ ?\ }$.

19. A trapezoid and its median are shown at the right above. Find the values of *x*, *y*, and *z*.

20. Join the midpoints of the sides of a rectangle. What special figure do you get? Write the key steps of a proof.

21. Join the midpoints of the sides of a rhombus. What special figure do you get? Write the key steps of a proof.

22. Given: $\angle C = 90°$; *T* is the midpoint of $\overline{CA}$; *S* is the midpoint of $\overline{BC}$; *P*, *Q*, and *R* divide $\overline{AB}$ into four equal segments.

Find: Perimeter of quad. *PRST* if $AB = 36$.

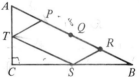

C 23. The solid shown is cut by a plane, not shown, that is parallel to the plane of $\triangle ABC$. The plane intersects $\overline{JA}$, $\overline{JB}$, and $\overline{JC}$ in points *X*, *Y*, and *Z*. If $AB = 12$, $BC = 8$, and $AC = 10$, find the perimeter of $\triangle XYZ$.

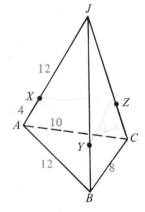

24. What can you prove about the diagonals of an isosceles trapezoid? Write a complete proof of the theorem you have discovered.

25. State and prove the converse of the theorem you discovered in Exercise 24. (*Hint:* Draw auxiliary lines as in Exercise 15.)

26. Join the midpoints of consecutive sides of an isosceles trapezoid. What special figure do you get? Prove it. (*Hint:* Draw the diagonals of the trapezoid.)

4-5 *Concurrent Lines*

When two or more lines intersect in one point, the lines are said to be **concurrent.** Below, you see two triangles and the bisectors of their angles. In both cases the bisectors are concurrent.

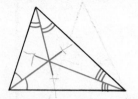
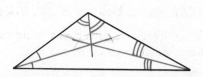

Theorem 4-16 The bisectors of the angles of a triangle intersect in a point which is equidistant from the three sides of the triangle.

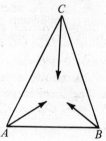

Given: $\triangle ABC$; the bisectors of $\angle A$, $\angle B$, and $\angle C$

Prove: The angle bisectors intersect in a point; that point is equidistant from $\overline{AB}$, $\overline{BC}$, and $\overline{AC}$.

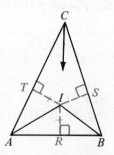

Key steps of proof:

1. Use I to label the point where the bisectors of $\angle A$ and $\angle B$ intersect.
2. Draw perpendiculars $\overline{IR}$, $\overline{IS}$, and $\overline{IT}$.
3. $IT = IR$; $IR = IS$ (Theorem 3-5)
4. $IT = IS$
5. I is on the bisector of $\angle C$. (Theorem 3-6)
6. The angle bisectors intersect in point I.
7. $IR = IS = IT$

Below, you see two triangles and the perpendicular bisectors of their sides. In both cases the perpendicular bisectors are concurrent.

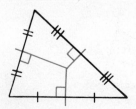
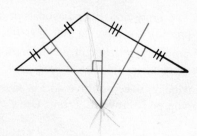

Theorem 4-17 The perpendicular bisectors of the sides of a triangle intersect in a point which is equidistant from the three vertices of the triangle.

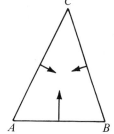

Given: $\triangle ABC$; the $\perp$ bisectors of $\overline{AB}$, $\overline{BC}$, and $\overline{AC}$

Prove: The $\perp$ bisectors intersect in a point; that point is equidistant from A, B, and C.

Key steps of proof:

1. Use O to label the point where the $\perp$ bisectors of $\overline{AC}$ and $\overline{BC}$ intersect.
2. Draw $\overline{OA}$, $\overline{OB}$, and $\overline{OC}$.
3. $OA = OC$; $OC = OB$ (Theorem 3-7)
4. $OA = OB$
5. O is on the $\perp$ bisector of $\overline{AB}$. (Theorem 3-8)
6. The $\perp$ bisectors intersect in point O.
7. $OA = OB = OC$

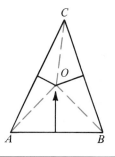

The following theorems will be proved in Chapter 11.

Theorem 4-18 The lines that contain the altitudes of a triangle intersect in a point.

Theorem 4-19 The medians of a triangle intersect in a point that is two thirds of the distance from each vertex to the midpoint of the opposite side.

According to Theorem 4-19, if $\overline{AM}$, $\overline{BN}$, and $\overline{CO}$ are medians of $\triangle ABC$, then

$$AX = \frac{2}{3}AM,$$

$$BX = \frac{2}{3}BN,$$

$$CX = \frac{2}{3}CO.$$

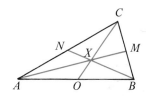

Classroom Exercises

1. Draw, if possible, a triangle in which the perpendicular bisectors of the sides intersect in the point described.
   a. A point inside the triangle   b. A point outside the triangle
   c. A point on the triangle

2. Repeat Exercise 1, but work with angle bisectors.

3. Is there some kind of triangle such that the perpendicular bisector of each side is also an angle bisector, a median, and an altitude?

4. $\triangle JAM$ is a right triangle.
   a. Is $\overline{JM}$ an altitude of $\triangle JAM$?
   b. Name another altitude shown.
   c. In what point do the three altitudes of $\triangle JAM$ meet?
   d. Where do the perpendicular bisectors of the sides of $\triangle JAM$ meet?
   e. Does your answer to (d) agree with Theorem 4-17?

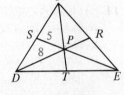

5. The medians of $\triangle DEF$ are shown. Find the lengths indicated.
   a. $EP = \underline{\ ?\ }$ b. $PR = \underline{\ ?\ }$
   c. If $FT = 9$, then $PT = \underline{\ ?\ }$ and $FP = \underline{\ ?\ }$.

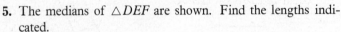

6. Given: $\overline{RJ}$ and $\overline{SK}$ are medians of $\triangle RST$; X and Y are the midpoints of $\overline{RG}$ and $\overline{SG}$.
   a. How are $\overline{XY}$ and $\overline{RS}$ related? Why?
   b. How are $\overline{KJ}$ and $\overline{RS}$ related? Why?
   c. What special kind of quadrilateral is $XYJK$? Why?
   d. Why does $XG = GJ$?
   e. Prove informally that $RG = \frac{2}{3}RJ$.

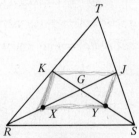

Written Exercises

Plot points A, B, and C on graph paper. Draw $\triangle ABC$. Draw the perpendicular bisectors of $\overline{AB}$, $\overline{BC}$, and $\overline{AC}$. Find the coordinates of the point where the perpendicular bisectors meet.

A   1. $A(0, 6)$ $B(0, 0)$ $C(8, 0)$ 2. $A(9, 12)$ $B(-3, 12)$ $C(3, -6)$
     3. $A(-4, -4)$ $B(8, 0)$ $C(2, 8)$ 4. $A(4, 1)$ $B(-2, 1)$ $C(0, 3)$

5. Draw a triangle such that the lines containing the three altitudes intersect in the point described.
   a. A point inside the triangle
   b. A point outside the triangle
   c. A point on the triangle

In Exercises 6–9, refer to the diagram.

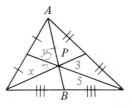

6. Find the values of x and y.

7. If $AB = 6$, then $BP = \underline{}$ and $AP = \underline{}$.

8. If $AB = 7$, then $BP = \underline{}$ and $AP = \underline{}$.

9. If $PB = 1.9$, then $AP = \underline{}$ and $AB = \underline{}$.

B 10. Use a ruler and a protractor to draw a pentagon with $108°$ angles and 5 cm sides. Draw the perpendicular bisectors of the five sides.

11. Repeat Exercise 10, but draw the five angle bisectors.

12. Three towns, located approximately as shown, plan to build one recreation center to serve all three towns. They decide that the fair thing to do is to build the hall equidistant from the three towns. Comment about the wisdom of the plan.

13. See Exercise 12. Draw three towns so located that it isn't possible to find a spot equidistant from the three towns.

14. In the figure, $\overline{AD}$ and $\overline{BE}$ are equal medians of $\triangle ABC$.
  a. Explain why $GD = GE$.
  b. $GA = \underline{}$
  c. Name three angles equal to $\angle GAB$.

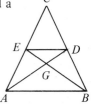

$\overline{AU}$, $\overline{BV}$, and $\overline{CW}$ are the medians of $\triangle ABC$.

15. If $AP = x^2$ and $PU = 2x$, then $x = \underline{}$.

16. If $BP = y^2 + 1$ and $PV = y + 2$, then $y = \underline{}$ or $\underline{}$.

C 17. If $CW = 2z^2 - 5z - 12$ and $CP = z^2 - 15$, then $z = \underline{}$ and $PW = \underline{}$.

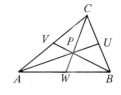

18. In the plane figure, point P is equidistant from R, S, and T. Describe the location of the following points in the plane.
  a. Points farther from both R and S than from T
  b. Points closer to both R and S than to T

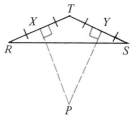

19. Prove: If two of the medians of a triangle are equal, then the triangle is isosceles.

Application ───────────────

Center of Gravity

The *center of gravity* of an object is a point where the weight of the object seems to be focused. If we lift or support an object, we do this most easily under its center of gravity. A seesaw on a playground has its center of gravity in the middle and will balance by itself when supported in the middle.

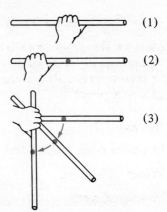

If an object is not supported under its center of gravity, it becomes unstable. Suppose you hold a heavy bar in one hand. If you support it near the center of gravity, it will be easy to hold (Figure 1). To support it at one end requires more effort (Figure 2), since the pole tends to turn until the center of gravity is directly below the point of support (Figure 3).

The center of gravity may be inside an object or outside of it. The center of gravity of an ice cube is in the middle of the ice, but the center of gravity of an automobile tire is not in a part of the tire itself.

Exercises

1. For this experiment, cut out a large, irregularly shaped piece of cardboard.

   a. Near the edge, poke a hole just large enough to allow the cardboard to rotate freely when pinned through the hole.

   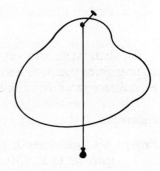

   b. Pin the cardboard through the hole to a suitable wall surface. The piece will position itself so that its center of gravity is as low as possible. This means that it will lie on a vertical line through the point of suspension. To find this line, tie a weighted string to the pin. Then draw on the cardboard the line determined by the string.

   c. Repeat (a) and (b) but use a different hole. The center of gravity of the cardboard ought to lie on both of the lines you have drawn and should therefore be their point of intersection. The cardboard should balance if supported at this point.

2. Cut out a piece of cardboard in the shape of a large scalene triangle.

   a. Follow the steps of Exercise 1 using three holes, one near each of the three vertices.

   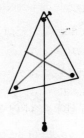

   b. If you worked carefully, all three lines drawn intersect in one point, the center of gravity of the cardboard. This point is also referred to as the *center of mass* or the *centroid* of the cardboard. Study the lines you have drawn and explain why in geometry the point of intersection of the medians of a triangle is called the *centroid of the triangle*.

3. Do you think that the center of gravity of a parallelogram is the point where the diagonals cross? Use the technique of Exercise 1 to test this idea.

4-6 *Indirect Proofs*

Most of the proofs you have seen or written have been direct proofs. Sometimes it is difficult or even impossible to find a direct proof, but easy to reason indirectly. Here is an example of the reasoning used in an **indirect proof.**

Example 1:

Given: x is an integer and x^2 is even.

Prove: x is even.

Proof (indirect):

Suppose x is *not* even. Then x is odd.
Thus, $x^2 = x \times x = \text{odd} \times \text{odd} = \text{odd}$.
But this contradicts the given fact that x^2 is even.
What we supposed, that x is *not* even, must be false.
It follows that x is even.

We shall write indirect proofs in *paragraph form* instead of the two-column form you have been using for direct proofs. Here is an example of an indirect proof in geometry.

Example 2:

Prove: If a line passes through the midpoint of one side of a triangle and is parallel to another side, then the line passes through the midpoint of the third side.

Given: $\triangle ABC$; M is the midpoint of $\overline{AC}$; $\overleftrightarrow{MX} \parallel \overline{AB}$

Prove: X is the midpoint of $\overline{BC}$.

Proof (indirect):

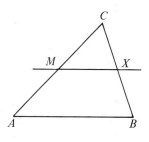

Suppose X is *not* the midpoint of $\overline{BC}$.
$\overline{BC}$ has a midpoint; call it N.
When we draw $\overleftrightarrow{MN}$, we have $\overleftrightarrow{MN} \parallel \overline{AB}$ (Theorem 4-15).
But this contradicts the given fact $\overleftrightarrow{MX} \parallel \overline{AB}$. (Through a point outside a line exactly *one* parallel can be drawn to the line.)
What we supposed to be true, that X is *not* the midpoint of $\overline{BC}$, must be false.
It follows that X *is* the midpoint of $\overline{BC}$.

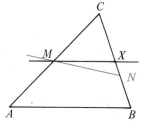

How to Write an Indirect Proof

1. Suppose that what you wish to prove true is not true.
2. Reason logically until you reach a contradiction of a known fact.
3. Conclude that what you supposed to be true is false and that what you wish to prove true is true.

Classroom Exercises ────────────────────────

1. An indirect proof is to be used to prove: If $a \perp b$, then $a \| d$.
   Which is the correct way to begin an indirect proof?
   Suppose a is not perpendicular to b.
   Suppose a is not parallel to d.

In Exercises 2–7 what is the first sentence of an indirect proof of each statement?

2. Today is a national holiday.

3. I did not rob the bank.

4. $\angle E$ is a right angle.

5. $\triangle ABC$ is equilateral.

6. M is the midpoint of $\overline{DE}$.

7. l, m, and k are not concurrent.

8. Leslie wishes to prove that $m \| k$. She can show it is false that $m \perp k$. Can she conclude that $m \| k$?

9. Brian wishes to prove that $w > z$. He has shown that the statement $w \leq z$ is false. Can he conclude that $w > z$?

10. Arrange sentences (a)–(e) in order so that they complete an indirect proof of the statement: In a plane, two lines parallel to a third line are parallel to each other.

    Given: Coplanar lines j, k, m; $j \| m$; $k \| m$

    Prove: $j \| k$

    Proof:

    (a) Therefore, it is false that j is not parallel to k, and it is true that $j \| k$.
    (b) Then j intersects k in some point X.
    (c) Suppose j is not parallel to k.
    (d) But this contradicts Theorem 2-8.
    (e) This means there are *two* lines, j and k, through X and parallel to m.

Challenge ──────────────────────────────

A baker, a carpenter, and a plumber are named Antonio Baker, Tricia Carpenter, and Alma Plumber, although no person's name matches his or her occupation. If Antonio drives a pick-up truck and the carpenter rides a bike, who drives a sports car and what is each person's occupation?

Written Exercises

Write indirect proofs in paragraph form.

A **1.** Given: Transversal t cuts lines g and h;
$\angle 1 \neq \angle 2$
Prove: g is not parallel to h.

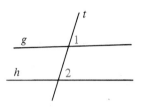

2. Given: Quad. $ABCD$
Prove: $\angle A$, $\angle B$, and $\angle C$ are not all 120° angles.

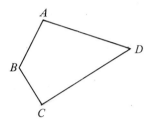

B **3.** Given: Scalene $\triangle REN$
Prove: $\angle R \neq \angle N$

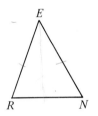

4. Given: $AC = RT$; $AB = RS$;
$\angle A \neq \angle R$
Prove: $BC \neq ST$

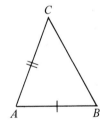 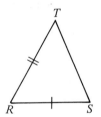

5. Given: $\overline{OA}$ is a median of $\triangle ABD$;
$\overline{OB}$ is not a median of $\triangle ABC$.
Prove: Quad. $ABCD$ is not a $\square$.

6. Given: Coplanar lines l, m, n;
n intersects l in P;
$l \parallel m$
Prove: n intersects m.

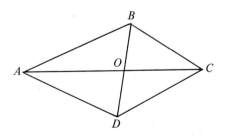 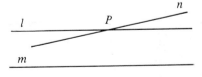

7. Prove that there is no smallest positive number.

8. Prove that a collection of nickels and dimes having the value 145¢ must have an odd number of nickels.

9. Prove that if x and y are integers and $3x + 12y = 450$, then x is even.

C **10.** Prove that the bisectors of two angles of a triangle are never perpendicular.

11. Prove that if a line passes through the midpoint of one leg of a trapezoid and is parallel to a base, then the line passes through the midpoint of the other leg. (*Hint:* See Example 2 on page 147.)

12. Prove that the diagonals of a trapezoid do not bisect each other.

13. Prove that if two lines are perpendicular to the same plane, then the lines do not intersect. (See diagram below.)

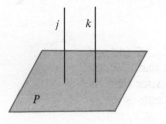

Ex. 13

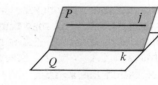

Ex. 14

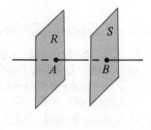

Ex. 15

14. Prove that if a line is parallel to a plane, and a plane containing that line intersects the first plane, then the line of intersection is parallel to the first line. (See diagram above.)

15. Prove that two planes perpendicular to the same line are parallel. (See diagram above.)

SELF-TEST

Exercises 1–3 refer to trapezoid *ABCD*.

1. Is $\overline{AD}$ parallel to $\overline{BC}$?

2. If *ABCD* is an isosceles trapezoid, which two pairs of angles are equal?

3. If $\overline{MN}$ is the median, then $MN = \underline{\ ?\ }$.

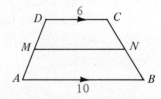

4. Name two properties of the segment that joins the midpoints of two sides of a triangle.

5. The perpendicular bisectors of the sides of a triangle intersect in a point and are said to be concurrent. Name three other kinds of concurrent lines (or rays or segments) associated with triangles.

6. You intend to write an indirect proof. It is given that $\angle 1 = \angle 2$, and you are to prove that line *j* bisects $\overline{XY}$. Which is a satisfactory start for your indirect proof, **a** or **b**?
   a. Suppose $\angle 1 \neq \angle 2$.   **b.** Suppose *j* does not bisect $\overline{XY}$.

7. In an indirect proof, you reason logically until you reach a $\underline{\ ?\ }$ of a known fact.

Inequalities

Objectives
1. **State and apply the inequality relations for one triangle.**
2. **State and apply the inequality relations for two triangles.**

4-7 *Inequalities for One Triangle*

Look at the figure at the left below and you can conclude that $\angle A$ and $\angle B$ are equal. Look at the figure at the right and you may think that $\angle X$ and $\angle Y$ are not equal, or perhaps that the measure of $\angle X$ is greater than the measure of $\angle Y$. We write this as $\angle X > \angle Y$. Statements such as $\angle X \neq \angle Y$ and $\angle X > \angle Y$ are called *inequalities.*

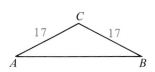
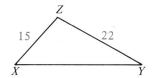

Some of the algebraic properties used in dealing with inequalities are listed below. Remember that any algebraic property can be used in geometry.

If $a > b$ and $c > d$, then $a + c > b + d$.
If $a > b$ and c is a positive number, then $ac > bc$.
If $a > b$ and $b > c$, then $a > c$.
If $a = b + c$ and c is a positive number, then $a > b$.

Here are three situations in geometry where the last property listed above can be used.

(1)

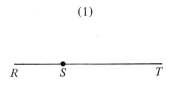

Since $RT = RS + ST$,
$RT > RS$

(2)

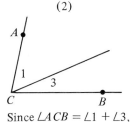

Since $\angle ACB = \angle 1 + \angle 3$,
$\angle ACB > \angle 1$.

(3)

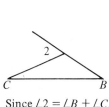

Since $\angle 2 = \angle B + \angle C$,
$\angle 2 > \angle B$.

Situations (2) and (3) occur in key steps in the proof of the following theorem. Look for them.

Theorem 4-20   If two sides of a triangle are unequal, then the angles opposite those sides are unequal in the same order.

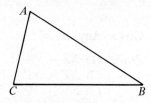

Given: △*ABC*; *AB* > *AC*

Prove: ∠*C* > ∠*B*

Key steps of proof:

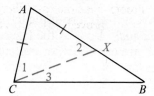

1. On $\overline{AB}$ choose point *X* so that *AX* = *AC*. Draw $\overline{CX}$.
2. ∠*ACB* > ∠1   (Why?)
3. ∠1 = ∠2   (Why?)
4. ∠2 > ∠*B*   (Why?)
5. ∠*ACB* > ∠*B*, or ∠*C* > ∠*B* (From Steps 2-4)

Theorem 4-21   If two angles of a triangle are unequal, then the sides opposite those angles are unequal in the same order.

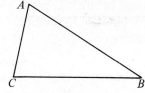

Given: △*ABC*; ∠*C* > ∠*B*

Prove: *AB* > *AC*

Proof (indirect):

Suppose *AB* ≯ *AC*. Then either *AB* = *AC* or *AB* < *AC*.
But if *AB* = *AC*, then ∠*C* = ∠*B*, and
   if *AB* < *AC*, then ∠*C* < ∠*B* by Theorem 4-20.
In either case we have a contradiction of the given fact that ∠*C* > ∠*B*. What we supposed to be true, *AB* ≯ *AC*, must be false. It follows that *AB* > *AC*.

Corollary 1   The perpendicular segment from a point to a line is the shortest segment from the point to the line.

Corollary 2   The perpendicular segment from a point to a plane is the shortest segment from the point to the plane.

See Classroom Exercises 15 and 16 for proofs of the corollaries.

The relationship stated in the following theorem is known as the *triangle inequality*.

Theorem 4-22   The sum of the lengths of any two sides of a triangle is greater than the length of the third side.

Given: Any $\triangle XYZ$

Prove: (1) $XZ + XY > YZ$
      (2) $XZ + YZ > XY$
      (3) $XY + YZ > XZ$

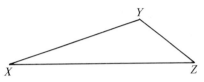

Proof: Take the case where $\overline{XZ}$ is the longest side. Then the first two statements to be proved are true. A proof of part (3) follows.

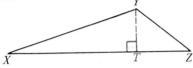

Statements	Reasons
1. Draw $\overline{YT} \perp \overline{XZ}$.	1. Through a point outside ___?___.
2. $XY > XT$; $YZ > TZ$	2. The $\perp$ segment from ___?___.
3. $XY + YZ > XT + TZ$, or	3. If $a > b$ and $c > d$, then
$XY + YZ > XZ$	$a + c > b + d$.

Classroom Exercises

Which is the largest angle and which is the smallest angle of each triangle?

1.

2.

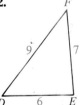

3.

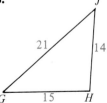

Which is the longest side and which is the shortest side of each triangle?

4.

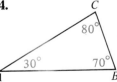

5.

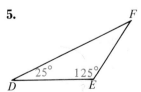

6.

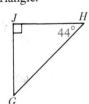

Is it possible to have a triangle whose sides have these lengths?

7. 5, 6, 7   **8.** 4, 5, 10   **9.** 8, 8, 0.1

10. 13, 8, 5   **11.** 3, 3, 8   **12.** 8, 10, 8

13. If the lengths of the two sides of a triangle are 6 and 9, the length of the third side must be greater than _?_, but it must be less than _?_.

14. Each leg of an isosceles triangle has length 10. The length of the base can be any number between _?_ and _?_.

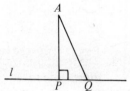

15. $\overline{AP} \perp l$, and $\overline{AQ}$ is any other segment from A to l. Explain how Corollary 1 follows from Theorem 4-21. That is, explain why $AP < AQ$.

16. $\overline{AP} \perp$ plane X, and $\overline{AQ}$ is any other segment from A to X. Draw a diagram and explain how Corollary 2 follows from Theorem 4-21 or from Corollary 1.

Written Exercises

The figures in Exercises 1–6 are not accurately drawn. If the figures were drawn as the measures indicate, which would be the longest side of each?

A **1.**

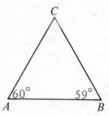

2.

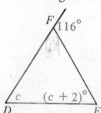

3.

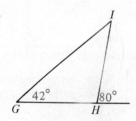

4.

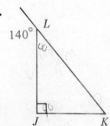

5.

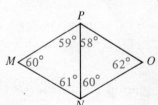

6.

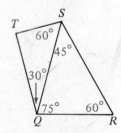

The lengths of two sides of $\triangle ABC$ are given. For each exercise write the numbers needed to complete this statement: CA must be greater than _?_ but less than _?_.

7. $AB = 5$; $BC = 7$ **8.** $AB = 10$; $BC = 7$

9. $AB = 12$; $BC = 12.2$ **10.** $AB = 93$; $BC = 1$

B 11. Given: Quadrilateral *ABCD*
Prove: $AB + BC + CD + DA > 2AC$

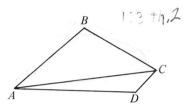

12. Given: $\angle A > \angle C$; $\angle B > \angle D$
Prove: $CD > AB$

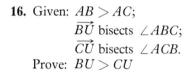

13. The lengths of three sides of a quadrilateral are 10, 4, and 2. The length of the fourth side must be greater than ? but less than ?. (*Hint:* Think of sticks 4 cm and 2 cm long fastened loosely at the ends of a stick 10 cm long.)

14. The lengths of four sides of a pentagon are 7, 8, 9, and 25. The length of the fifth side must be greater than ?, but less than ?.

15. Given: Rect. *DEFG*;
$DE > EF$
Prove: $\angle 2 > \angle 1$

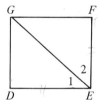

16. Given: $AB > AC$;
$\overrightarrow{BU}$ bisects $\angle ABC$;
$\overrightarrow{CU}$ bisects $\angle ACB$.
Prove: $BU > CU$

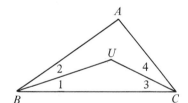

17. Given: *P* is a point on the base $\overline{JK}$ of isosceles $\triangle JKL$.
Prove: $LP < LJ$

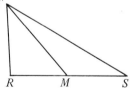

C 18. Given: $\overline{TM}$ is a median of $\triangle RST$.
Prove: $TR + TS > 2TM$
(*Hint:* Draw a segment from *M* to the midpoint of $\overline{TS}$.)

19. Prove that the sum of the lengths of the medians of a triangle is greater than half the perimeter of the triangle.

20. If you replace "medians" with "altitudes" in Exercise 19, can you prove the resulting statement? Explain.

21. Given: *P* lies inside $\triangle ABC$.
Prove: $AB + AC > PB + PC$
(*Hint:* Draw an auxiliary line.)

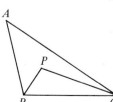

4-8 *Inequalities for Two Triangles*

Suppose you have two identical compasses. Open one of them more than the other so that $\angle A > \angle X$, and you have $BC > YZ$. The theorem suggested by this simple example is not easy to prove.

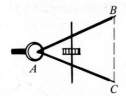 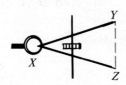

Theorem 4-23 If two sides of one triangle are equal to two sides of another triangle, but the included angle of the first triangle is larger than the included angle of the second, then the third side of the first triangle is longer than the third side of the second. (SAS Inequality)

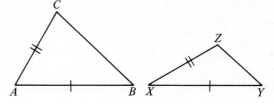

Given: $AB = XY$;
　　　$AC = XZ$;
　　　$\angle A > \angle X$

Prove: $BC > YZ$

Key steps of proof:

First draw $\overrightarrow{AT}$ so that $\angle BAT = \angle X$. On $\overrightarrow{AT}$ choose point K so that $AK = XZ$. There are then two cases to consider in the proof.

Case 1: K is on $\overline{BC}$.

1. $BC > BK$
2. $\triangle ABK \cong \triangle XYZ$ (SAS Postulate)
3. $BK = YZ$ (Why?)
4. $BC > YZ$ (From Steps 1 and 3)

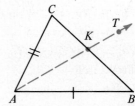

Case 2: K is not on $\overline{BC}$.

1. Draw $\overrightarrow{AP}$, the bisector of $\angle CAK$.
2. Draw $\overline{PK}$ and $\overline{KB}$.
3. $AC = XZ = AK$ (Why?)
4. Since $\triangle CAP \cong \triangle KAP$ (SAS), $PC = PK$.
5. $BP + PK > BK$ (Why?)
6. $BP + PC > BK$ (Why?), or $BC > BK$
7. Since $\triangle KAB \cong \triangle ZXY$ (SAS), $YZ = BK$.
8. $BC > YZ$ (From Steps 6 and 7)

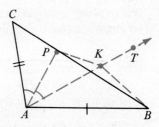

Theorem 4-24 If two sides of one triangle are equal to two sides of another triangle, but the third side of the first triangle is longer than the third side of the second, then the included angle of the first triangle is larger than the included angle of the second. (SSS Inequality)

Given: $AB = XY$;
$\qquad AC = XZ$;
$\qquad BC > YZ$

Prove: $\angle A > \angle X$

Proof (indirect):

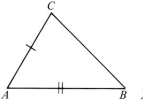 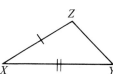

Suppose $\angle A \not> \angle X$. Then either $\angle A = \angle X$ or $\angle X > \angle A$.
But if $\angle A = \angle X$, then $\triangle BAC \cong \triangle YXZ$, and $BC = YZ$;
$\qquad$ if $\angle X > \angle A$, then $YZ > BC$ by Theorem 4-23.
In either case we have a contradiction of the given fact that $BC > YZ$. What we supposed to be true, that $\angle A \not> \angle X$, must be false. It follows that $\angle A > \angle X$.

Classroom Exercises

In each exercise, some facts are given. What can you conclude?

1. $DE = RS$; $DF = RT$; $\angle D > \angle R$
2. $DE = RS$; $DF = RT$; $\angle D = \angle R$
3. $ED = SR$; $EF = ST$; $DF > RT$
4. $FE = TS$; $FD = TR$; $\angle F < \angle T$
5. $DF > DE$
6. $\angle S > \angle T$

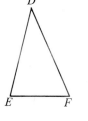 

7.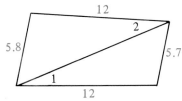

8.

9. State the theorems that were applied in Exercises 7 and 8.

Written Exercises

Choose from the words *always, sometimes,* and *never* to complete the sentence in the best way.

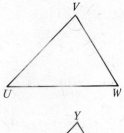

A **1.** If $UV = XY$ and $UW = XZ$, then $\angle U$ is ___ equal to $\angle X$.

2. If $UV = XY$ and $UW = XZ$ and $\angle U > \angle X$, then VW is ___ greater than YZ.

3. If $VU = YX$ and $VW = YZ$ and $\angle V = \angle Y$, then UW is ___ greater than XZ.

4. If $WV = ZY$ and $WU = ZX$ and $UV > XY$, then $\angle W$ is ___ greater than $\angle Z$.

5. If $\angle X > \angle Z$, then YZ is ___ greater than YX.

6. If $\angle U > \angle X$ and $\angle W > \angle Z$, then UW is ___ greater than XZ.

Which is longer, $\overline{AJ}$ or $\overline{AK}$?

7.

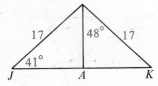

8.

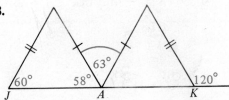

Which is larger, $\angle 1$ or $\angle 2$? (*Note:* The diagrams show figures in space.)

9.

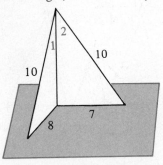

10.

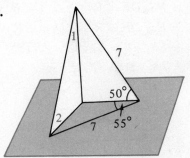

B **11.** Given: O is the midpoint of the hypotenuse of rt. $\triangle ABC$; $\angle BOC = 100°$

Prove: $CB > AB$

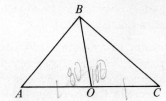

12. Given: Plane *P* bisects $\overline{XZ}$ at *Y*;
 $XV > ZV$
Use the given information to prove
something about the diagram.

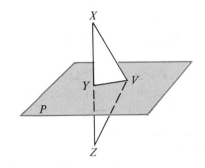

13. Given: $RE = RS = ST$
Prove: $ET > ER$

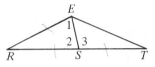

C 14. Given: $AB > AC$; $BD = EC$
Prove: $BE > CD$

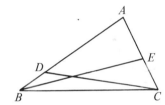

15. Given: $\square RSTQ$; $RT > SQ$
Prove: $\angle RST$ is an obtuse angle.

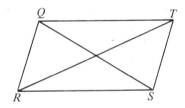

16. Given: The solid shown;
 $VA = VB$;
 $\angle BVC > \angle AVC$
Prove: $\angle BAC > \angle ABC$

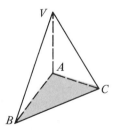

17. Given: The solid shown;
 $GH = GK$;
 $XH = NK$
Prove: $\angle HGN > \angle KGN$

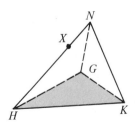

18. Draw a figure, list what is given and what is to be proved, and then write the key steps of a proof for the statement:

If two segments are drawn to a plane from a point in a line that is perpendicular to the plane, and one segment meets the plane at a greater distance from the foot of the perpendicular than the other segment, then the first segment is the longer.

SELF-TEST

Complete the statements.

1. In a $\triangle ABC$, $AC > BC$. Then $\angle\,\underline{\ ?\ } > \angle\,\underline{\ ?\ }$.
2. In a $\triangle TED$, $TE = 10$ and $ED = 19$. Then $TD > \underline{\ ?\ }$.
3. In a $\triangle XYZ$, $XY = 25$ and $YZ = 70$. Then $XZ < \underline{\ ?\ }$.

Complete the statements with the correct symbol ($<, =, >$).

4. If $ST = TW$ and $\angle 3 < \angle 4$, then $RS\,\underline{\ ?\ }\,RW$.
5. If $ST = TW$ and $RW > RS$, then $\angle 4\,\underline{\ ?\ }\,\angle 3$.
6. If $RS = RW$ and $TS = TW$, then $\angle 1\,\underline{\ ?\ }\,\angle 2$.

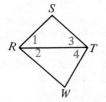

COMPUTER KEY-IN

If you break a stick into three pieces, do you think it is always possible to join the pieces end-to-end to form a triangle?

It's easy to see that if the sum of the lengths of any two of the pieces is less than or equal to that of the third, a triangle can't be formed.

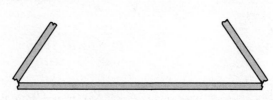

By an experiment, your class can estimate the probability that three pieces of broken stick will form a triangle. Suppose everyone in your class has a stick 1 unit long and breaks it into 3 pieces. If there are 30 people in your class and 8 people are able to form a triangle with their pieces, we estimate that the probability of forming a triangle is about $\dfrac{8}{30}$.

Of course, this experiment is not very practical. You can get much better results by having a computer simulate the breaking of many, many sticks, as in the program in BASIC on the opposite page.

In lines 30 and 40 of the following program, you tell the computer how many sticks you want to break. Each stick is 1 unit long, and the computer breaks each stick by choosing two random numbers x and y between 0 and 1. These numbers divide the stick into 3 lengths, a, b, and c.

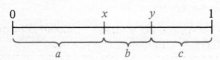

The computer then keeps count of the number of sticks (N) which form a triangle when broken.

Notice that RND is used in lines 70 and 80. Since usage of RND varies, check this with the manual for your computer and make any necessary changes. The computer print-outs shown in this text use capital letters. The *x*, *y*, *a*, *b*, and *c* used in the discussion above appear as X, Y, A, B, C.

```
10   PRINT "SIMULATION--BREAKING STICKS TO MAKE TRIANGLES"
20   PRINT
30   PRINT "HOW MANY STICKS DO YOU WANT TO BREAK";
40   INPUT D
50   LET N=0
60   FOR I=1 TO D
70   LET X=RND(0)
80   LET Y=RND(0)
90   IF X>Y THEN 120
100    LET A=X
110    GOTO 130
120    LET A=Y
130    LET B=ABS(X-Y)
140    LET C=1-A-B
150    IF A+B <= C THEN 210
160    IF B+C <= A THEN 210
170    IF C+A <= B THEN 210
180    PRINT
190    PRINT TAB(5);A;B;C
200    LET N=N+1
210    NEXT I
220    LET P=N/D
230    PRINT
240    PRINT "THE EXPERIMENTAL PROBABILITY THAT"
250    PRINT "A BROKEN STICK CAN MAKE A TRIANGLE IS";P;"."
260    END
```

Line Number	Explanation
60–140	These lines simulate the breaking of each stick. When I = 10, for example, the computer is "breaking" the tenth stick.
150–170	Here the computer tests to see whether the pieces of the broken stick can form a triangle. If not, the computer goes on to the next stick (line 210) and the value of N is not affected.
200	If the broken stick has survived the tests of the previous three steps, then the pieces can form a triangle and the value of N is increased by 1 here.
210	Lines 60–210 form a loop which is repeated D times. When I = D, the probability P is calculated and printed.

Exercises

1. Pick any two numbers x and y between 0 and 1 with $x < y$. With paper and pencil, carry out the instructions in Lines 90 through 140 of the program to see how the computer finds a, b, and c and tests to see whether the values can be the lengths of the sides of a triangle. Do the same for a pair x and y with $x > y$.

2. If your computer uses a language other than BASIC, write a similar program for your computer.

3. Run the program several times for D = 25.

4. Delete the print statements in Lines 180 and 190 and then run the program for large values of D, say 100, 400, 800, and compare your results with those of some classmates. What appears to be the probability that the pieces of a broken stick can form a triangle?

Chapter Summary

1. A parallelogram has these properties:
   a. Opposite sides are parallel.
   b. Opposite sides are equal.
   c. Opposite angles are equal.
   d. Diagonals bisect each other.

2. You can prove that a quadrilateral is a parallelogram if you can show any one of the following:
   a. Both pairs of opposite sides are parallel.
   b. One pair of opposite sides are equal and parallel.
   c. Both pairs of opposite sides are equal.
   d. Both pairs of opposite angles are equal.
   e. The diagonals bisect each other.

3. If three parallel lines cut off equal segments on one transversal, then they cut off equal segments on every transversal.

4. Rectangles, rhombuses, and squares are special parallelograms.

5. The median of a trapezoid is parallel to the bases and its length is half the sum of the lengths of the bases.

6. The segment that joins the midpoints of two sides of a triangle is parallel to the third side and its length is half the length of the third side.

7. a. The bisectors of the angles of a triangle intersect in a point equidistant from the three sides.
   b. The perpendicular bisectors of the sides intersect in a point equidistant from the three vertices.
   c. The lines containing the altitudes intersect in a point.
   d. The medians intersect in a point which is two thirds of the distance from each vertex to the midpoint of the opposite side.

8. You begin an indirect proof by supposing that what you wish to prove true is *not* true. If this supposition leads to a contradiction of a known fact, then what you wish to prove true must be true.

9. In △*ABC*, if *AB* > *AC*, then ∠*C* > ∠*B*. The converse is also true.

10. The sum of any two sides of a triangle is greater than the third side.

11. The SSS and SAS Inequality Theorems provide methods for comparing angles and sides of two triangles.

Chapter Review ———————————

4-1 *Properties of Parallelograms*

In parallelogram *ABCD*, ∠*ABC* = 115°.

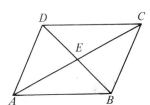

1. If *AC* = 15, then *EC* = _?_ .
2. ∠*DCB* = _?_ °
3. If ∠*ABD* = 42°, then ∠*ADB* = _?_ °.
4. If *AB* = 3*x* + 5 and *CD* = 7*x* − 3, then *x* = _?_ .

4-2 *Ways to Prove that Quadrilaterals are Parallelograms*

In each exercise you could prove that quad. *MATH* is a parallelogram if one more equation were given. State the equation.

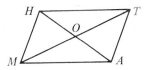

5. *MH* = 3; *AT* = 3; *HT* = 5
6. $\overline{MA} \parallel \overline{HT}$; *MA* = 8
7. *MO* = *OT*
8. ∠*HMA* = ∠*ATH*

4-3 *Special Parallelograms*

Complete with the word *always*, *sometimes*, or *never*.

9. If a quadrilateral is a rhombus, then it is _?_ a square.
10. If a quadrilateral is a square, then it is _?_ a rectangle.
11. If a rectangle is a rhombus, then it is _?_ a square.
12. If a quadrilateral is a rhombus, then it is _?_ a regular polygon.

4-4 *Trapezoids*

Quad. *ABCD* is a trapezoid with median $\overline{EF}$.

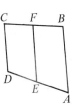

13. Name the bases of the trapezoid.
14. Name the legs of the trapezoid.
15. If *AB* = 23 and *CD* = 16, then *EF* = _?_ .
16. If ∠*DEF* = 67°, then ∠*EDC* = _?_ .

4-5 *Concurrent Lines*

17. The perpendicular bisectors of the sides of a triangle intersect in a point which is equidistant from the three ___?___ of the triangle.
18. The bisectors of the angles of a triangle intersect in a point which is equidistant from the three ___?___ of the triangle.

Exercises 19 and 20 refer to the diagram. The medians of △*ABC* are shown.

19. If $DG = 7$, then $DB = $ _?_ .
20. If $AF = 18$, then $AD = $ _?_ .

4-6 *Indirect Proofs*

Arrange sentences 21–24 to form an indirect proof of the statement: If $x^2 - 3 = 7$, then $x \neq 5$.

21. But this contradicts the fact that $x^2 - 3 = 7$.
22. It follows that $x \neq 5$.
23. Then $x^2 - 3 = 22$.
24. Suppose $x = 5$.

4-7 *Inequalities for One Triangle*

25. In △*ABC*, if $\angle A > \angle B$, then $BC > $ _?_ .
26. In △*RST*, if $RS > ST$, then $\angle$_?_ $> \angle$_?_ .
27. Two sides of a triangle have lengths 4 and 9. The length of the third side must be greater than _?_ .
28. Two sides of a triangle have lengths of 8 and 11. The length of the third side must be less than _?_ .

4-8 *Inequalities for Two Triangles*

Complete with the symbol $<$, $=$, or $>$.

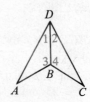

29. If $AB = CB$ and $\angle 3 > \angle 4$, then CD _?_ AD.
30. If $AD = CD$ and $\angle 1 = \angle 2$, then AB _?_ BC.
31. If $AB = CB$ and $AD > CD$, then $\angle 3$ _?_ $\angle 4$.
32. If $AD = CD$ and $\angle 1 > \angle 2$, then AB _?_ CB.

Chapter Test

Classify each statement as true or false

1. All squares are rectangles.
2. All rhombuses are squares.
3. All squares are rhombuses.
4. Some rhombuses are rectangles.
5. Some parallelograms are rhombuses.
6. Some trapezoids are parallelograms.
7. It is possible to form a triangle having sides 15 cm, 19 cm, and 35 cm.
8. A diagonal of a rhombus bisects a pair of opposite angles.
9. The diagonals of a rhombus bisect each other.
10. In triangle ABC, if $BC > AB$, then $\angle C > \angle A$.

Exercises 11–13 refer to trapezoid $ABCD$ with median $\overline{EF}$.

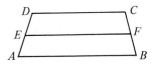

11. If $AB = 21$ and $CD = 13$, then $EF = \underline{?}$.
12. If $AB = 16$ and $EF = 12$, then $CD = \underline{?}$.
13. If $AB = 3x + 1$, $CD = 2x$, and $EF = 2x + 3$, find the value of x.

In Exercises 14–17, state the definition or theorem that allows you to conclude that quad. $HJKL$ is a parallelogram.

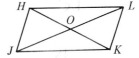

14. $HL = JK$; $\overline{HL} \| \overline{JK}$
15. $\overline{HJ} \| \overline{LK}$; $\overline{HL} \| \overline{JK}$
16. $HO = OK$; $JO = OL$
17. $HL = JK$; $HJ = LK$

18. If two sides of a triangle are 5 and 8, then the third side must be less than $\underline{?}$.
19. If two sides of a triangle are 11 and 6, then the third side must be greater than $\underline{?}$.
20. Is it possible to have a triangle with sides 39 cm, 78 cm, and 1.2 m?

Exercises 21–23 refer to $\triangle PQR$ with medians $\overline{PS}$ and $\overline{RT}$.

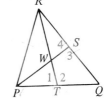

21. If $RT = 21$, then $WT = \underline{?}$.

22. If $\angle 1 < \angle 2$, then $PR \underline{\quad ? \quad} QR$.
$$(<, =, >)$$

23. If $PR < PQ$, then $\angle 3 \underline{\quad ? \quad} \angle 4$.
$$(<, =, >)$$

24. Given: $\square WXYZ$; $\angle 1 = \angle 2$
Prove: Quad. $WXTS$ is a parallelogram.

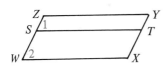

25. Given: $\square ABCD$; $\angle 3 = \angle 4$
Prove: $AD = EC$

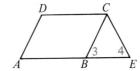

5

Similar Polygons

Definitions and Properties

Objectives
1. Express a ratio in simplest form.
2. Solve for an unknown term in a given proportion.
3. Transform a proportion into an equivalent equation.
4. State and apply the properties of similar figures.

5-1 *Ratio and Proportion*

The **ratio** of one number to another is the quotient when the first number is divided by the second.

$$\text{The ratio of 12 to 16 equals } \frac{12}{16}, \text{ or } \frac{3}{4}.$$

Since we cannot divide by zero, we never talk about the ratio of some number to 0. When an expression such as $\frac{a}{b}$ appears in this book, you know that $b \neq 0$.

A ratio is a number. When we speak of the ratio of two segments or two angles, we mean the ratio of the number of units in each, when each has been measured in terms of the same unit.

Example 1: **a.** Find the ratio of the longest side to the shortest side of the triangle.

          b. Find the ratio of the smallest angle to the largest angle of the triangle.

Solution: **a.** $\dfrac{\text{longest side}}{\text{shortest side}} = \dfrac{12}{6} = \dfrac{2}{1}$ **b.** $\dfrac{\text{smallest angle}}{\text{largest angle}} = \dfrac{30}{90} = \dfrac{1}{3}$

           The ratio is 2 to 1.   The ratio is 1 to 3.

Example 2: A stair rug is 2 m long and 60 cm wide. Find the ratio of its width to its length.

Solution:

Method 1	Method 2
Convert to centimeters.	Convert to meters.
2 m is 200 cm.	60 cm is 0.6 m.
$\dfrac{w}{l} = \dfrac{60}{200} = \dfrac{3}{10}$	$\dfrac{w}{l} = \dfrac{0.6}{2} = \dfrac{6}{20} = \dfrac{3}{10}$

Example 2 illustrates that the unit of measure used does not affect the ratio of two quantities.

Sometimes the ratio of a to b is written as $a:b$. This form can be used for comparing more than two numbers also. The statement that three numbers are in the ratio $h:j:k$ (read h to j to k) means the following:

(1) The ratio of the first two numbers is $h:j$.
(2) The ratio of the last two numbers is $j:k$.
(3) The ratio of the first and last is $h:k$.

Ratios are usually expressed in *simplest form* as in these examples:

$$\frac{8}{4} = \frac{2}{1} \qquad \frac{16\,m^2n}{12\,mn^2} = \frac{4m}{3n} \qquad 15:10:5 = 3:2:1$$

Example 3: The three angles of a triangle are in the ratio $4:3:8$. Find the measure of each angle.

Solution: Let $4x$, $3x$, and $8x$ represent the measures.
$$4x + 3x + 8x = 180$$
$$15x = 180$$
$$x = 12$$
Therefore $4x = 48$, $3x = 36$, and $8x = 96$.
The angles are $48°$, $36°$, and $96°$.

A **proportion** is an equation stating that two ratios are equal. We can write a proportion in either of these forms:

$$\frac{a}{b} = \frac{c}{d} \qquad a:b = c:d$$

Either form of the proportion may be read "a is to b as c is to d." The number a is called the first *term* of the proportion. The numbers b, c, and d are the second, third, and fourth terms.

When you want to state that three or more ratios are equal, you can write an *extended proportion:* $\dfrac{a}{b} = \dfrac{c}{d} = \dfrac{e}{f}$.

Classroom Exercises

Express each ratio in simplest form.

1. $\dfrac{20}{30}$ **2.** $\dfrac{2k}{3k}$ **3.** $\dfrac{n^2}{n^5}$ **4.** $\dfrac{n^5}{n^2}$

5. $AB:BC$ **6.** $BC:AB$

7. $AB:AC$ **8.** $AC:BC$

9. $AC:CD$ **10.** $AB:BC:CD$

Exs. 5-10

11. If the ratio of two segments is $2:3$ when measured in centimeters, what is their ratio when measured in meters?

12. If three numbers are in the ratio $2:3:4$, what is the ratio of the first two numbers?

13. If three numbers are in the ratio $2:3:4$, *must* the numbers be 2, 3, and 4?

14. What is the third term of the proportion $\dfrac{w}{x} = \dfrac{y+z}{z}$?

15. Is the equation $\dfrac{w}{x} = \dfrac{y}{z} + 1$ a proportion? Explain.

Written Exercises

In Exercises 1-9, $a = 10$, $b = 6$, and $c = 20$. Express each ratio in simplest form.

A   **1.** a to b **2.** c to a **3.** $\dfrac{a+b}{c}$

    4. $\dfrac{b}{a+c}$ **5.** $b:a$ **6.** $(a+b+c):(a+b)$

    7. $a:b:c$ **8.** $c:a:b$ **9.** $a:(a+b):(a+b+c)$

Exercises 10-13 refer to a rectangle. For each, express the ratio of length to width in simplest form.

	10.	11.	12.	13.
Length	3 m	18 cm	2 m	1 m
Width	2 m	15 cm	15 cm	0.7 cm

Write the ratios in simplest form.

14. $\dfrac{2ab}{3ab}$ **15.** $\dfrac{12xy}{15y^2}$ **16.** $\dfrac{5(a+b)}{7(a+b)}$

17. $\dfrac{c(x+7)}{c(x-y)}$ **18.** $\dfrac{(a+7)(a-y)}{2(a-y)}$ **19.** $\dfrac{6c+9}{3(2c+3)}$

Find the value of x.

Example: $\dfrac{4}{7} = \dfrac{6}{x}$

Solution: $7x\left(\dfrac{4}{7}\right) = 7x\left(\dfrac{6}{x}\right)$ (Multiplication Postulate)

$$4x = 42$$

$$x = \dfrac{42}{4} = \dfrac{21}{2}$$

B 20. $\dfrac{x}{8} = \dfrac{5}{12}$ **21.** $\dfrac{6}{x} = \dfrac{3}{14}$ **22.** $\dfrac{5x}{6} = \dfrac{4}{3}$

23. $\dfrac{2x + 1}{3} = \dfrac{4}{5}$ **24.** $\dfrac{15}{2x + 1} = \dfrac{5}{3}$ **25.** $\dfrac{x + 2}{5} = \dfrac{x + 1}{4}$

Find the measure of each angle in Exercises 26–29.

26. The ratio of two complementary angles is $5:7$.

27. The ratio of two supplementary angles is $2:3$.

28. The three angles of a triangle are in the ratio $1:2:3$.

29. The three angles of a triangle are in the ratio $2:2:5$.

30. A pentagon has a $90°$ angle and a $100°$ angle. Find the measures of the remaining three angles if the ratio is $8:12:15$.

31. The ratio of the exterior angles of a 7-gon is $2:4:4:4:5:5:6$. Find the measure of its largest interior angle.

32. This year a baseball star has a batting average of .310. This means that the ratio of the number of hits to the number of times at bat is .310. The player has had 155 hits. How many times has the player been at bat?

33. A tennis player decides to invest her bonus in stock with a price-to-earnings ratio no greater than 12 to 1. A certain stock priced at $85 had earnings of $6 last year. Does the stock meet her requirement?

C 34. How many sides does a regular polygon have if the ratio of an interior angle to an exterior angle is $3:1$?

35. In a polygon the ratio of one exterior angle to its adjacent interior angle is $4:41$. Is it possible that the polygon is a regular polygon? Explain.

36. The angles of pentagon *ABCDE* are in the ratio $4:5:6:4:8$.
    a. Find the measure of each angle.
    b. Draw the pentagon.
    c. Explain why two sides of the pentagon are parallel.

5-2 *Properties of Proportions*

The first and last terms of a proportion are called the *extremes*. The middle terms are called the *means*. In the proportions below, the extremes are shown in color.

extremes

$$a:b = c:d \qquad 6:8 = 3:4 \qquad \frac{6}{8} = \frac{3}{4}$$

means

Note above that $6 \cdot 4 = 8 \cdot 3$. This is a property of all proportions:

The product of the extremes equals the product of the means.

$$\frac{a}{b} = \frac{c}{d} \text{ is equivalent to } ad = bc.$$

The equations are equivalent since we can change either equation into the other by multiplying (or dividing) each member by bd.

It is often helpful to change a proportion into an equivalent proportion. The following properties will help you in your work. When you use any of them in a proof, you may use the reason "A property of proportions."

Properties of Proportions

1. $\dfrac{a}{b} = \dfrac{c}{d}$ is equivalent to $ad = bc$.

2. $\dfrac{a}{b} = \dfrac{c}{d}$ is equivalent to $\dfrac{a}{c} = \dfrac{b}{d}$.

3. $\dfrac{a}{b} = \dfrac{c}{d}$ is equivalent to $\dfrac{b}{a} = \dfrac{d}{c}$.

4. $\dfrac{a}{b} = \dfrac{c}{d}$ is equivalent to $\dfrac{a + b}{b} = \dfrac{c + d}{d}$.

5. If $\dfrac{a}{b} = \dfrac{c}{d} = \dfrac{e}{f} = \cdots$, then $\dfrac{a + c + e + \cdots}{b + d + f + \cdots} = \dfrac{a}{b} = \dfrac{c}{d} = \cdots$.

Classroom Exercises

1. Which proportions are equivalent to $\dfrac{x}{12} = \dfrac{2}{3}$?

   a. $\dfrac{x}{3} = \dfrac{12}{2}$

   b. $\dfrac{x}{2} = \dfrac{12}{3}$

   c. $\dfrac{x + 12}{12} = \dfrac{5}{3}$

Complete each statement.

2. If $\dfrac{y}{z} = \dfrac{3}{5}$, then $5y = $ __?__ .

3. If $\dfrac{w}{x} = \dfrac{4}{7}$, then $\dfrac{w}{4} = \dfrac{?}{?}$.

4. If $\dfrac{u}{v} = \dfrac{10}{3}$, then $\dfrac{v}{u} = \dfrac{?}{?}$.

5. If $\dfrac{s}{t} = \dfrac{4}{5}$, then $\dfrac{s+t}{t} = \dfrac{?}{?}$.

6. If $\dfrac{a}{b} = \dfrac{c}{d} = \dfrac{3}{5}$, then $\dfrac{a+c+3}{b+d+5} = \dfrac{?}{?}$.

7. a. Apply the fact that the product of the extremes equals the product of the means to the proportion $\dfrac{r}{t} = \dfrac{3}{5}$. You get $5r = $ __?__ .

b. Apply the fact to $\dfrac{5}{t} = \dfrac{3}{r}$. You get __?__ = __?__ .

c. Are the proportions $\dfrac{r}{t} = \dfrac{3}{5}$ and $\dfrac{5}{t} = \dfrac{3}{r}$ equivalent?

8. Explain how you can quickly tell that $\dfrac{u}{7} = \dfrac{3}{10}$ and $\dfrac{u}{3} = \dfrac{10}{7}$ are not equivalent.

9. Given the proportions $\dfrac{a}{b} = \dfrac{c}{d}$ and $\dfrac{a}{c} = \dfrac{b}{t}$, what can you conclude? Show that your conclusion is correct.

Written Exercises

Complete each statement.

A **1.** If $\dfrac{a}{7} = \dfrac{3}{2}$, $2a = $ __?__ .

2. If $b:3 = 4:9$, $9b = $ __?__ .

3. If $\dfrac{11}{6} = \dfrac{5}{c}$, $11c = $ __?__ .

4. If $\dfrac{13}{2} = \dfrac{d}{10}$, $2d = $ __?__ .

5. If $\dfrac{e}{5} = \dfrac{f}{6}$, $\dfrac{e}{f} = \dfrac{?}{?}$.

6. If $\dfrac{g}{h} = \dfrac{8}{11}$, $\dfrac{h}{g} = \dfrac{?}{?}$.

7. If $2:9 = j:k$, $k:j = $ __?__ : __?__ .

8. If $\dfrac{m}{n} = \dfrac{13}{4}$, $\dfrac{13}{m} = \dfrac{?}{?}$.

9. If $\dfrac{p}{5} = \dfrac{q}{12}$, $\dfrac{p+5}{5} = \dfrac{?}{?}$.

10. If $\dfrac{r-s}{s} = \dfrac{2}{7}$, $\dfrac{r}{s} = \dfrac{?}{?}$.

Find the value of x.

11. $\dfrac{x}{6} = \dfrac{2}{5}$

12. $\dfrac{x+4}{3} = \dfrac{15}{2}$

13. $\dfrac{15}{x} = \dfrac{6}{5}$

14. $9:11 = 4:x$

15. $\dfrac{x+5}{3} = \dfrac{x+9}{4}$

16. $\dfrac{12}{x} = \dfrac{8}{x-2}$

For the figure shown it is given that $\dfrac{CD}{DA} = \dfrac{CE}{EB}$. Copy and complete the table.

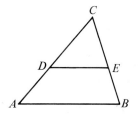

	CD	DA	CA	CE	EB	CB
17.	6	9	?	8	?	?
18.	3	?	8	6	?	?
19.	?	15	?	16	20	?
20.	2	8/3	?	3	?	7
B 21.	?	?	15	16	?	24
22.	7	8	?	?	?	15
23.	?	?	18	9	12	?
24.	?	?	21	12	?	28

(*Hint for Ex. 21:* Let $CD = x$; then $DA = 15 - x$.)

25. Given the proportions $\dfrac{a + b}{b} = \dfrac{c + d}{d}$ and $\dfrac{e}{a + b} = \dfrac{d}{b}$, what can you conclude? Show that your conclusion is correct.

26. Show that $\dfrac{a}{b} = \dfrac{c}{d}$ and $\dfrac{a + c}{a - c} = \dfrac{b + d}{b - d}$ are equivalent.

Find the value of x.

27. $\dfrac{x + 1}{x - 2} = \dfrac{x + 5}{x - 6}$

28. $\dfrac{x - 3}{x + 2} = \dfrac{x - 5}{x + 7}$

C 29. $\dfrac{x^2 - 6x + 20}{x} = \dfrac{x + 1}{2}$

30. $\dfrac{x + 7}{x - 1} = \dfrac{2x - 1}{x - 2}$

Find the values of x and y for each pair of equations.

31. $\dfrac{3x - 2y}{6} = 1$

$\dfrac{x - 5}{y} = \dfrac{1}{6}$

32. $\dfrac{1}{2} = \dfrac{6}{2x - 3y}$

$\dfrac{5x + 3y}{17} = \dfrac{3}{1}$

33. a. Show that if $\dfrac{a}{b} = \dfrac{c}{d} = \dfrac{e}{f}$, then $\dfrac{a + c + e}{b + d + f} = \dfrac{a}{b}$. (*Hint:* Let $\dfrac{a}{b} = r$. Then $a = br$, $c = dr$, and $e = \underline{\ ?\ }$.)

b. Explain how your proof can be extended to prove the last property of proportions listed on page 171.

34. When $\dfrac{a}{b} = \dfrac{c}{d} = \dfrac{e}{f}$, the ratio $\dfrac{a - c + e}{b - d + f}$ can be expressed in much simpler ways. Find one way and show that you are correct.

5-3 *Similar Figures*

Maps, scale drawings, and photographic enlargements are all used to show a given shape in a different size. Two figures, such as those below, which have the same shape are called *similar*.

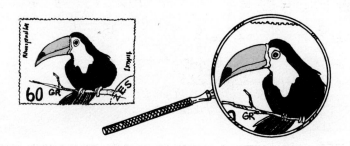

Two polygons are **similar** if their vertices can be paired so that:

(1) Corresponding angles are equal.

(2) Corresponding sides are in proportion (have the same ratio).

When referring to similar polygons, we name their corresponding vertices in the same order and use the symbol $\sim$ for "is similar to." Below, polygon *ABCDE* $\sim$ polygon *UVWXY*.

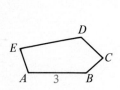
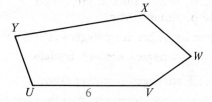

Thus (1) $\angle A = \angle U,$ $\angle B = \angle V,$ $\angle C = \angle W,$
$\angle D = \angle X,$ $\angle E = \angle Y$

and (2) $\dfrac{AB}{UV} = \dfrac{BC}{VW} = \dfrac{CD}{WX} = \dfrac{DE}{XY} = \dfrac{EA}{YU}.$

The ratio of two corresponding sides is called the **scale factor**. Since $\dfrac{AB}{UV} = \dfrac{3}{6} = \dfrac{1}{2}$, the scale factor of polygon *ABCDE* to polygon *UVWXY* is $\dfrac{1}{2}$.

In diagrams of two similar polygons, it is often helpful to label two corresponding vertices *Q* and *Q'* (read *Q prime*), *R* and *R'*, *S* and *S'*, and so on.

Example: Quad. $QRST \sim$ quad $Q'R'S'T'$
Find (a) the scale factor
and (b) the values of x, y, and z.

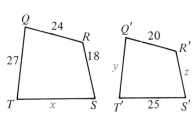

Solution: (a) Scale factor $= \dfrac{24}{20} = \dfrac{6}{5}$

(b) $\dfrac{x}{25} = \dfrac{6}{5}$ $\dfrac{27}{y} = \dfrac{6}{5}$ $\dfrac{18}{z} = \dfrac{6}{5}$

$5x = 150$ $6y = 135$ $6z = 90$

$x = 30$ $y = 22.5$ $z = 15$

Classroom Exercises

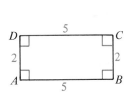 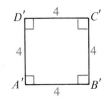 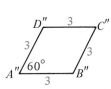

1. Use polygons $ABCD$ and $A'B'C'D'$.
   a. Are corresponding angles equal?
   b. Are corresponding sides in proportion?
   c. Are the polygons similar?
2. Repeat Exercise 1, but use polygons $A'B'C'D'$ and $A''B''C''D''$.
3. Are all regular octagons similar? Explain.

Tell whether each statement is true or false.

4. If two polygons are congruent, then they are similar. T
5. If two polygons are similar, then they are congruent. F
6. Every polygon is similar to itself. T
7. All rectangles are similar. F
8. All squares are similar. T
9. If the scale factor of two similar triangles is $1:1$, then the triangles are congruent. T
10. Given the lengths and angle measures shown in the diagram, what can you conclude? Explain.

 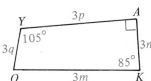

For Exercises 11 and 12, recall that the perimeter of a figure is the sum of the lengths of the sides.

11. The two triangles shown are similar.
    a. What is the scale factor?
    b. Find the perimeter of each triangle.
    c. Find the ratio of the perimeters.

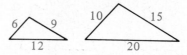

12. The two pentagons are similar.
    a. What is the scale factor?
    b. Find the perimeter of each pentagon.
    c. Find the ratio of the perimeters.

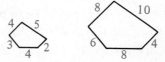

Written Exercises

Tell whether the two polygons are *always, sometimes,* or *never* similar.

A 1. An acute triangle and an obtuse triangle
   2. A scalene triangle and an isosceles triangle
   3. Two squares   4. Two rectangles
   5. Two equiangular pentagons   6. Two equilateral pentagons
   7. Two regular pentagons   8. Two regular polygons

Given: Quad. *RSTU* ~ quad. *R′S′T′U′*

9. $\angle R' = \underline{\ ?\ }°$

10. If $\angle T = 100°$, then $\angle T' = \underline{\ ?\ }°$.

11. The scale factor of quad. *RSTU* to quad. *R′S′T′U′* is ?.

12. $U'R' = \underline{\ ?\ }$ 13. $TS = \underline{\ ?\ }$

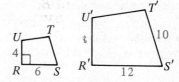

B 14. You can estimate the height of a flagpole by placing a mirror on level ground so that you see the top of the flagpole in it.

   The girl shown is 172 cm tall. Her eyes are about 12 cm from the top of her head. By measurement, *AM* is about 120 cm and *A′M* is about 4.5 m.

   From physics it is known that $\angle 1 = \angle 2$ and that the triangles are similar. Find the approximate height of the pole.

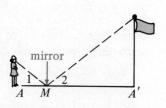

Two similar polygons are shown. Find the values of *x*, *y*, and *z*.

15.

16.

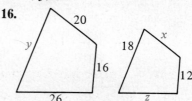

17.

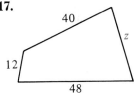

18.

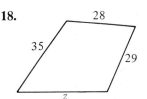

Plot the given points on graph paper. Draw pentagon $ABCDE$ and $\overline{A'B'}$. Locate points C', D', and E' so that $A'B'C'D'E'$ is a similar pentagon.

19. $A(0, 0)$, $B(4, 0)$, $C(5, 3)$, $D(3, 6)$, $E(0, 3)$, $A'(5, 5)$, $B'(13, 5)$

20. $A(0, 0)$, $B(4, 0)$, $C(5, 3)$, $D(3, 6)$, $E(0, 3)$, $A'(-1, 0)$, $B'(-9, 0)$

21. Given the lengths and angle measures shown in the diagram, what can you prove? Write the key steps of a proof.

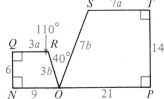

22. Given: Regular pentagon $ABCDE$;
$$\frac{AV}{AB} = \frac{BW}{BC} = \frac{CX}{CD} = \frac{DY}{DE} = \frac{EZ}{EA} = \frac{1}{4}$$
What can you prove? Write the key steps of a proof.

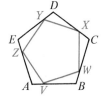

C 23. When a certain card 20 cm long is cut in half as shown, each part has the same shape as the original card. How wide was the original card?

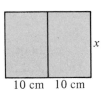

10 cm   10 cm

24. The large rectangle shown below is a *golden rectangle* if when a square is cut away, the rectangle that remains is similar to the original rectangle.

a. Write a proportion and solve for x.
(*Hint:* For the equation $ax^2 + bx + c = 0$,
$$x = \frac{-b \pm \sqrt{b^2 - 4ac}}{2a}.)$$

b. The ratio $\dfrac{20}{x}$ is called the *golden ratio*. Use your result in (a) to write the golden ratio in simplified radical form.

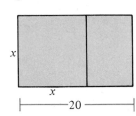

SELF-TEST

A rectangle has length 12 cm, width 9 cm, and diagonal 15 cm. Express each ratio in simplest form.

1. width to length

2. diagonal to width.

3. Two complementary angles are in the ratio $2:3$. Find the measures of the two angles.

Is the equation equivalent to the proportion $\dfrac{a}{b} = \dfrac{k}{n}$?

4. $\dfrac{a}{n} = \dfrac{b}{k}$

5. $\dfrac{a+b}{b} = \dfrac{k+n}{n}$

6. $an = bk$

7. $k:a = n:b$

Given: Pentagon $ABCDE \sim$ pentagon $RSTUV$

8. $\angle D = \angle \underline{\ ?\ }$

9. $\dfrac{BC}{ST} = \dfrac{EA}{?}$

10. If $AB = 8$, $BC = 16$, and $ST = 12$, then $RS = \underline{\ ?\ }$.

Application

Scale Drawings

When you use a ruler to measure the distance between Calgary and Spokane on the map below, you get about 2.5 cm. A map is an example of a *scale drawing*. The scale given on the map tells you that a *map distance* of 1 cm represents a *true distance* of 190 km. The true distance between Calgary and Spokane is therefore about 2.5 · 190, or 475 km.

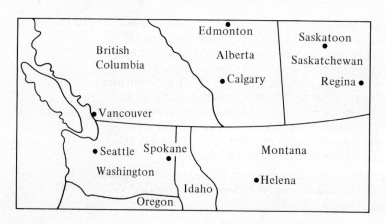

Scale 1 cm = 190 km

Suppose you wish to locate a town which you know is less than 250 km from Helena. The first step is to find the map distance corresponding to 250 km.

$$\frac{\text{True distance (km)}}{\text{Map distance (cm)}} = \frac{190}{1}$$

$$\frac{250}{x} = \frac{190}{1}$$

$$x = 1.3 \text{ (cm)}$$

The places which are less than 250 km from Helena can be pictured on the map as the region inside a circle of radius 1.3 cm centered at Helena.

Exercises

Use the map on page 178. Find the map distance and then the true distance, to the nearest 10 km, between the cities named.

1. Saskatoon and Seattle

2. Edmonton and Regina

3. How long would it take an airplane with a ground speed of 300 km per hour to fly from Helena to Vancouver?

4. Trace the map on a separate sheet of paper. Then shade in the region of the map which shows places less than 300 km from Vancouver.

A scale drawing of a house plan is sketched.

5. What does the architect mean when she prints $\frac{3}{8}'' = 10'$?

6. Measure the width of the garage in the plan. What is the width of the real garage?

7. The bedroom has a door $2\frac{1}{2}$ ft wide. The opening on the diagram should be _?_ in. Check by measuring.

8. Measure the length and width of the bathroom in the plan. What are its dimensions in the house itself?

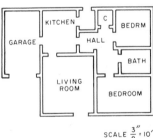

SCALE $\frac{3''}{8} = 10'$

9. Make a drawing of a regulation college football field to a scale of 1 unit = 40 in. The goal lines are 100 yd apart. The side lines are 160 ft apart. Include the ten-yard lines; the inbounds lines, 53 ft 4 in. in from the side lines; the end zones, which extend 10 yd beyond the goal lines; and points marking the goal post uprights, 23 ft 4 in. apart and centered 10 yd beyond the goal lines. (Hint for choosing a convenient unit: Convert all given true measurements to inches and make a table listing the corresponding drawing measurements in terms of "units." Base your choice of unit on the size you want your drawing to be. Probably you will choose a small metric unit or some fraction of an inch.)

10. Make a scale drawing of your school gymnasium basketball court. Measure the principal markings on the court and include them in your drawing. Choose a convenient scale.

Working with Similar Triangles

Objectives
1. State and apply the AA Similarity Postulate.
2. State and apply the SAS Similarity Theorem and the SSS Similarity Theorem.
3. State and apply the Triangle Proportionality Theorem and its corollary.
4. State and apply the theorem about the bisector of an angle of a triangle.

5-4 *A Postulate for Similar Triangles*

We can prove that two triangles are similar by showing that they satisfy the definition of similar polygons. The following experiment will suggest a simple way to prove that two triangles are similar.

1. Draw $\overline{AB}$ 6 cm long and $\overline{DE}$ 10 cm long. Note that $\dfrac{AB}{DE} = \dfrac{6}{10}$, or 0.6.

2. Draw any angle at A and an equal angle at D. Draw any angle at B and an equal angle at E. Label points C and F as indicated. Does $\angle ACB = \angle DFE$?

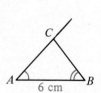
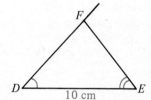

3. Measure $\overline{BC}$ and $\overline{EF}$ on your drawing. Is $\dfrac{BC}{EF}$ about 0.6?

4. Measure $\overline{AC}$ and $\overline{DF}$ on your drawing. Is $\dfrac{AC}{DF}$ about 0.6?

If you worked carefully, your answers to the questions in Steps 2, 3, and 4 were all Yes. In $\triangle ABC$ and $\triangle DEF$, (1) corresponding angles are equal, and (2) corresponding sides are in proportion. Thus, $\triangle ABC \sim \triangle DEF$ by the definition of similar polygons.

When you draw two triangles such that two angles of one triangle are equal to two angles of the other, it turns out that the third angles are equal and that corresponding sides are in proportion.

Postulate 13   If two angles of one triangle are equal to two angles of another triangle, then the triangles are similar. (AA Similarity Postulate)

Example:

Given: $\angle R = \angle U$

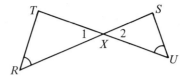

Prove: $\dfrac{RT}{US} = \dfrac{TX}{SX}$

Proof:

Statements	Reasons
1. $\angle 1 = \angle 2$	1. Vertical $\angle$s are $=$.
2. $\angle R = \angle U$	2. Given
3. $\triangle RXT \sim \triangle UXS$	3. AA Similarity Postulate
4. $\dfrac{RT}{US} = \dfrac{TX}{SX}$	4. Corr. sides of $\sim\triangle$ are in proportion.

Classroom Exercises

Tell whether the statements are true or false.

Exercises 1–7 refer to the diagram, in which $\triangle ABC \sim \triangle XYZ$.

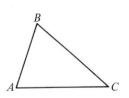

1. If $\angle C = 40°$, then $\angle Z = 40°$.

2. If $BC = 5$, then $YZ = 5$.

3. $\dfrac{AC}{XZ} = \dfrac{YZ}{BC}$

4. If $AB = 12$ and $XY = 16$, the scale factor of $\triangle ABC$ to $\triangle XYZ$ equals $3:4$.

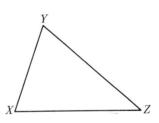

5. If $\dfrac{AC}{XZ} = \dfrac{7}{10}$, then $\dfrac{BC}{YZ} = \dfrac{7}{10}$.

6. If $AC:XZ = 7:10$, then $\angle A:\angle X = 7:10$.

7. If the scale factor of $\triangle ABC$ to $\triangle XYZ$ is $2:3$, and if $YZ = 30$, then $BC = 20$.

8. All equilateral triangles are similar.

9. All right triangles are similar.

10. All right triangles with a $42°$ angle are similar.

11. Suppose $\triangle RUN \sim \triangle JOG$. Which of the following are correct?

   a. $\overline{RU}$ and $\overline{JG}$ are corresponding sides.

   b. $\angle N$ and $\angle G$ are corresponding angles.

   c. $\triangle RNU \sim \triangle JGO$

   d. $\dfrac{RU}{JO} = \dfrac{UN}{OG}$ **e.** $\dfrac{RU}{JO} = \dfrac{RN}{JG}$ **f.** $\dfrac{RN}{JG} = \dfrac{OG}{UN}$

12. Suppose you wish to prove the equation $AB \cdot XZ = AC \cdot XY$. Complete each proportion so that it will lead to the desired equation.

   a. $\dfrac{AB}{XY} = \dfrac{?}{?}$ **b.** $\dfrac{AB}{AC} = \dfrac{?}{?}$

Written Exercises

State whether the triangles are or are not similar.

A 1.
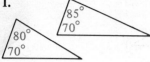

2.
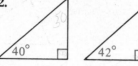

3.

4.
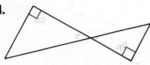

5.

6.

7.
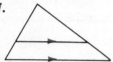

8.

9.
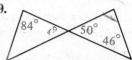

Complete the statements.

10. a. $\triangle CAB \sim \triangle\ \underline{?}$

 b. $\dfrac{AB}{?} = \dfrac{BC}{?} = \dfrac{CA}{?}$

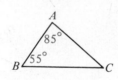 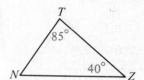

11. a. $\triangle DMF \sim \triangle\ \underline{?}$

 b. $\dfrac{?}{MG} = \dfrac{?}{GE} = \dfrac{?}{EM}$

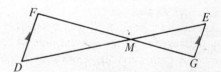

12. a. $\triangle RUT \sim \triangle\ \underline{?}$

 b. $\dfrac{x}{?} = \dfrac{4}{?}$ and $\dfrac{y}{?} = \dfrac{4}{?}$

 c. $x = \underline{?}$ and $y = \underline{?}$

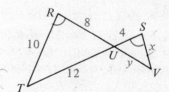

13. a. $\triangle ABC \sim \triangle\ \underline{?}$

 b. $\dfrac{4}{x} = \dfrac{8}{?}$ and $\dfrac{10}{y + 10} = \dfrac{?}{12}$

 c. $x = \underline{?}$ and $y = \underline{?}$

14. What is the scale factor of $\triangle ABC$ to $\triangle AMN$? Is it $\frac{2}{1}$, $\frac{1}{2}$, or $\frac{2}{3}$?

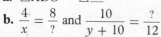 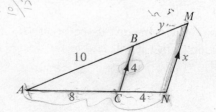

B **15.** Given: $\overline{EX} \parallel \overline{GY}$; $\angle 1 = \angle 2$
Prove: $\triangle DEX \sim \triangle FGY$

16. Given: $\triangle DEX \sim \triangle FGY$
Prove: $DE \cdot GY = EX \cdot FG$

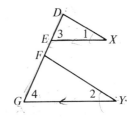

17. Given: $\angle R = \angle S$
Prove: $\dfrac{JM}{KM} = \dfrac{RM}{SM}$

18. Given: $\angle J = \angle K$
Prove: $JM \cdot SK = KM \cdot RJ$

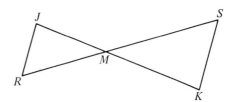

19. Write the key steps of a proof of the statement: If two similar triangles have the scale factor k, then corresponding altitudes have the scale factor k.

Given: $\triangle ABC \sim \triangle XYZ$; $\dfrac{AC}{XZ} = k$;

$\overline{CR}$ and $\overline{ZS}$ are altitudes.

Prove: $\dfrac{CR}{ZS} = k$

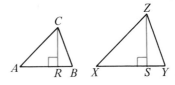

20. In $\square ABCD$, $DB = 5$.
Find DE and BE.

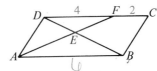

21. a. Draw two right triangles, one with legs 3 cm and 4 cm and the other with legs 6 cm and 8 cm.
b. With a protractor, measure the acute angles of each. What do you notice?
c. With a ruler, measure the longest side of each triangle. How do they compare? Do you think the triangles are similar?

22. Draw two triangles, one with sides 3 cm and 4 cm and an included angle of 115°, and the other with sides 6 cm and 8 cm and an included angle of 115°. Measure with ruler and protractor and decide whether the triangles are similar.

23. On the basis of Exercises 21 and 22, what do you guess about two triangles having two pairs of sides in proportion and the included angles equal? Is your guess based on deductive or inductive reasoning?

Find the values of x and y for each diagram.

24. Given: $\square RETS$

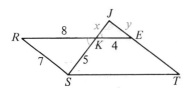

25. Given: $\square LMNO$

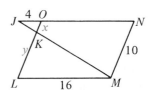

C 26. Given: Regular pentagon *ABCDE*

  a. Make a large copy of the diagram.

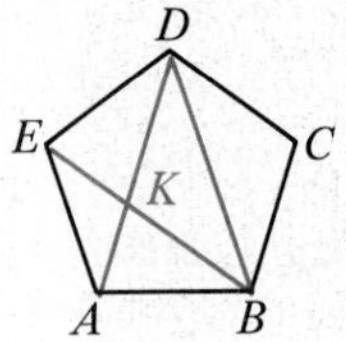

  b. Write the angle measures on your diagram.

  c. Write the key steps of a proof that $\dfrac{DA}{DK} = \dfrac{DK}{AK}$.

  d. Show that $\dfrac{DA}{DK} = \dfrac{1 + \sqrt{5}}{2}$, the golden ratio.

  (*Hint:* Let $AB = 1$ and $DA = x$.)

27. Given: $\overline{QN}$ and $\overline{RM}$ are medians of $\triangle QRS$.

  a. Write the key steps of a proof that $\dfrac{QP}{PN} = \dfrac{2}{1}$ and $\dfrac{RP}{PM} = \dfrac{2}{1}$.

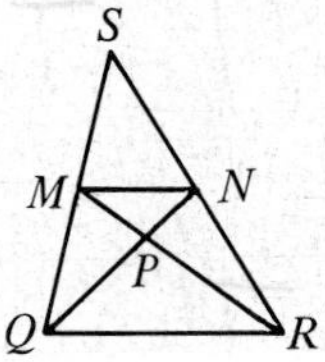

  Base your proof on similar triangles and not on Theorem 4-19 on page 143.

  b. Explain how you can extend your proof to prove Theorem 4-19.

28. The diagram shows an object $\overline{AB}$ and its image $\overline{A'B'}$ in a lens. The lens, $\overline{AB}$, and $\overline{A'B'}$ are all perpendicular to $\overline{BB'}$.

$\overline{CA'}$ intersects $\overline{BB'}$ in *F*, the focus of the lens.

OF is called the focal distance of the lens.

Prove: $\dfrac{1}{OB} + \dfrac{1}{OB'} = \dfrac{1}{OF}$

$\left(\dfrac{1}{\text{object distance}} + \dfrac{1}{\text{image distance}} = \dfrac{1}{\text{focal distance}} \right)$

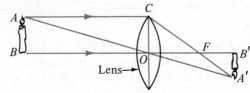

Lens→

CALCULATOR KEY-IN

To the ancient Greeks and many artists of the Renaissance, a rectangle was most pleasing to the eye when the length *l* and width *w* satisfied the equation

$\dfrac{l}{w} = \dfrac{l + w}{l}$. A rectangle with this property is said to be a *golden rectangle*,

and the ratio $\dfrac{l}{w}$ is called the *golden ratio*.

  Modern architects and artists have also used the golden rectangle in their designs. The dimensions of the home in suburban Paris shown below neatly fit within a golden rectangle. The architect was Le Corbusier.

Exercises

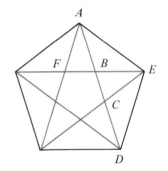

1. A regular pentagon is shown. It happens to be true that $\dfrac{AC}{AB}$, $\dfrac{AB}{BC}$, and $\dfrac{AD}{AC}$ all equal the golden ratio. Measure the appropriate lengths to the nearest millimeter and compute the ratios with a calculator.

2. From the equation $\dfrac{l}{w} = \dfrac{l + w}{l}$ it can be shown that the numerical value of $\dfrac{l}{w}$ is $\dfrac{1 + \sqrt{5}}{2}$. Express the value of $\dfrac{l}{w}$, the golden ratio, as a decimal.

3. The sequence 1, 1, 2, 3, 5, 8, 13, 21, . . . is called a *Fibonacci sequence* after its discoverer, Leonardo Fibonacci, a 13th century mathematician. The first two terms are 1 and 1. You then add two consecutive terms to get the next term.

1st term	+	2nd term	=	3rd term		2nd term	+	3rd term	=	4th term
1	+	1	=	2		1	+	2	=	3
3rd	+	4th	=	5th		4th	+	5th	=	6th
2	+	3	=	5		3	+	5	=	8

  a. Find the first twenty terms of the Fibonacci sequence.

  b. Express the ratios of several consecutive terms as decimals. That is, compute $\dfrac{1}{1} = \underline{\;?\;}$, $\dfrac{2}{1} = \underline{\;?\;}$, $\dfrac{3}{2} = \underline{\;?\;}$, ... , $\dfrac{6765}{4181} = \underline{\;?\;}$.

  c. What happens to your ratios as you move along in the sequence?

4. Sometimes the golden ratio is expressed as $\dfrac{w}{l}$ rather than $\dfrac{l}{w}$. From Exercise 2 you see that $\dfrac{w}{l} = \dfrac{2}{1 + \sqrt{5}}$. Express $\dfrac{w}{l}$ as a decimal.

5-5 *Similarity Theorems*

When a line is parallel to one side of a triangle as shown at the left below, the top triangle is similar to the larger triangle. Do you see why?

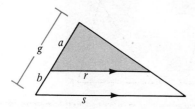 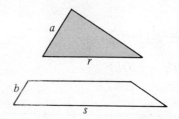

This proportion is correct: $\dfrac{a}{g} = \dfrac{r}{s}$

The two parts into which the large triangle is divided are *not* similar:

$$\frac{a}{b} \neq \frac{r}{s}$$

At this point you can show that two triangles are similar by definition or by the AA Postulate, as in the case above. The next two theorems provide other ways to prove that two triangles are similar. Proofs of the theorems are rather difficult. See Exercises 18 and 19.

Theorem 5-1

If an angle of one triangle is equal to an angle of another triangle and the sides including those angles are in proportion, then the triangles are similar. (SAS Similarity Theorem)

Given: $\angle A = \angle D$;
$$\frac{AB}{DE} = \frac{AC}{DF}$$

Prove: $\triangle ABC \sim \triangle DEF$

Analysis: We can choose X on $\overline{DE}$ so that $DX = AB$, and through X draw a line parallel to $\overline{EF}$. We can get $\dfrac{DX}{DE} = \dfrac{DY}{DF}$.
Using the given proportion and the fact that $DX = AB$, we can show that $DY = AC$. Then $\triangle ABC \cong \triangle DXY$ and $\angle B = \angle 1$. Since $\angle 1 = \angle E$, we have $\angle B = \angle E$. Since $\angle A = \angle D$, we can use the AA Postulate to prove $\triangle ABC \sim \triangle DEF$.

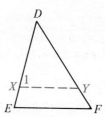

Theorem 5-2   If the sides of two triangles are in proportion, then the triangles are similar. (SSS Similarity Theorem)

Given: $\dfrac{AB}{DE} = \dfrac{BC}{EF} = \dfrac{AC}{DF}$

Prove: $\triangle ABC \sim \triangle DEF$

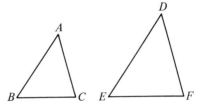

Classroom Exercises

For each diagram in Exercises 1–3, complete the proportion $\dfrac{r}{s} = \dfrac{?}{?}$.

1.

2.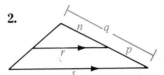

3.

4. In this exercise you will establish an unexpected result: If two posts are set upright and a rope is stretched from the top of each post to the foot of the other post, then the ropes cross at a height above the ground that is always the same no matter how far apart the posts are set.

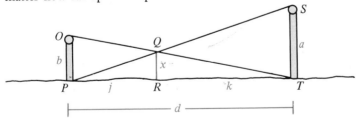

a. $\dfrac{j}{d} = \dfrac{x}{a}$ because $\triangle PRQ \sim \triangle\underline{\,?\,}$.

b. $\dfrac{k}{d} = \dfrac{x}{b}$ because $\triangle TRQ \sim \triangle\underline{\,?\,}$.

c. Using the addition property of equality and equations (a) and (b), we have
$\dfrac{j}{d} + \dfrac{k}{d} = \dfrac{x}{a} + \dfrac{x}{b}$. Show that this equation simplifies to $1 = \dfrac{x}{a} + \dfrac{x}{b}$.

d. Solve equation (c) for x.

e. Use equation (d) to explain that the height at which the ropes in the diagram cross is the same regardless of the distance between the two poles.

Written Exercises

Find the value of *x*.

A 1.
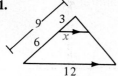

2.
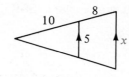

3.

4.
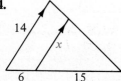

5.
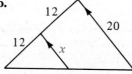

6.
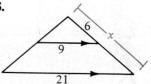

In Exercises 7–10, name two similar triangles. State a theorem that supports your answer.

7.
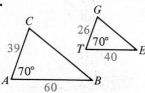

8.
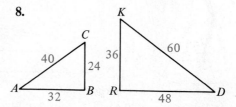

9.
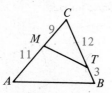

10.
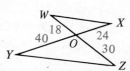

11. One triangle has sides 6 cm, 7 cm, and 8 cm long. Another has sides 8 cm, 9 cm, and 10 cm long. Are the triangles similar? Why or why not?

12. One triangle has sides 6 cm, 9 cm, and 12 cm long. Another has sides 9 cm, 13.5 cm, and 18 cm long. Are the triangles similar? Why or why not?

B 13. In $\triangle ABC$, $AB = 8$, $BC = 5$, and $CA = 6$. $\overline{AB}$ is extended through B to D so that $BD = 16$, and $\overline{AC}$ is extended through C to E so that $CE = 26$. $\overline{DE}$ is drawn.
  a. Name two similar triangles.
  b. What theorem supports your conclusion in part (a)?
  c. Find DE.

14. In the diagram at the top of the next page, $\triangle DEF$ and $\triangle DBC$ are coplanar but are not coplanar with $\triangle ABC$. State at least two things you can prove. (No proofs are required.)

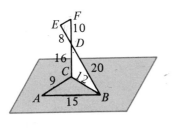

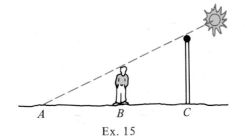

Ex. 14   Ex. 15

15. To find the height of a pole, John made use of shadows as suggested by the diagram above. By measuring, he found that $AB = 3.4$ m and $AC = 6.3$ m. If John is 1.7 m tall, how tall is the pole?

16. The schematic drawing, *not* drawn to scale, shows a film being projected on a screen. $PA = 5$ cm and $PB = 5$ m. The image of the person on the screen is 60 cm tall. How tall is the image of the person on the film?

Lamp

Film image

Screen image

17. To draw a polygon $A'B'C'D'$ similar to given polygon $ABCD$, Sarah chose any point O inside polygon $ABCD$. She located point A' so that $OA' = 3 \cdot OA$. Copy the diagram and locate B' so that $OB' = 3 \cdot OB$. Locate C' and D' in like manner. Explain why $A'B'C'D' \sim ABCD$.

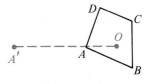

18. Prove Theorem 5-1 on page 186.

C 19. Prove Theorem 5-2 on page 187. (*Hint:* Begin as for Theorem 5-1. Get the proportion $\dfrac{DX}{DE} = \dfrac{XY}{EF} = \dfrac{DY}{DF}$. In this proportion, substitute AB for DX. Compare the new proportion with the given proportion to show that $BC = XY$ and $AC = DY$.

20. Given: $\square ABCD$; P is any point on $\overline{AB}$; Q is any point on $\overline{CD}$; $\overline{PQ}$ intersects $\overline{AC}$ at X; a line through X intersects $\overline{BC}$ at R and $\overline{AD}$ at S.
Prove: $\triangle PXS \sim \triangle QXR$
(*Hint:* Draw a good figure. Look for two sets of similar triangles that will permit you to show that $\dfrac{PX}{QX}$ and $\dfrac{SX}{RX}$ are equal to the same ratio.)

5-6 *Sides Divided Proportionally*

Suppose X and Y are points on $\overline{GH}$ and $\overline{JK}$.

$\overline{GH}$ and $\overline{JK}$ are said to be **divided proportionally** when $\dfrac{GX}{GH} = \dfrac{JY}{JK}$. For this definition, you could write any proportion equivalent to the one given.

Theorem 5-3   If a line is parallel to one side of a triangle and intersects the other two sides, then it divides them proportionally. (Triangle Proportionality Theorem)

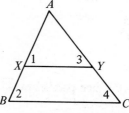

Given: $\triangle ABC$; $\overleftrightarrow{XY} \parallel \overline{BC}$

Prove: $\dfrac{AX}{AB} = \dfrac{AY}{AC}$

Proof:

Statements	Reasons
1. $\overleftrightarrow{XY} \parallel \overline{BC}$	1. Given
2. $\angle 1 = \angle 2$; $\angle 3 = \angle 4$	2. If two parallel lines are ___?___.
3. $\triangle AXY \sim \triangle ABC$	3. AA Similarity Postulate
4. $\dfrac{AX}{AB} = \dfrac{AY}{AC}$	4. Corr. sides of $\sim \triangle$ are in proportion.

The proportion shown in Theorem 5-3 can be expressed in other ways. One common way is $\dfrac{AX}{XB} = \dfrac{AY}{YC}$. (See Exercise 16.) In this book you may use the theorem to support *any* proportion that is equivalent to $\dfrac{AX}{AB} = \dfrac{AY}{AC}$.

For the diagram at the right, here are some of the proportions you may write now that Theorem 5-3 has been proved.

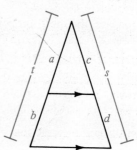

$$\frac{a}{b} = \frac{c}{d} \qquad \frac{b}{a} = \frac{d}{c} \qquad \frac{a}{c} = \frac{b}{d}$$

$$\frac{a}{t} = \frac{c}{s} \qquad \frac{b}{t} = \frac{d}{s} \qquad \frac{b}{d} = \frac{t}{s}$$

Corollary If three parallel lines intersect two transversals, then they divide the transversals proportionally.

In the diagram, $\dfrac{a}{b} = \dfrac{c}{d}$.

For a proof, see Exercise 15.

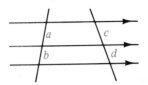

Theorem 5-4 If a ray bisects an angle of a triangle, it divides the opposite side into segments proportional to the other two sides.

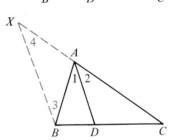

Given: $\triangle ABC$; $\angle 1 = \angle 2$

Prove: $\dfrac{BD}{DC} = \dfrac{AB}{AC}$

Key steps of proof:

1. Through B draw a line parallel to $\overline{AD}$.
2. $\angle 3 = \angle 1 = \angle 2 = \angle 4$
3. Because $\angle 3 = \angle 4$, $XA = AB$
4. $\dfrac{BD}{DC} = \dfrac{XA}{AC}$ (Theorem 5-3)
5. $\dfrac{BD}{DC} = \dfrac{AB}{AC}$ (Substitution)

Classroom Exercises

1. For the diagram shown, state proportions informally in this way:

$$\frac{\text{upper left}}{\text{lower left}} = \frac{\text{upper right}}{\text{lower right}},$$

$$\frac{\text{whole left}}{\text{lower left}} = \frac{?}{?}, \ldots$$

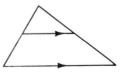

In each exercise, decide whether the three lines are parallel.

2.

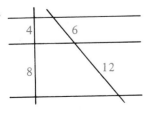

3.

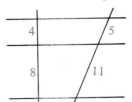

4.

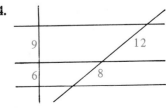

The diagram for Exercises 5 and 6 shows a segment parallel to one side of a triangle.

5. In the two similar triangles, three pairs of corresponding lengths are r and $\underline{\ ?\ }$; v and $\underline{\ ?\ }$; $\underline{\ ?\ }$ and u.

6. Tell whether the proportions are correct.

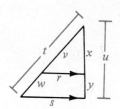

a. $\dfrac{v}{w} = \dfrac{x}{y}$ **b.** $\dfrac{v}{w} = \dfrac{r}{s}$ **c.** $\dfrac{v}{t} = \dfrac{x}{u}$

d. $\dfrac{r}{s} = \dfrac{v}{t}$ **e.** $\dfrac{r}{s} = \dfrac{x}{u}$ **f.** $\dfrac{r}{s} = \dfrac{v}{u}$

7. Three different approaches to the problem of finding JK are suggested in (a)–(c). Show each solution.

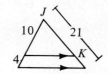

a.

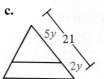

b.   **c.**

8. Refer to the diagram for Exercise 7(c) and explain why the two lengths can be labeled $5y$ and $2y$.

9. Given the diagram shown with $DO = EO = 12$, find CO and FO. What can you prove about quad. *CEFD*? Explain.

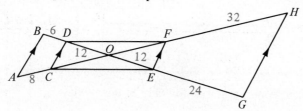

Written Exercises ——————————————————————

A   **1.** Tell whether the proportions are correct.

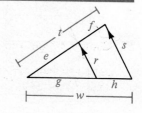

a. $\dfrac{r}{s} = \dfrac{t}{e}$ **b.** $\dfrac{e}{f} = \dfrac{h}{g}$ **c.** $\dfrac{h}{e} = \dfrac{f}{g}$

d. $\dfrac{r}{s} = \dfrac{g}{w}$ **e.** $\dfrac{f}{e} = \dfrac{h}{g}$ **f.** $\dfrac{t}{f} = \dfrac{w}{h}$

2. Tell whether the proportions are correct.

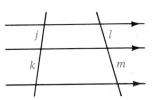

a. $\dfrac{j}{k} = \dfrac{l}{m}$ **b.** $\dfrac{j}{m} = \dfrac{k}{l}$

c. $\dfrac{j}{m} = \dfrac{l}{k}$ **d.** $\dfrac{m}{l} = \dfrac{k}{j}$

Find the value of *x*.

3.

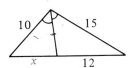

4.

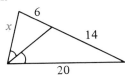

5.

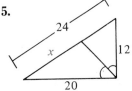

6.

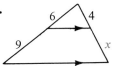

7.

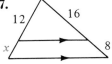

8.

9.

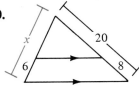

10.

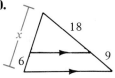

11.

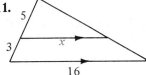

12.

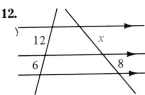

13.

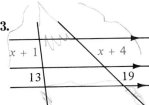

14.

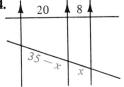

B 15. Write a complete proof of the corollary of Theorem 5-3.

Given: $j \parallel k \parallel l$; transversals *m* and *n*

Prove: $\dfrac{a}{b} = \dfrac{c}{d}$

(*Hint:* First draw the auxiliary line shown.)

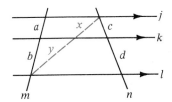

16. Show that the proportions $\dfrac{AX}{AB} = \dfrac{AY}{AC}$ and $\dfrac{AX}{XB} = \dfrac{AY}{YC}$

are equivalent. (*Hint:* Replace *AB* with $AX + XB$ and *AC* with $AY + YC$.)

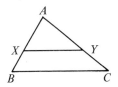

In Exercises 17–22, $\overline{PJ} \parallel \overline{QK}$. Copy the table and fill in as many blanks as possible. *Suggestion:* Draw a new figure for each exercise and label lengths as you find them.

	OP	PQ	OQ	OJ	JK	OK	PJ	QK
17.	8	5	?	10	?	?	?	3
18.	?	?	12	?	?	?	14	21
19.	6	?	?	8	8	?	10	?
20.	?	?	?	36	24	?	?	50
21.	10	8	?	?	?	27	15	?
22.	15	?	?	20	16	?	?	?

23. Find *JK*.

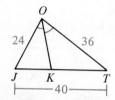

24. Find *PQ*.

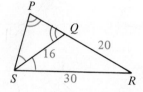

C 25. Three lots extend from Scudder Street to Tishler Road. The side boundaries are perpendicular to Scudder Street. Find the frontage of each lot on Tishler Road.

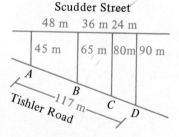

26. Prove: If two transversals cut three parallel planes, the transversals are divided proportionally.

Suggestion: Draw $\overline{DC}$, then four segments that permit you to apply Theorem 5-3 twice.

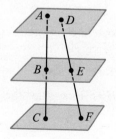

27. Prove: If a line divides two sides of a triangle proportionally, then the line is parallel to the third side of the triangle.

28. Given: $\overline{JL} \parallel \overline{MN} \parallel \overline{PR}$; ·
   $PM = 12$; $MK = 7$
   What can you prove about MN and JL?
   Write the key steps of a proof.

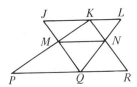

29. Prove Ceva's Theorem: If P is any point inside $\triangle ABC$,
   then $\dfrac{AX}{XB} \cdot \dfrac{BY}{YC} \cdot \dfrac{CZ}{ZA} = 1$.

   (*Hint:* Draw lines parallel to $\overline{CX}$ through A and B. In
   $\triangle ABM$ apply Theorem 5-3. Show $\triangle APN \sim \triangle MPB$,
   $\triangle BYM \sim \triangle CYP$, $\triangle CZP \sim \triangle AZN$.

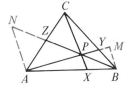

SELF-TEST

State the postulate or theorem you can use to prove that each pair of triangles is similar.

1.   **2.** 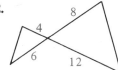   **3.**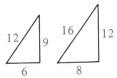

Given $\overline{AB} \parallel \overline{CD}$, complete the proportions.

4. $\dfrac{u}{v} = \dfrac{?}{y}$ **5.** $\dfrac{u}{w} = \dfrac{x}{?}$

6. $\dfrac{u}{x} = \dfrac{w}{?}$ **7.** $\dfrac{AB}{CD} = \dfrac{u}{?}$

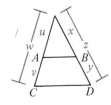

Find the value of x for each diagram.

8. 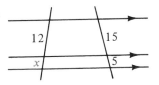   **9.**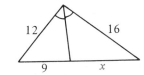

Extra Topology

In the geometry we have been studying, our interest has been in congruent figures and similar figures, that is, figures with the same size and shape, or at least the same shape. If we were studying another branch of geometry, called *topology,* our interest would be in figures that can be stretched, bent, or molded into the same shape without cutting or puncturing. Such figures are called *topologically equivalent.* For example, imagine taking a rubber band and stretching it into all kinds of figures.

  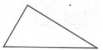

These figures are all topologically equivalent to each other. In the study of topology, circles, squares, and triangles are all equivalent because each one can be stretched into the other. Likewise, the straight line segment and wiggly curves below are equivalent.

The following plane figures are also equivalent to each other but not to any of the figures above.

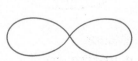 

A person who studies topology is called a *topologist.* Such a person would be interested in solid figures as well as figures in a plane. For example, the topologist would consider a ball, block, and potato equivalent to each other.

Solid ball

Solid block

Potato

In fact, a doughnut is topologically equivalent to a coffee cup! (See diagrams below.) For this reason, a topologist has been humorously described as a mathematician who can't tell the difference between a doughnut and a coffee cup!

Think of the objects as made of modeling clay.

Push thumb into clay to make room for coffee.

Exercises

In each exercise, tell which figure is *not* topologically equivalent to the rest. Exercises 1 and 2 show plane figures.

1. **a.**  **b.**  **c.**  **d.**

2. **a.**  **b.**  **c.**  **d.**

3. **a.** solid ball   **b.** hollow ball   **c.** teaspoon   **d.** comb
4. **a.** saucer   **b.** car key   **c.** coffee cup   **d.** wedding ring
5. **a.** hammer   **b.** screwdriver   **c.** thimble   **d.** sewing needle

6. Group the block numbers shown into three groups such that the numbers in each group are topologically equivalent to each other.

7. Make a series of drawings showing that the items in each pair are topologically equivalent to each other.
   a. a drinking glass and a dollar bill
   b. a tack and a paper clip

Chapter Summary

1. The ratio of a to b is the quotient $\dfrac{a}{b}$ (b cannot be 0). The ratio may be written $a:b$.

2. A proportion is an equation stating that two ratios are equal, such as $\dfrac{a}{b} = \dfrac{c}{d}$.

3. The properties of proportions (see page 171) are used to change proportions into equivalent equations. For example, the product of the extremes equals the product of the means.

4. Similar figures have the same shape. Two polygons are similar if corresponding angles are equal and corresponding sides are in proportion.

5. Ways to prove two triangles similar:

   AA Similarity Postulate   SAS Similarity Theorem   SSS Similarity Theorem

6. Ways to show that segments are proportional:
   a. If segments are corresponding sides of similar triangles, then they are in proportion.
   b. If a line is parallel to one side of a triangle and intersects the other two sides, then it divides them proportionally.
   c. If three parallel lines intersect two transversals, they divide the transversals proportionally.
   d. If a ray bisects an angle of a triangle, then it divides the opposite side into segments proportional to the other two sides.

Chapter Review

5-1 *Ratio and Proportion*

Write the ratios in simplest form.

1. $42x^2y : 24xy^2$ 2. $36:24:48$

3. If $\dfrac{x}{4} = \dfrac{7}{5}$, $x = \underline{\ ?\ }$.

4. The ratio of two supplementary angles is $1:4$. What are the measures of the angles?

5-2 *Properties of Proportions*

5. If $\dfrac{a}{b} = \dfrac{4}{9}$, then $4b = \underline{\ ?\ }$.

6. If $\dfrac{h}{k} = \dfrac{c}{d}$, then $\dfrac{h}{c} = \dfrac{?}{?}$.

7. If $\dfrac{x}{y} = \dfrac{2}{3}$, then $\dfrac{x+y}{y} = \dfrac{?}{?}$.

8. If $\dfrac{2}{m} = \dfrac{7}{n}$, then $\dfrac{m}{2} = \dfrac{?}{?}$.

5-3 *Similar Figures*

9. If $\triangle TRI \sim \triangle ANG$, then $\dfrac{TR}{AN} = \dfrac{IT}{?}$.

10. If quad. $ABCD \sim$ quad. $JKLM$, then $\angle DAB = \angle\,\underline{?}$.

11. If $\triangle ABC \sim \triangle PQR$, the scale factor of $\triangle ABC$ to $\triangle PQR$ is $\dfrac{2}{3}$, and $PR = 24$, then $AC = \underline{?}$.

12. The quadrilaterals shown are similar. Find the value of x.

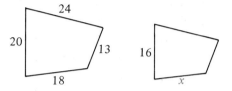

5-4 *A Postulate for Similar Triangles*

13. If $\angle A = \angle E$, is $\triangle ABC \sim \triangle EDC$?

14. If $\triangle ABC \sim \triangle EDC$, then $\dfrac{AB}{?} = \dfrac{AC}{?} = \dfrac{BC}{?}$.

15. If $\triangle ABC \sim \triangle EDC$, then $AB \cdot EC = \underline{?} \cdot ED$.

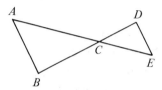

5-5 *Similarity Theorems*

16. If $\dfrac{AD}{AB} = \dfrac{AE}{AC}$, is $\triangle ADE \sim \triangle ABC$?

17. If $AD = 5$, $AB = 8$, $AE = 6$, and $AC = 9$, is $\triangle ADE \sim \triangle ABC$?

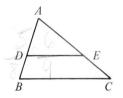

Are the two triangles similar?

18.

 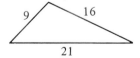

19.

5-6 *Sides Divided Proportionally*

20. If $\angle 1 = \angle 2$, then $\dfrac{CE}{EB} = \dfrac{AC}{?}$.

21. If $\overline{DE} \| \overline{AB}$, $AC = 24$, $CE = 10$, and $EB = 8$, find CD.

22. If $\angle 1 = \angle 2$, $CE = 9$, $EB = 6$, and $AB = 10$, find AC.

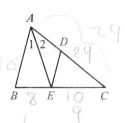

Chapter Test

1. If two sides of a rectangle measure 54 cm and 81 cm, what is the ratio of the longer side to the shorter side in simplest form?

2. If $\triangle ABC \sim \triangle A'B'C'$, then $\angle A = \angle \underline{\ ?\ }$ and $\dfrac{AC}{AB} = \dfrac{?}{?}$.

3. If $x:y:z = 2:3:4$ and $x = 6$, then $y = \underline{\ ?\ }$ and $z = \underline{\ ?\ }$.

4. If $\dfrac{12}{x} = \dfrac{5}{7}$, then $x = \underline{\ ?\ }$.

5. If $\dfrac{x}{y} = \dfrac{s}{t}$, then $\dfrac{x}{s} = \dfrac{?}{?}$.

6. $\triangle ABX \sim \triangle ACB$ by what postulate?

7. $\dfrac{AB}{AC} = \dfrac{BX}{?}$

8. If $AB = 28$, $BC = 24$, and $AC = 40$, the scale factor of $\triangle ABX$ to $\triangle ACB$ is $\underline{\ ?\ }$.

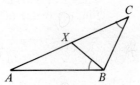

9. Supply the reasons.

Given: $\overline{DE} \parallel \overline{AB}$

Prove: $DB \cdot AC = AE \cdot BC$

Proof:

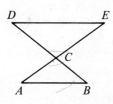

Statements	Reasons
1. $\overline{DE} \parallel \overline{AB}$	1. ___?___
2. $\angle EDC = \angle ABC$	2. ___?___
3. $\angle DCE = \angle BCA$	3. ___?___
4. $\triangle DCE \sim \triangle BCA$	4. ___?___
5. $\dfrac{DC}{BC} = \dfrac{EC}{AC}$	5. ___?___
6. $\dfrac{DC + BC}{BC} = \dfrac{EC + AC}{AC}$, or $\dfrac{DB}{BC} = \dfrac{AE}{AC}$	6. ___?___
7. $DB \cdot AC = AE \cdot BC$	7. ___?___

Find the value of x for each diagram.

10.

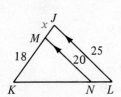

11.

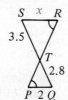

12.

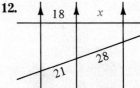

13. Given: $\square ABCD$
    a. Name two triangles similar to $\triangle GFB$.
    b. $\triangle AEF \sim \triangle\,\underline{?}$ **c.** $\triangle AED \sim \triangle\,\underline{?}$

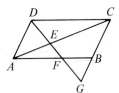

14. Given: $\overline{WX} \parallel \overline{PQ}$; $\overline{TS} \parallel \overline{PR}$
    Prove: $\triangle SYX \sim \triangle RPQ$

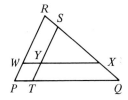

Algebra Review

Solve each equation by the method of factoring.

Example: $x^2 + 2x - 8 = 0$

Solution: $(x - 2)(x + 4) = 0$
             $x - 2 = 0$ or $x + 4 = 0$
                $x = 2$ or $x = -4$

1. $x^2 - 4x + 3 = 0$ **2.** $x^2 - 5x - 6 = 0$

3. $x^2 - 9x + 8 = 0$ **4.** $x^2 - 8x + 15 = 0$

5. $t^2 + 3t = 54$ **6.** $r^2 - 32 = 4r$

7. $z^2 = 6z - 9$ **8.** $2v + 24 = v^2$

9. $2x^2 - 5x + 2 = 0$ **10.** $3y^2 - 10y + 3 = 0$

11. $6p^2 - 7p - 3 = 0$ **12.** $13n = 20n^2 - 15$

Solve each equation by using the quadratic formula.

Example: $x^2 + 3x - 2 = 0$

Solution: For $ax^2 + bx + c = 0$, $x = \dfrac{-b \pm \sqrt{b^2 - 4ac}}{2a}$.

           For $x^2 + 3x - 2 = 0$, $x = \dfrac{-3 \pm \sqrt{3^2 - 4(1)(-2)}}{2(1)}$

                                  $= \dfrac{-3 \pm \sqrt{9 + 8}}{2}$.

       Thus, $x = \dfrac{-3 + \sqrt{17}}{2}$ or $x = \dfrac{-3 - \sqrt{17}}{2}$.

13. $x^2 + 2x - 3 = 0$ **14.** $x^2 + 4x + 3 = 0$

15. $x^2 + 7x + 3 = 0$ **16.** $x^2 - 7x + 3 = 0$

17. $2x^2 + 5x + 1 = 0$ **18.** $3x^2 - 3x - 10 = 0$

19. $3x^2 + 9x + 2 = 0$ **20.** $8x^2 - 3x - 1 = 0$

6

Right Triangles

The Pythagorean Theorem

Objectives
1. Determine the geometric mean between two numbers.
2. State and apply the relationships that exist when the altitude is drawn to the hypotenuse of a right triangle.
3. State and apply the Pythagorean Theorem.

6-1 *Geometric Means*

Suppose a, x, and d are positive numbers.

If $\dfrac{a}{x} = \dfrac{x}{d}$, then x is the **geometric mean** between a and d.

Example 1: Find the geometric mean between the given numbers.

a. 3 and 12   **b.** 15 and 3

Solution:
a. $\dfrac{3}{x} = \dfrac{x}{12}$

$x^2 = 36$

$x = \sqrt{36} = 6$

b. $\dfrac{15}{x} = \dfrac{x}{3}$

$x^2 = 45$

$x = \sqrt{45} = \sqrt{9 \cdot 5} = \sqrt{9} \cdot \sqrt{5} = 3\sqrt{5}$

In (b) above, the *radical* $\sqrt{45}$ could be simplified, since the *radicand* 45 had the factor 9, a perfect square. In the rest of this book you will be expected to write expressions involving radicals in **simplest form,** that is:

(1) No radicand may have a factor, other than 1, which is a perfect square.
(2) No radicand may be a fraction.
(3) No denominator of a fraction may contain a radical.

The symbol $\sqrt{}$ will always indicate the positive square root of a number. Do not use decimal approximations unless requested to do so.

Example 2:   Simplify the expressions.

a. $\sqrt{\dfrac{5}{8}}$
b. $\dfrac{3}{4\sqrt{5}}$

Solution: a. $\sqrt{\dfrac{5}{8}} = \sqrt{\dfrac{5}{8} \cdot \dfrac{2}{2}}$

$$= \sqrt{\dfrac{10}{16}} = \dfrac{\sqrt{10}}{4}$$

b. $\dfrac{3}{4\sqrt{5}} = \dfrac{3}{4\sqrt{5}} \cdot \dfrac{\sqrt{5}}{\sqrt{5}}$

$$= \dfrac{3\sqrt{5}}{4 \times 5} = \dfrac{3\sqrt{5}}{20}$$

The corollaries of Theorem 6-1, below, involve geometric means.

Theorem 6-1   If the altitude is drawn to the hypotenuse of a right triangle, then the two triangles formed are similar to the original triangle and to each other.

Given: Rt. $\triangle ABC$ with rt. $\angle ACB$;
$\overline{CD} \perp \overline{AB}$

Prove: $\triangle ADC \sim \triangle ACB \sim \triangle CDB$

Key steps of proof:

1. $\angle A = \angle 2$ (Both angles are complementary to $\angle 1$.)
2. $\triangle ADC$, $\triangle ACB$, and $\triangle CDB$ have these equal angles:
   $\angle ADC = \angle ACB = \angle CDB$ (Why?)
   $\angle A \ \ = \ \ \angle A \ \ = \ \ \angle 2$
3. $\triangle ADC \sim \triangle ACB \sim \triangle CDB$ (AA Postulate)

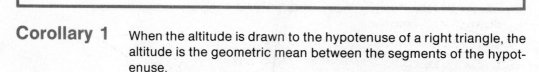

Corollary 1   When the altitude is drawn to the hypotenuse of a right triangle, the altitude is the geometric mean between the segments of the hypotenuse.

Corollary 2   When the altitude is drawn to the hypotenuse of a right triangle, either leg is the geometric mean between the hypotenuse and the segment of the hypotenuse that is adjacent to that leg.

For proofs of the corollaries, see Classroom Exercises 5-9.

Example 3:   Find the values of h, b, and a if $\angle ACB$ is a right angle.

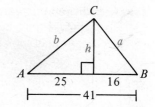

Solution:

$$\frac{25}{h} = \frac{h}{16}$$
$$h^2 = 25 \times 16$$
$$h = 5 \times 4$$
$$h = 20$$

$$\frac{41}{b} = \frac{b}{25}$$
$$b^2 = 25 \times 41$$
$$b = 5\sqrt{41}$$

$$\frac{41}{a} = \frac{a}{16}$$
$$a^2 = 16 \times 41$$
$$a = 4\sqrt{41}$$

Classroom Exercises

$\angle RTS$ is a right angle and $\overline{TN} \perp \overline{RS}$.

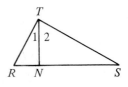

1. If $\angle R = 70°$, then
   $\angle 1 = \underline{\ ?\ }°$, $\angle 2 = \underline{\ ?\ }°$, $\angle S = \underline{\ ?\ }°$.

2. If $\angle R = k°$, then
   $\angle 1 = \underline{\ ?\ }°$, $\angle 2 = \underline{\ ?\ }°$, $\angle S = \underline{\ ?\ }°$.

3. Name two triangles similar to $\triangle RTS$.

4. Name two triangles similar to $\triangle TNR$.

The triangles in Exercises 5–9 are similar by Theorem 6-1. Complete the proportions.

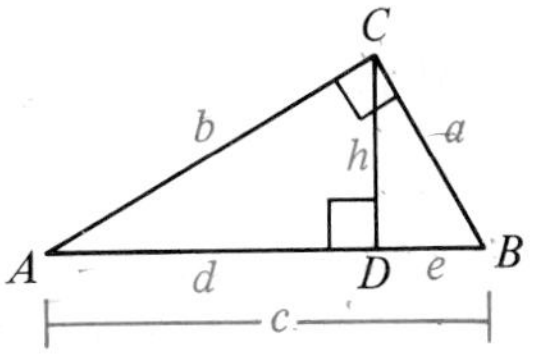

5. Since $\triangle ADC \sim \triangle CDB$, $\dfrac{d}{h} = \dfrac{h}{?}$.

6. Since $\triangle ACB \sim \triangle ADC$, $\dfrac{c}{b} = \dfrac{b}{?}$.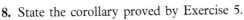

7. Since $\triangle ACB \sim \triangle CDB$, $\dfrac{c}{a} = \dfrac{?}{e}$.

8. State the corollary proved by Exercise 5.

9. State the corollary proved by Exercises 6 and 7.

10. The diagram shows a right triangle with the altitude drawn to the hypotenuse.

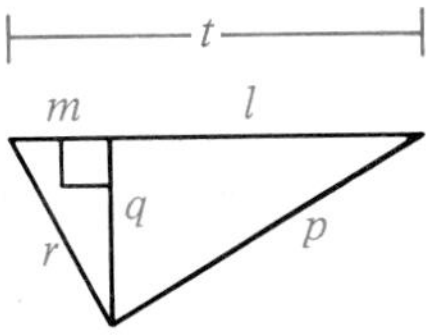

    a. p is the geometric mean between $\underline{\ ?\ }$ and $\underline{\ ?\ }$.

    b. q is the geometric mean between $\underline{\ ?\ }$ and $\underline{\ ?\ }$.

    c. r is the geometric mean between $\underline{\ ?\ }$ and $\underline{\ ?\ }$.

11. Which number does $\sqrt{18}$ equal, $2\sqrt{3}$ or $3\sqrt{2}$?

12. Find the geometric mean between the given numbers.

    a. 4 and 9   b. 2 and 8

    c. 2 and 5   d. positive numbers a and b

Simplify the expressions.

13. a. $\sqrt{36 + 64}$
    b. $\sqrt{36} + \sqrt{64}$

14. a. $\sqrt{2^2 + 3^2}$
    b. $\sqrt{2^2} + \sqrt{3^2}$

15. a. $\sqrt{4 \cdot 9}$
    b. $\sqrt{4} \cdot \sqrt{9}$

16. a. $\sqrt{25 \cdot 4}$
    b. $\sqrt{25} \cdot \sqrt{4}$

17. $7\sqrt{81}$

18. $\sqrt{50}$

19. $\sqrt{\dfrac{4}{5}}$

20. $\dfrac{7}{\sqrt{3}}$

Written Exercises

Simplify the expressions.

A **1.** $\sqrt{36}$ **2.** $5\sqrt{9}$ **3.** $\sqrt{\dfrac{4}{9}}$

4. $\dfrac{3}{5}\sqrt{25}$ **5.** $\sqrt{20}$ **6.** $\sqrt{12}$

7. $5\sqrt{12}$ **8.** $\dfrac{1}{2}\sqrt{300}$ **9.** $\sqrt{\dfrac{1}{3}}$

10. $\sqrt{\dfrac{2}{5}}$ **11.** $\dfrac{3}{\sqrt{7}}$ **12.** $\dfrac{5}{\sqrt{8}}$

Find the geometric mean between the given numbers.

13. 3 and 12   **14.** 50 and 2   **15.** 7 and 36
16. 5 and 7   **17.** 6 and 14   **18.** 5 and 15

Each diagram shows a right triangle with the altitude drawn to the hypotenuse.
Find the values of x, y, and z.

B **19.**   **20.**   **21.**

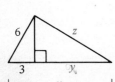

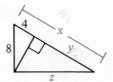

22.   **23.**   **24.**

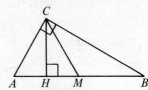

C **25.** Prove: In a right triangle, the product of the hypotenuse and the altitude
drawn to the hypotenuse is equal to the product of the two legs.

26. The arithmetic mean between two numbers r and s is defined to be the number $\dfrac{r+s}{2}$.

a. $\overline{CM}$ is the median and $\overline{CH}$ is the altitude to the hypotenuse of right $\triangle ABC$. Show that CM is the arithmetic mean between AH and BH and that CH is the geometric mean between AH and BH. Then use the diagram to show that the arithmetic mean is greater than the geometric mean.

b. Show algebraically that the arithmetic mean between two different numbers r and s is greater than the geometric mean. (*Hint:* The geometric mean is $\sqrt{rs}$. Work from $\dfrac{r+s}{2} > \sqrt{rs}$ to $(r-s)^2 > 0$ and then reverse the steps.)

6-2 *The Pythagorean Theorem*

One of the best known and most useful theorems in all of mathematics is the Pythagorean Theorem. It is believed that Pythagoras, a Greek mathematician and philosopher, proved this theorem about twenty-five hundred years ago. Many different proofs have been discovered since then.

Theorem 6-2   In a right triangle, the square of the hypotenuse is equal to the sum of the squares of the legs. (Pythagorean Theorem)

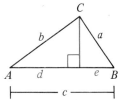

Given: Rt. $\triangle ABC$; $\angle C$ is a rt. $\angle$.

Prove: $c^2 = a^2 + b^2$

Proof:

Statements	Reasons
1. Draw a perpendicular from C to $\overline{AB}$.	1. Through a point outside a line, exactly one perpendicular ___?___.
2. $\dfrac{c}{a} = \dfrac{a}{e}$; $\dfrac{c}{b} = \dfrac{b}{d}$	2. When the altitude is drawn to the hypotenuse of a rt. $\triangle$, either leg is the geometric mean between ___?___.
3. $ce = a^2$; $cd = b^2$	3. A property of proportions
4. $ce + cd = a^2 + b^2$	4. Addition Postulate
5. $c(e + d) = a^2 + b^2$	5. Distributive Property
6. $c^2 = a^2 + b^2$	6. Substitution Property

Example: Find the value of x. (Recall that the length of a segment is defined to be a positive number.)

a.

b.

Solution: a. $x^2 = 5^2 + 6^2$
$x^2 = 25 + 36$
$x^2 = 61$
$x = \sqrt{61}$

b. $x^2 + (x + 3)^2 = 15^2$
$x^2 + x^2 + 6x + 9 = 225$
$2x^2 + 6x - 216 = 0$
$x^2 + 3x - 108 = 0$
$(x + 12)(x - 9) = 0$
$\cancel{x = -12}, x = 9$

Classroom Exercises

1. The early Greeks thought of the Pythagorean Theorem in this form: *The area of the square on the hypotenuse of a right triangle is equal to the sum of the areas of the squares on the legs.* Draw a figure to illustrate this interpretation.

Tell whether each equation is correct or incorrect.

2. $j^2 + k^2 = m^2$

3. $m = j + k$

4. $k^2 = m^2 - j^2$

5. $m^2 + j^2 = k^2$

6. $m^2 - k^2 = j^2$

7. $j^2 = (m + k)(m - k)$

State an equation you could use to find the value of x.

8.

9.

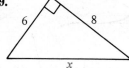

10.

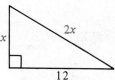

11.

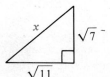

12.

13.

14. Simplify each of the following.

    a. $(\sqrt{6})^2$ **b.** $(2\sqrt{5})^2$ **c.** $(5\sqrt{2})^2$ **d.** $(3\sqrt{3})^2$

15. In the figures shown, $\angle 1 > 90°$ and $\angle 2 = 90°$. Notice that two sides of the obtuse triangle are equal to two sides of the right triangle. Write a reason for each statement.

    a. $t > n$

    b. $t^2 > n^2$

    c. $n^2 = a^2 + b^2$

    d. $t^2 > a^2 + b^2$

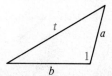

16. Exercise 15 proves a theorem informally. Complete a statement of the theorem: In an obtuse triangle, the square of the side opposite the obtuse angle ___?___ .

Written Exercises

Copy and complete the table.

A

	1.	**2.**	**3.**	**4.**	**5.**	**6.**	**7.**	**8.**
a	3	?	6	$\frac{3}{7}$	8	11	4	?
b	?	12	8	$\frac{4}{7}$	15	?	$4\sqrt{3}$	$5\sqrt{3}$
c	5	13	?	?	?	61	?	10

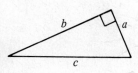

A rectangle and one diagonal are shown for each exercise. Find the value of *x*.

9.

10.

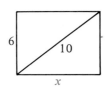

11.

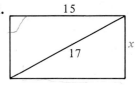

A square and one diagonal are shown for each exercise. Find the value of *x*.

12.

13.

14.

Find the length of the altitude to the base of the isosceles triangles.

B 15.

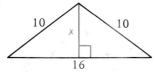

16.

Find the values of *x* and *h*.

17.

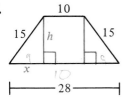

18.

19. Pierre walks 2 km east, then 3 km north, then 2 km east. How far is he from his starting point?

20. Sofia walks 6 km west, 1 km north, then 2 km east. How far is she from her starting point?

In Exercises 21–24, quad. *ABCD* is a rhombus.

21. If $AB = 10$ and $AC = 12$, $BD = \underline{\ ?\ }$.

22. If $AC = 10$ and $BD = 24$, $AB = \underline{\ ?\ }$.

23. If $AC = 4$ and $BD = 4\sqrt{3}$, $AB = \underline{\ ?\ }$.

24. If $BD = 14$ and $AD = 7\sqrt{2}$, $AC = \underline{\ ?\ }$.

Find the value of *x* in each three-dimensional figure.

25.

26.

In Exercises 27–30, find the measure of a diagonal of a rectangular box with the given dimensions.

Example: Dimensions 8, 5, and 3

Solution:

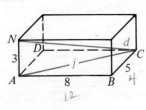

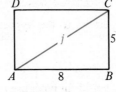

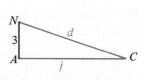

$$j^2 = (8)^2 + (5)^2$$
$$j^2 = 64 + 25$$
$$j^2 = 89$$

$$d^2 = j^2 + (3)^2$$
$$d^2 = 89 + 9 = 98$$
$$d = \sqrt{98} = \sqrt{49 \cdot 2} = 7\sqrt{2}$$

27. Dimensions 3, 4, 12

28. Dimensions 5, 5, 3

29. Dimensions 7, 6, 5

30. Dimensions l, w, h

Find the value of h.

C 31.

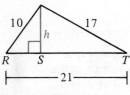

(*Hint:* Let $RS = x$.)

32.

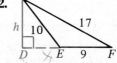

(*Hint:* Let $DE = x$.)

33. Find the values of r, s, t, u, and v for the rectangle shown.

34. Each of the six edges of the solid shown is 10 cm long. P and Q are midpoints. Find PQ.

35. Find the value of z.

36. The segment with length m is a median. Find a formula for m in terms of a, b, and c.

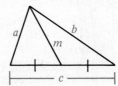

SELF-TEST

1. Which number does $\sqrt{\dfrac{3}{8}}$ equal, $8\sqrt{24}$ or $\dfrac{\sqrt{6}}{4}$?

2. What is the geometric mean between 2 and 50?

The diagram shows an altitude drawn to the hypotenuse of a right triangle.

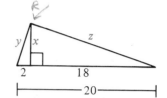

3. $x = \underline{\ ?\ }$ 4. $y = \underline{\ ?\ }$ 5. $z = \underline{\ ?\ }$

The table refers to a right triangle with hypotenuse t and legs r and s. Complete the table.

	6.	7.	8.	9.	10.
r	6	?	3	?	5
s	8	12	3	24	$5\sqrt{3}$
t	?	13	?	25	?

Challenge

A room is 10 meters long, 4 meters wide, and 4 meters high. A spider is at the middle of an end wall, $\frac{1}{3}$ meter from the floor. A fly is at the middle of the other end wall, $\frac{1}{3}$ meter from the ceiling, too frightened to move. The spider crawls to the fly. What is the shortest distance? (14 meters is not the correct answer.)

Biographical Note Charlotte Angas Scott

Charlotte Angas Scott (1858–1931) was born in Lincoln, England. She studied mathematics at Cambridge University and in 1885 received her doctorate of science from the University of London. That same year the newly founded Bryn Mawr College invited her to the United States to inaugurate a mathematics program. The only woman on the original faculty of six, she quickly established a reputation as an outstanding lecturer and research mathematician, publishing over thirty papers in the developing field of algebraic geometry.

Professor Scott was an active supporter of the New York Mathematical Society, contributing articles to its *Bulletin* from the time of its founding in 1891. The group reorganized as the American Mathematical Society in 1894, and Professor Scott served as its Vice President in 1906.

Right Triangles

Objectives

1. State and apply the converse of the Pythagorean Theorem, and related theorems regarding obtuse and acute triangles.
2. Determine the length of two sides of a 45-45-90 or a 30-60-90 triangle when the length of the third side is known.

6-3 *Converse of the Pythagorean Theorem*

The converse of the Pythagorean Theorem is also true.

Theorem 6-3   If the square of one side of a triangle is equal to the sum of the squares of the other two sides, then the triangle is a right triangle.

Given: $\triangle ABC$ with $c^2 = a^2 + b^2$

Prove: $\triangle ABC$ is a rt. $\triangle$.

Key steps of proof:

1. Using the given lengths a and b, draw rt. $\triangle RST$ as shown. Call the hypotenuse x.
2. $x^2 = a^2 + b^2$ (Why?)
3. $c^2 = a^2 + b^2$ (Given)
4. Thus $x^2 = c^2$ and $x = c$.
5. $\triangle ABC \cong \triangle RST$ (SSS Postulate)
6. Because $\angle T$ is a rt. $\angle$, $\angle C$ must be a rt. $\angle$ (why?), and $\triangle ABC$ is a rt. $\triangle$.

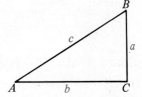

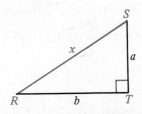

You can see that 3, 4, and 5 satisfy the equation $a^2 + b^2 = c^2$. Thus a triangle with sides 3 cm, 4 cm, and 5 cm is a right triangle. Some right triangles that appear often in geometry are listed below.

Some Common Right Triangle Lengths

3, 4, 5	**5, 12, 13**	**8, 15, 17**	**7, 24, 25**
6, 8, 10	**10, 24, 26**		
9, 12, 15			
12, 16, 20			
15, 20, 25			

Theorems 6-4 and 6-5 won't surprise you. The proofs are left as Exercises 18 and 19.

Theorem 6-4 If the square of the longest side of a triangle is greater than the sum of the squares of the other two sides, then the triangle is an obtuse triangle.

Theorem 6-5 If the square of the longest side of a triangle is less than the sum of the squares of the other two sides, then the triangle is an acute triangle.

Classroom Exercises

The lengths of the sides of a triangle are given in each exercise. Is the triangle acute, right, or obtuse?

1. 3, 4, 5
2. 4, 5, 6
3. 2, 3, 4
4. 3, 7, $\sqrt{57}$
5. 2, $\sqrt{5}$, 3
6. $\sqrt{13}$, $\sqrt{23}$, 6

7. Choose a value for c so that each statement is true.

    a. $\angle C$ is a right angle.
    b. $\angle C$ is an obtuse angle.
    c. $\angle C$ is an acute angle.
    d. There isn't any triangle with sides 6, 8, and c.

8. Try to find x mentally.

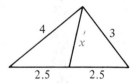

Written Exercises

Tell whether the triangle formed with the given sides is acute, right, or obtuse. If a triangle can't be formed, write *No triangle*.

A **1.** 5, 12, 13
2. 8, 15, 18
3. 9, 12, 14
4. 9, 13, 23
5. 0.7, 0.7, 1
6. 1.1, 6.0, 6.1
7. $\frac{3}{5}$, $\frac{4}{5}$, 1
8. 5, 6, $2\sqrt{15}$
9. $\sqrt{7}$, $\sqrt{8}$, $\sqrt{15}$

How many right triangles would each diagram have if it were drawn to scale? (The diagrams in Exercises 14 and 15 indicate three dimensions.)

10.

11.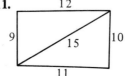

12.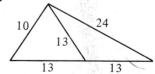

B 13.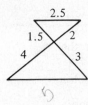

14.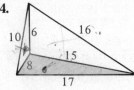

15.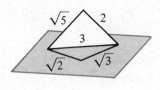

16. Sketch $\square YEUM$ with $YE = 30$, $EU = 11$, and $YU = 32$. Which diagonal is longer, $\overline{YU}$ or $\overline{ME}$?

17. Sketch $\square QUIZ$ with $QU = 5$, $QI = 8$ and $ZU = 6$. What special kind of parallelogram is $QUIZ$?

18. Write the key steps of a proof of Theorem 6-4.
Given: $t^2 > a^2 + b^2$
Prove: $\angle J$ is an obtuse angle.
(*Hint:* Draw a rt. $\triangle$ with legs a and b. Call the hypotenuse c. Compare t with c.)

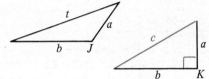

19. Write the key steps of a proof of Theorem 6-5.

20. a. Complete the table below.

	$a = 2x$	$b = x^2 - 1$	$c = x^2 + 1$
$x = 2$	4	3	5
$x = 3$	?	?	?
$x = 4$	?	?	?
$x = 5$	?	?	?

b. For each value of x, there is something special about the corresponding values of a, b, and c. What is it?

c. Prove that a triangle with sides $a = 2x$, $b = x^2 - 1$, and $c = x^2 + 1$ is a right triangle when $x > 1$.

C 21. Each side of square $ABCD$ is 25 cm long. Point P, inside the square, is 9 cm from $\overline{AB}$ and 12 cm from $\overline{BC}$. Draw $\overline{PA}$, $\overline{PB}$, $\overline{PC}$, and $\overline{PD}$. Consider $\triangle APB$, $\triangle BPC$, $\triangle CPD$, and $\triangle DPA$. Are they acute, right, or obtuse triangles?

22. Suppose a, b, and c are the lengths of the sides of a right triangle. Explain why a triangle whose sides have lengths $a + 1$, $b + 1$, and $c + 1$ must be an acute triangle.

23. Find the values of x and y.

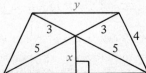

COMPUTER KEY-IN

Suppose a, b, and c are positive integers such that $a^2 + b^2 = c^2$. Then the converse of the Pythagorean Theorem guarantees that a, b, and c are the lengths of the sides of a right triangle. Because of this, any such triple of integers is called a **Pythagorean triple.**

For example, 3, 4, 5 is a Pythagorean triple since $3^2 + 4^2 = 5^2$. Another triple is 6, 8, 10, since $6^2 + 8^2 = 10^2$. 3, 4, 5 is called a *primitive* Pythagorean triple because no factor (other than 1) is common to all three integers. 6, 8, 10 is *not* a primitive triple.

The following program in BASIC lists all Pythagorean triples where a and b are less than or equal to 20.

```
10 FOR A=1 TO 20
20 FOR B=A TO 20
30 LET C=SQR(A*A+B*B)
40 IF C<>INT(C) THEN 60
50 PRINT A;B;C
60 NEXT B
70 NEXT A
80 END
RUN

3 4 5
5 12 13
6 8 10
8 15 17
9 12 15
12 16 20
15 20 25

END
```

Exercises

In Exercises 1–3, consider what the computer does the *first* time it runs through lines 10 through 40.

1. $a = 1$ and $b = \underline{?}$

2. In line 30 the computer finds c. $c = \sqrt{\underline{?}}$

3. In line 40 the computer checks the calculated value of c and finds that c is not an integer. What does the computer do next?

Look at the print-out of the RUN of the program.

4. When $a = 1$, is there any value of b from 1 to 20 which results in a Pythagorean triple?

5. How can you adjust the program so that the RUN will include the triple 20, 21, 29?

6. Which triples shown in the print-out are primitive Pythagorean triples?

7. Type and run the program. (If your computer uses a language other than BASIC, write and run a similar program for your computer.)

6-4 *Special Right Triangles*

An isosceles right triangle is often called a 45-45-90 triangle.

Theorem 6-6 In a 45-45-90 triangle, the hypotenuse is $\sqrt{2}$ times as long as a leg. (45-45-90 Theorem)

Given: A 45-45-90 triangle

Prove: hypotenuse = leg $\times \sqrt{2}$

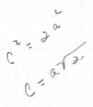

Key steps of proof:

1. $c^2 = a^2 + a^2$ (Why?)
2. $c = \sqrt{2a^2} = a\sqrt{2}$ (By algebra)

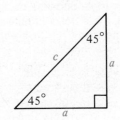

Example 1: Find the value of x in each diagram.

a.   **b.**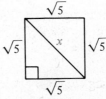

Solution: All the triangles in the diagrams are 45-45-90 triangles.

a. hyp. = leg $\times \sqrt{2}$
$$x = 7\sqrt{2}$$

b. hyp. = leg $\times \sqrt{2}$
$$x = \sqrt{5} \cdot \sqrt{2}$$
$$x = \sqrt{10}$$

Another special triangle is a 30-60-90 triangle.

Theorem 6-7 In a 30-60-90 triangle, the hypotenuse is twice as long as the shorter leg, and the longer leg is $\sqrt{3}$ times as long as the shorter leg. (30-60-90 Theorem)

Given: A 30-60-90 triangle

Prove: hypotenuse = short leg $\times$ 2;
long leg = short leg $\times \sqrt{3}$

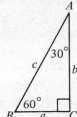

Key steps of proof:

1. Build on the given triangle as shown, forming two ≅ △. (SAS)
2. Since △ABB′ is equiangular (why?), it is equilateral, and $c = a + a$, or $c = 2a$.
3. $a^2 + b^2 = c^2$ (Why?)
4. $a^2 + b^2 = 4a^2$ (Substitution)
   $$b^2 = 3a^2$$
   $$b = a\sqrt{3}$$ (By algebra)

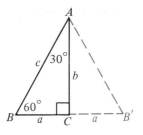

Example 2: Find the values of x and y in each diagram.

a.

b.

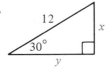

Solution: Each triangle is a 30-60-90 triangle.

a. hyp. = short leg × 2
$$x = 5 \times 2$$
$$x = 10$$

long leg = short leg × $\sqrt{3}$
$$y = 5\sqrt{3}$$

b. hyp. = short leg × 2
$$12 = x \cdot 2$$
$$6 = x$$

long leg = short leg × $\sqrt{3}$
$$y = 6\sqrt{3}$$

Classroom Exercises

1. The angles of the triangle are in the ratio 30:60:90, or 1:2:3. Are the sides also in the ratio 1:2:3?

2. The angles of the triangle are in the ratio 45:45:90, or 1:1:2. Are the sides also in the ratio 1:1:2?

Find the value of x.

3.

4.

5.

6.

7.

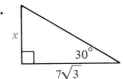

8.

Written Exercises

Complete the tables.

A

	1.	2.	3.	4.	5.	6.
RT	5	?	?	?	?	?
ST	?	$\frac{3}{7}$	$\sqrt{13}$	?	?	?
RS	?	?	?	$13\sqrt{2}$	15	$\sqrt{10}$

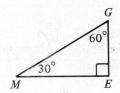

	7.	8.	9.	10.	11.	12.
GE	5	$\frac{7}{3}$	?	?	?	?
ME	?	?	?	?	$8\sqrt{3}$	5
GM	?	?	20	19	?	?

13. Draw an equilateral triangle, each side 8 units long. Draw an altitude. Note that the altitude bisects the base.

   a. Use Theorem 6-7 to find the altitude.

   b. Use the Pythagorean Theorem to find the altitude.

14. Draw a square, each side 10 units long. Draw a diagonal.

   a. Use Theorem 6-6 to find the diagonal.

   b. Use the Pythagorean Theorem to find the diagonal.

Draw a new figure for each exercise. Label lengths as you find them.

	RS	ST	RT	RU	TU	SU
B 15.	?	5	?	?	?	?
16.	?	?	?	10	?	?
17.	$8\sqrt{3}$	?	?	?	?	?
18.	?	?	?	?	4	?

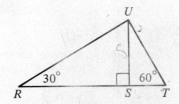

Find the lengths of as many segments as possible in each solid.

19.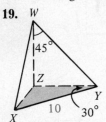

The three angles at *Z* are rt. ∠s.

20.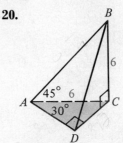

21. A rhombus has a 60° angle and a 12 cm side. How long are the diagonals?

22. Each side of a regular octagon is 4 cm long. *A* and *B* are midpoints of two directly opposite sides. Find *AB*.

C 23. If the wrench just fits the nut, what is the value of *x*?

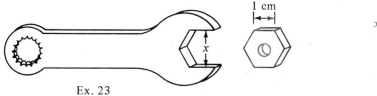

Ex. 23   Ex. 24

24. The corners are cut off a 6 cm square to form a regular octagon. Find the value of *x*.

In Exercises 25 and 26 the figures are trapezoids.

25. Find the perimeter.   **26.** Find *WX*.

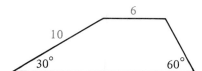

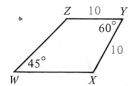

SELF-TEST

Three sides of a triangle are given. Is it an acute, right, or obtuse triangle?

1. 7, 8, 9   **2.** 2, 3, $\sqrt{17}$ **3.** $\sqrt{5}$, $\sqrt{6}$, $\sqrt{11}$

4. If $a = 9$, then $b = \underline{\ ?\ }$ and $c = \underline{\ ?\ }$.

5. If $c = 12\sqrt{2}$, then $a = \underline{\ ?\ }$ and $b = \underline{\ ?\ }$.

6. If $r = 7$, then $t = \underline{\ ?\ }$ and $s = \underline{\ ?\ }$.

7. If $t = 20$, then $r = \underline{\ ?\ }$ and $s = \underline{\ ?\ }$.

8. If $s = \dfrac{5\sqrt{3}}{2}$, then $r = \underline{\ ?\ }$ and $t = \underline{\ ?\ }$.

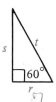

Trigonometry

Objectives
1. Define the tangent, sine, and cosine ratios for an acute angle.
2. Solve right triangle problems by correct selection and use of the tangent, sine, and cosine ratios.

6-5 *The Tangent Ratio*

The word *trigonometry* comes from Greek words that mean "triangle measurement." Study in this book will be limited to the trigonometry of right triangles. In the right triangle shown, one acute angle is marked. The leg opposite this angle and the leg adjacent to this angle are labeled.

The following ratio of the lengths of the legs is called the *tangent ratio*.

tangent of $\angle A = \dfrac{\text{leg opposite } \angle A}{\text{leg adjacent to } \angle A}$

In abbreviated form: $\tan A = \dfrac{\text{opposite}}{\text{adjacent}}$

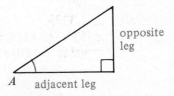

Example 1: Find $\tan X$ and $\tan Y$.

Solution: $\tan X = \dfrac{\text{leg opposite } \angle X}{\text{leg adjacent to } \angle X} = \dfrac{12}{5}$

$\tan Y = \dfrac{\text{leg opposite } \angle Y}{\text{leg adjacent to } \angle Y} = \dfrac{5}{12}$

In the right triangles shown below, $\angle A = \angle R$. Then by the AA Postulate, the triangles are similar. We can write these proportions:

$\dfrac{a}{r} = \dfrac{b}{s}$ (Why?)

$\dfrac{a}{b} = \dfrac{r}{s}$ (A property of proportions)

$\tan A = \tan R$ (Def. of tangent ratio)

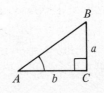
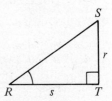

We have shown that if $\angle A = \angle R$, then $\tan A = \tan R$. Thus we have shown that the value of the tangent of an angle depends only on the size of the angle, not on the size of the right triangle. It is also true that if $\tan A = \tan R$, then $\angle A = \angle R$.

The table on page 225 lists the values of the tangents of some acute angles. Most of the values are approximations, rounded to four decimal places. Locate an angle of 43° in the table. Read across to .9325 in the column labeled "Tangent." We write tan 43° ≈ 0.9325. The symbol ≈ means "is approximately equal to."

Example 2: Find the value of x to the nearest tenth.

Solution: $\tan 43° = \dfrac{x}{24}$

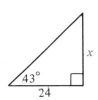

$0.9325 \approx \dfrac{x}{24}$ (From the table)

$x \approx 24 \cdot 0.9325$

$x \approx 22.3800$, or 22.4

Example 3: Find the value of y to the nearest degree.

Solution: $\tan y° = \dfrac{8}{6}$

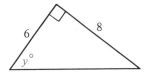

$\tan y° \approx 1.3333$

$y° \approx 53°$ (1.3333 is closer to table value 1.3270 than to 1.3764.)

Classroom Exercises

In Exercises 1–3, find tan A.

1. B **2.** B **3.**

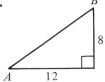   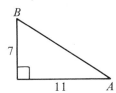   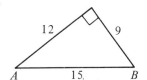

4–6. Find tan B for each triangle above.

Use the table on page 225 to complete the statements.

7. tan 10° ≈ _?_   **8.** tan 42° ≈ _?_   **9.** tan 77° ≈ _?_

10. tan _?_ ° ≈ 2.2460   **11.** tan _?_ ° ≈ 0.3249   **12.** 0.9004 ≈ tan _?_ °

13. Three 45-45-90 triangles are shown.

a. In each triangle find tan 45°.

b. Is the value of tan 45° in the table on page 225 approximate or exact?

14. Three 30-60-90 triangles are shown.

 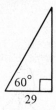

    a. In each triangle, find the length of the longer leg.

    b. In each triangle, find tan 60°.

    c. You should have found that $\tan 60° = \dfrac{\sqrt{3}}{1}$. Since $\sqrt{3} = 1.732051$ correct
to six decimal places, tan 60° = 1.732051 correct to six decimal places.
Compare this value with the one listed on page 225.

15. Compare the two ways of finding the value of x, correct to the nearest integer.

    Using tan A: Using tan B:

$$\tan 36° = \frac{120}{x}$$
$$\tan 54° = \frac{x}{120}$$

$$0.7265 \approx \frac{120}{x}$$
$$1.3764 \approx \frac{x}{120}$$

$$0.7265x \approx 120$$
$$120 \times 1.3764 \approx x$$

$$x \approx \frac{120}{0.7265}$$
$$165 \approx x$$

$$x \approx 165$$

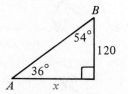

If you did your own arithmetic, which method would you prefer?
If you used a hand calculator, which method would you prefer?

Written Exercises

A Find tan A and tan B.

  1.

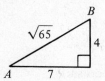

  2.

In Exercises 3–12, use the table on page 225 to find the values of x and y. Find
lengths correct to the nearest tenth and angle measures to the nearest degree.

  3.   **4.**   **5.**

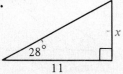 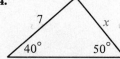

6.

7.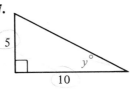

8.

B **9.**

10.

11.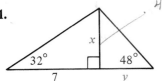

12.

13. The sides of a rectangle are 10 and 16. Find, to the nearest degree, one of the smaller angles formed at the intersection of the diagonals.

14. The diagonals of a rhombus are 12 and 20. Find the angles of the rhombus to the nearest degree.

15. The purpose of this exercise is to investigate the statement
"$\tan A + \tan B = \tan (A + B)$."

  a. $\tan 32° \approx$ _?_   **b.** $\tan 18° \approx$ _?_

  c. $\tan 32° + \tan 18° \approx$ _?_

  d. $\tan (32° + 18°) = \tan 50° \approx$ _?_

  e. Is "$\tan A + \tan B = \tan (A + B)$" correct?

16. Investigate the correctness of the statement "$\tan A - \tan B = \tan (A - B)$."

In Exercises 17–20, find lengths correct to the nearest tenth. Use the table on page 225.

C **17.** Point O is equidistant from the five vertices of the regular pentagon shown. Find OT.

18. Point P is equidistant from the ten vertices of a regular decagon. The distance between P and a side of the decagon is 12. Find the perimeter of the decagon.

19. $BC = 9$; $CD \approx$ _?_

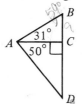

20. $RT = 12$; $ST \approx$ _?_

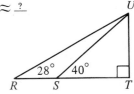

Career Notes

Computer
Programmer

The logical, step-by-step type of reasoning used in writing geometry proofs is the daily business of a computer programmer. Problems for the computer may vary from keeping records of students in school to tracking spacecraft. Before writing a program to solve a particular problem, the computer programmer often makes a flow chart, outlining the logical sequence of steps the computer will take. Some programs take an hour to write, others may require a year of work.

Some programmers specialize, writing programs only for scientific research or only for business. Different computer languages are appropriate for different types of problems, so a computer programmer's training generally includes learning more than one computer language. Employment often requires a college degree, though some firms will train high school graduates to be programmers.

TABLE OF TRIGONOMETRIC RATIOS

Angle	Sine	Cosine	Tangent	Angle	Sine	Cosine	Tangent
1°	.0175	.9998	.0175	46°	.7193	.6947	1.0355
2°	.0349	.9994	.0349	47°	.7314	.6820	1.0724
3°	.0523	.9986	.0524	48°	.7431	.6691	1.1106
4°	.0698	.9976	.0699	49°	.7547	.6561	1.1504
5°	.0872	.9962	.0875	50°	.7660	.6428	1.1918
6°	.1045	.9945	.1051	51°	.7771	.6293	1.2349
7°	.1219	.9925	.1228	52°	.7880	.6157	1.2799
8°	.1392	.9903	.1405	53°	.7986	.6018	1.3270
9°	.1564	.9877	.1584	54°	.8090	.5878	1.3764
10°	.1736	.9848	.1763	55°	.8192	.5736	1.4281
11°	.1908	.9816	.1944	56°	.8290	.5592	1.4826
12°	.2079	.9781	.2126	57°	.8387	.5446	1.5399
13°	.2250	.9744	.2309	58°	.8480	.5299	1.6003
14°	.2419	.9703	.2493	59°	.8572	.5150	1.6643
15°	.2588	.9659	.2679	60°	.8660	.5000	1.7321
16°	.2756	.9613	.2867	61°	.8746	.4848	1.8040
17°	.2924	.9563	.3057	62°	.8829	.4695	1.8807
18°	.3090	.9511	.3249	63°	.8910	.4540	1.9626
19°	.3256	.9455	.3443	64°	.8988	.4384	2.0503
20°	.3420	.9397	.3640	65°	.9063	.4226	2.1445
21°	.3584	.9336	.3839	66°	.9135	.4067	2.2460
22°	.3746	.9272	.4040	67°	.9205	.3907	2.3559
23°	.3907	.9205	.4245	68°	.9272	.3746	2.4751
24°	.4067	.9135	.4452	69°	.9336	.3584	2.6051
25°	.4226	.9063	.4663	70°	.9397	.3420	2.7475
26°	.4384	.8988	.4877	71°	.9455	.3256	2.9042
27°	.4540	.8910	.5095	72°	.9511	.3090	3.0777
28°	.4695	.8829	.5317	73°	.9563	.2924	3.2709
29°	.4848	.8746	.5543	74°	.9613	.2756	3.4874
30°	.5000	.8660	.5774	75°	.9659	.2588	3.7321
31°	.5150	.8572	.6009	76°	.9703	.2419	4.0108
32°	.5299	.8480	.6249	77°	.9744	.2250	4.3315
33°	.5446	.8387	.6494	78°	.9781	.2079	4.7046
34°	.5592	.8290	.6745	79°	.9816	.1908	5.1446
35°	.5736	.8192	.7002	80°	.9848	.1736	5.6713
36°	.5878	.8090	.7265	81°	.9877	.1564	6.3138
37°	.6018	.7986	.7536	82°	.9903	.1392	7.1154
38°	.6157	.7880	.7813	83°	.9925	.1219	8.1443
39°	.6293	.7771	.8098	84°	.9945	.1045	9.5144
40°	.6428	.7660	.8391	85°	.9962	.0872	11.4301
41°	.6561	.7547	.8693	86°	.9976	.0698	14.3007
42°	.6691	.7431	.9004	87°	.9986	.0523	19.0811
43°	.6820	.7314	.9325	88°	.9994	.0349	28.6363
44°	.6947	.7193	.9657	89°	.9998	.0175	57.2900
45°	.7071	.7071	1.0000				

6-6 *The Sine and Cosine Ratios*

Can you find the values of x or y in the diagram by using
tan 64°? by using tan 26°? The answer is No, because the
only side known, the hypotenuse, is not involved in the
definition of tangent. Two ratios that do relate the hypotenuse to the legs are the *sine* and *cosine*.

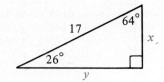

$$\text{sine of } \angle A = \frac{\text{leg opposite } \angle A}{\text{hypotenuse}}$$

$$\text{cosine of } \angle A = \frac{\text{leg adjacent to } \angle A}{\text{hypotenuse}}$$

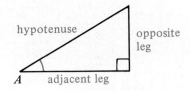

We now have three useful trigonometric ratios:

$$\tan A = \frac{\text{opposite}}{\text{adjacent}}$$

$$\sin A = \frac{\text{opposite}}{\text{hypotenuse}}$$

$$\cos A = \frac{\text{adjacent}}{\text{hypotenuse}}$$

Example 1:   Find x and y correct to the nearest integer. Use the table on page 225.

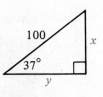

Solution:

$$\sin 37° = \frac{x}{100} \qquad \cos 37° = \frac{y}{100}$$

$$0.6018 \approx \frac{x}{100} \qquad 0.7986 \approx \frac{y}{100}$$

$$100 \times 0.6018 \approx x \qquad 100 \times 0.7986 \approx y$$

$$60.18 \approx x \qquad 79.86 \approx y$$

$$x \approx 60 \qquad y \approx 80$$

Example 2:   Find t correct to the nearest degree.

Solution: $\sin t° = \dfrac{17}{20}$

$\sin t° = 0.8500$

Note that 0.8500 is closer to the table value .8480 than
to table value .8572. Therefore, $t \approx 58$ to the nearest
degree.

Classroom Exercises

In Exercises 1–3, find sin A, cos A, and tan A.

1.
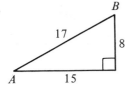

2.
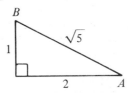

3.
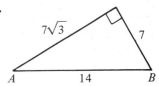

4–6. For each triangle above, find sin B, cos B, and tan B.

For each triangle, state two different equations you could use to find the value of x.

7.

8.

9.

10. a. Expressed as a common fraction, sin $A = \underline{\ ?\ }$.

    b. Expressed as a decimal, sin $A = \underline{\ ?\ }$.

    c. From the table on page 225, you find $\angle A = \underline{\ ?\ }°$.

    d. Explain how you could have found the measure of $\angle A$ in this case without using tables.

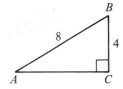

Written Exercises

In all exercises, use the table on page 225. Find lengths correct to the nearest tenth and angle measures to the nearest degree.

A   **1.** sin $5° \approx \underline{\ ?\ }$ **2.** sin $72° \approx \underline{\ ?\ }$ **3.** cos $48° \approx \underline{\ ?\ }$

    4. cos $A \approx 0.2079$; $\angle A = \underline{\ ?\ }°$.   **5.** sin $A \approx 0.9550$; $\angle A = \underline{\ ?\ }°$.

In Exercises 6–11, find the value of x.

6.
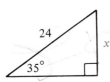

7.

8.
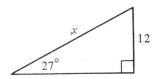

9.
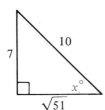

10.
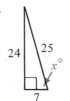

11.

B 12. A 5 m ladder leans against a building as shown.

    a. How far up the side of the building does the ladder reach?

    b. How far from the building is the bottom of the ladder?

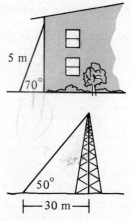

13. A guy wire is to make a 50° angle with the ground at a point 30 m from the foot of the tower. Allowing one extra meter at each end, how long a wire is needed?

In Exercises 14–17, it will help to sketch the figure described.

14. The equal sides of an isosceles triangle are 24 units long. Each base angle is a 75° angle. Find the altitude.

15. An isosceles triangle has a 38° base angle and 14 cm legs. Find the base. (*Hint:* Draw an altitude and find half the base.)

16. An isosceles triangle has a 54° vertex angle and a 36 cm base. How long are the legs?

17. Each side of a regular decagon is 20 mm long. Find the length of a segment that joins the midpoints of two consecutive sides.

In Exercises 18–20, find the value of x.

C 18.

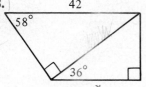

19.

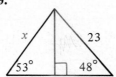

20.

Challenge

Start with a right triangle. Build a square on each side. Locate the center of the square on the longer leg. Through the center, draw a parallel to the hypotenuse and a perpendicular to the hypotenuse.

    Cut out the pieces numbered 1–5. Can you arrange the five pieces to cover exactly the square built on the hypotenuse?

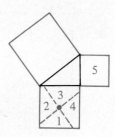

6-7 *Using Trigonometric Ratios*

Suppose that an operator at the top of the lighthouse can sight the sailboat by lowering a telescope through an angle of 3° from the horizontal. This angle, shown in the diagram, is called an **angle of depression.**

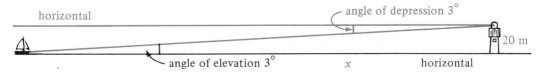

At the same time, a person in the boat can sight the top of the lighthouse by raising a telescope through a 3° angle from the horizontal. This angle, shown in the diagram above, is called an **angle of elevation.** Note that the angle of elevation equals the angle of depression.

If the top of the lighthouse is known to be 20 m above sea level, the distance, x, between the boat and the lighthouse can be found by these two methods:

Method 1

$$\tan 3° = \frac{20}{x}$$

$$0.0524 \approx \frac{20}{x}$$

$$0.0524x \approx 20$$

$$x \approx 20 \div 0.0524$$

$$x \approx 381.6794$$

Method 2

$$\tan 87° = \frac{x}{20}$$

$$19.0811 \approx \frac{x}{20}$$

$$20 \times 19.0811 \approx x$$

$$381.6220 \approx x$$

Both methods tell us that the sailboat is roughly 400 m from the lighthouse. The arithmetic is easier in Method 2 because x is in the numerator of the fraction instead of in the denominator.

Classroom Exercises

In Exercises 1–4, lines l and m are horizontal. Name the angles described.

1. The angle of elevation from A to E.

2. The angle of depression from D to B.

3. The angle of elevation from B to F.

4. The angle of depression from E to C.

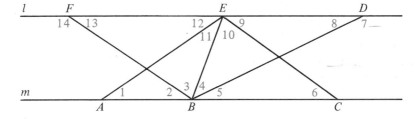

In Exercises 5–7, state the equation you would use to find the value of x and tell whether the arithmetic involved in each would require division or multiplication.

5.

6.

7.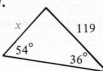

Written Exercises

A 1. A person at R sights S. Which angle at R is the angle of elevation, $\angle 1$ or $\angle 2$?

2. A person at S sights R. Which angle at S is the angle of depression, $\angle 3$ or $\angle 4$?

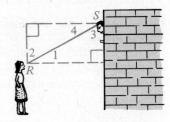

In Exercises 3–10, find lengths correct to the nearest meter and angle measures to the nearest degree. Use the table on page 225.

3. When the sun's elevation is 38°, a building casts a shadow 21 m long. How high is the building?

4. At a certain time, an upright meter stick casts a shadow 3 m long. What is the angle of elevation of the sun?

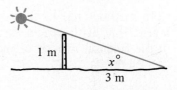

B 5. Engineers at G wish to reach P, 15 m below ground level, by tunneling at a 40° angle. How long will the tunnel be?

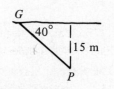

6. The pilot of a small boat measures and determines that the angle of elevation of the top of a lighthouse is 9°. The pilot knows that the lighthouse reaches 21 m above sea level.

a. Make a sketch.   **b.** Find the distance between the boat and the lighthouse.

7. A cliff is known to be 27 m high. A person on the cliff sights a steamer at an angle of depression of 4°.

a. Make a sketch.   **b.** Find the distance between the steamer and the cliff.

8. A pilot flying over level ground at an altitude of 650 m sights a marker. The angle of depression measures 9°.

   a. Make a sketch.

   b. Find the distance between the marker and the plane.

   c. Find the ground distance between the marker and the point directly below the plane.

9. A beam of light along $\overrightarrow{LC}$ illuminates the clouds at a point C directly over a light detector at D. The angle of the beam is 40° and the detector is 400 m from the light source as shown. Find h, the height of the clouds.

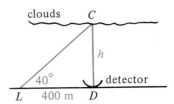

10. A rhombus has a 5 m side and an 80° angle. Find the diagonals of the rhombus.

Given the value of one trigonometric ratio of $\angle A$, find the other two ratios.

Example: $\sin A = \dfrac{4}{7}$

Solution: (1) Sketch a rt. $\triangle$ with a leg and hypotenuse in the ratio $\dfrac{4}{7}$.

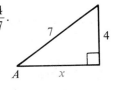

      (2) Use the Pythagorean Theorem to find the third side, x.

$$x^2 + 16 = 49$$
$$x^2 = 33$$
$$x = \sqrt{33}$$

      (3) Find $\cos A$ and $\tan A$.

$$\cos A = \frac{\sqrt{33}}{7} \qquad \tan A = \frac{4}{\sqrt{33}}, \text{ or } \frac{4\sqrt{33}}{33}$$

11. $\cos A = \dfrac{3}{8}$ **12.** $\sin A = \dfrac{3}{8}$ **13.** $\tan A = \dfrac{5}{9}$ **14.** $\sin A = \dfrac{11}{61}$

C 15. A regular pentagon is shown. Find the length of $\overline{JK}$ correct to the nearest integer.

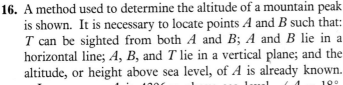

16. A method used to determine the altitude of a mountain peak is shown. It is necessary to locate points A and B such that: T can be sighted from both A and B; A and B lie in a horizontal line; A, B, and T lie in a vertical plane; and the altitude, or height above sea level, of A is already known.

    In one case, A is 4306 m above sea level, $\angle A = 18°$, $\angle TBC = 24°$, and $AB = 1672$ m. How far above sea level is T?

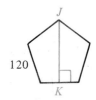

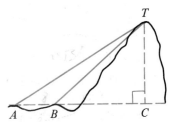

SELF-TEST

Exercises 1–5 refer to the diagram.

1. $\sin R = \dfrac{?}{?}$

2. $\sin T = \dfrac{?}{?}$

3. $\tan R = \dfrac{?}{?}$

4. $\cos R = \dfrac{?}{?}$

5. To the nearest degree, $\angle R = \underline{?}°$.
   (Use the table on page 225.)

Find the value of x correct to tenths. (Use the table on page 225.)

6.

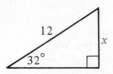

7.

8.

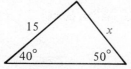

9.

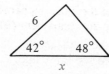

Extra Non-Euclidean Geometries _____

When you develop a geometry, you have some choice as to which statements you are going to postulate and which you are going to prove. For example, consider these two statements:

(A) If two parallel lines are cut by a transversal, then corresponding angles are equal.

(B) Through a point outside a line, exactly one parallel can be drawn to the line.

In this book, statement (A) is Postulate 8 and statement (B) is Theorem 2-8. In some books, statement (B) is a postulate and statement (A) is a theorem. In still other developments, both of these statements are proved on the basis of some third statement chosen as a postulate.

A natural question to raise is, Do we have to assume either statement, or an equivalent, at all? Couldn't each of them be proved on the basis of the other postulates? The answer, which wasn't known until the nineteenth century, is that some assumption about parallels *must* be made.

For historical reasons, statement (B) is often called the *Parallel Postulate.* A geometry that provides for a unique parallel to a line through a point not on the line is called *Euclidean,* so this book is a book on Euclidean geometry. Geometries that do not provide for a unique parallel are called *non-Euclidean.* Such geometries, developed in the nineteenth century, aren't mere curiosities; for example, Einstein's Theory of Relativity is based on non-Euclidean geometry. The statements below show the key differences between Euclidean geometry and two types of non-Euclidean geometry.

Euclidean geometry	Through a point outside a line, *exactly one* parallel can be drawn to the line.
Hyperbolic geometry	Through a point outside a line, *more than one* parallel can be drawn to the line. (Bolyai, Lobachevsky)
Elliptic geometry	Through a point outside a line, *no* parallel can be drawn to the line. (Riemann)

To see a model of a no-parallel geometry, visualize the surface of a sphere. Think of a *line* as being a great circle of the sphere, that is, the intersection of the sphere and a plane that passes through the center of the sphere. On the sphere, through a point outside a line, there isn't any line parallel to the line. All lines, as defined, intersect.

Finally, we return to the opening sentences of this Extra and use the postulates of this book to prove statement (B).

Given: Point P outside line k

Prove: (1) There is a line through P parallel to k.
     (2) There is only one line through P parallel to k.

Proof of (1), Key Steps:

1. Draw a line through P and some point Q on k. (Postulates 1 and 2)
2. Draw line m so that $\angle 2$ and $\angle 1$ are corresponding angles and $\angle 2 = \angle 1$. (Protractor Postulate)
3. $m \| k$, and there is a line through P parallel to k. (Postulate 9)

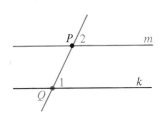

Proof of (2), Indirect Proof:

Suppose there are at least two lines, x and y, through P parallel to k. Draw a line through P and some point R on k. $\angle 4 = \angle 3$ and $\angle 5 = \angle 3$ by Postulate 8, so $\angle 5 = \angle 4$. But $\angle 5 > \angle 4$. This is impossible, our supposition must be false, and it follows that there is only one line through P parallel to k.

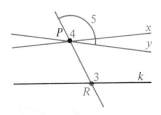

Notice the crucial role played by Postulate 8 in the second part of the proof above. Without Postulate 8 or an equivalent, there couldn't be a proof. For over two thousand years people didn't know that this was the case and tried, without success, to find a proof.

Chapter Summary

1. When $\dfrac{a}{x} = \dfrac{x}{d}$, x is the geometric mean between a and d.

2. A right triangle is shown with the altitude drawn to the hypotenuse.
   a. The two triangles formed are similar to the original triangle and to each other.

   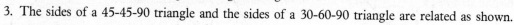

   $$\dfrac{x}{h} = \dfrac{h}{y} \qquad\qquad \dfrac{c}{b} = \dfrac{b}{x} \qquad\qquad \dfrac{c}{a} = \dfrac{a}{y}$$

   b. Pythagorean Theorem: $c^2 = a^2 + b^2$
   c. Converse of Pythagorean Theorem:
      If $c^2 = a^2 + b^2$, the triangle is a right triangle.

3. The sides of a 45-45-90 triangle and the sides of a 30-60-90 triangle are related as shown.

4. In the right triangle shown:

   $$\tan A = \dfrac{a}{b} \qquad\qquad \sin A = \dfrac{a}{c} \qquad\qquad \cos A = \dfrac{b}{c}$$

   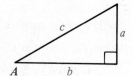

Chapter Review

6-1 *Geometric Means*

1. Simplify $\sqrt{\dfrac{20}{9}}$.

   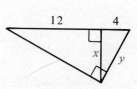

2. Find the geometric mean between 3 and 21.
3. $x = \underline{\ ?\ }$
4. $y = \underline{\ ?\ }$

6-2 *The Pythagorean Theorem*

5. A right triangle has legs 2 and 6. What is the length of the hypotenuse?
6. A rectangle has sides 4 and 8. What is the length of a diagonal?
7. A square has side 10. How long is the diagonal?
8. An isosceles triangle has sides 17, 17, and 16. How long is the altitude to the base?

6-3 *Converse of the Pythagorean Theorem*

Tell whether the triangle formed with the given sides is acute, right, or obtuse. If a triangle can't be formed, write *No triangle.*

9. 2, 3, 4   **10.** $2\sqrt{3}, \sqrt{3}, \sqrt{15}$ **11.** 9, 12, 22   **12.** 5, 7, 8

6-4 *Special Right Triangles*

In Exercises 13–15, find the value of *x*.

13.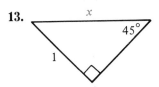

14.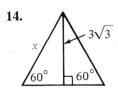

15.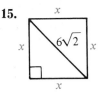

6-5 *The Tangent Ratio*

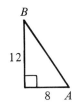

16. $\tan A = \underline{\ ?\ }$

17. $\tan B = \underline{\ ?\ }$

Use the table on page 225 for Exercises 18 and 19.

18. $\tan 21° \approx \underline{\ ?\ }$ **19.** $\tan \underline{\ ?\ }° \approx 1.5399$

6-6 *The Sine and Cosine Ratios*

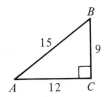

20. $\sin A = \dfrac{?}{?}$ $\cos A = \dfrac{?}{?}$

$\sin B = \dfrac{?}{?}$ $\cos B = \dfrac{?}{?}$

Use the table on page 225 for Exercises 21 and 22.

21. $\sin 67° \approx \underline{\ ?\ }$; $\cos 67° \approx \underline{\ ?\ }$ **22.** $\cos \underline{\ ?\ } \approx 0.7314$

6-7 *Using Trigonometric Ratios*

In Exercises 23–25, write the equation you would use to find the value of *x*. Do not use the table and do not find the numerical value of *x*.

23.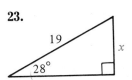

24.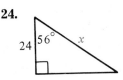

25.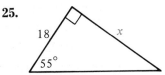

Chapter Test

Find the geometric mean between the given numbers.

1. 2 and 32   **2.** 3 and 27   **3.** 9 and 25

Simplify the expressions.

4. $\sqrt{27}$ **5.** $\sqrt{120}$ **6.** $\sqrt{\dfrac{8}{9}}$

The diagram for Exercises 7–10 shows right $\triangle ABC$ with altitude $\overline{CD}$ drawn to the hypotenuse.

7. Name two triangles similar to $\triangle ABC$.

8. CD is the geometric mean between AD and _?_.

9. AC is the geometric mean between _?_ and _?_.

10. If $AC = 12$ and $BC = 16$, then $AB = $ _?_.

Tell whether the triangle formed with the given sides is acute, right, or obtuse. If a triangle can't be formed, write *No triangle*.

11. 5, 12, 13   **12.** 9, 8, 17   **13.** 2, 2, $2\sqrt{2}$

14. 2, 2, 3   **15.** 4, 5, 6   **16.** 1.5, 2, 2.5

In Exercises 17 and 18, find the value of x.

17.

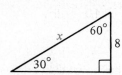

18.

Find each value of x to the nearest integer. Use the table of trigonometric ratios on page 225.

19.

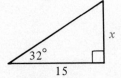

20.

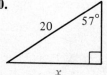

21.

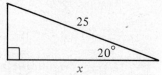

22. Given: Rt. $\triangle ABC$; rt. $\triangle EFG$;
$\angle B = \angle E$; $\overline{CD} \perp \overline{AB}$
Prove: $\triangle ADC \sim \triangle FGE$

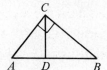

Cumulative Review: Chapters 1–6

True-False Exercises

Write T or F to indicate your answer.

A   **1.** Theorems are statements which we accept without proof.

  2. The sum of the interior angles of a convex quadrilateral is $360°$.

  3. The sum of the exterior angles of a convex quadrilateral is $360°$.

  4. Opposite angles of a parallelogram are equal.

  5. Every square is a rectangle.

  6. A parallelogram is a trapezoid.

  7. In $\triangle ABC$, if $\angle A > \angle B$, then $AC > BC$.

  8. If two triangles are similar, then their corresponding sides must be equal.

  9. If the shorter leg of a 30-60-90 triangle is 25 cm, then the longer leg is 50 cm.

  10. If $\dfrac{a}{b} = \dfrac{c}{d}$, then $\dfrac{a}{c} = \dfrac{b}{d}$.

  11. If $\overleftrightarrow{AB}$ and $\overleftrightarrow{CD}$ do not intersect, then $\overleftrightarrow{AB}$ and $\overleftrightarrow{CD}$ must be parallel.

  12. It is possible to have a triangle with sides 27 cm, 57 cm, and 36 cm.

B **13.** If D lies on $\overline{BC}$ and $\overline{AD}$ bisects $\angle A$ of $\triangle ABC$, then $\dfrac{BD}{DC} = \dfrac{AB}{AC}$.

  14. If line l is perpendicular to line m, and line m is perpendicular to line n, then line l must be perpendicular to line n.

  15. In rt. $\triangle ABC$ if D lies on hypotenuse $\overline{AB}$ and $\overline{CD} \perp \overline{AB}$, then $\dfrac{CD}{AD} = \dfrac{BD}{BC}$.

  16. If $\triangle ABC \sim \triangle DEF$, and $\triangle DEF \cong \triangle JKL$, then $\dfrac{AB}{DE} = \dfrac{AC}{JL}$.

  17. If quad. $ABCD$ has $\angle A$ and $\angle B$ supplementary and $\angle C$ and $\angle D$ also supplementary, then quad. $ABCD$ must be a parallelogram.

  18. If x, y, and z are the lengths of the sides of a triangle, then $z > y - x$.

Algebraic Exercises

In Exercises 1–20 find the value of x.

A   **1.** An angle and its complement measure $(2x)°$ and $(4x - 12)°$.

  2. An angle and its supplement measure $(3x - 24)°$ and $(4x - 34)°$.

  3. The angles of a triangle measure $x°$, $(4x - 3)°$, and $(7x + 15)°$.

  4. The legs of an isosceles triangle are $5x + 2$ and $9x - 10$.

  5. Two vertical angles measure $(17x - 41)°$ and $(25x - 65)°$.

  6. The angles of a quadrilateral are $4x°$, $8x°$, $7x°$, and $5x°$.

7. A line segment along the number line has endpoints at -13 and 41. The midpoint is at $4x$.

8. A transversal cuts a pair of parallel lines. The interior angles on the same side of the transversal measure $(5x - 9)°$ and $(9x + 7)°$.

9. A line segment is bisected, making two segments of lengths $13x$ and $5x + 152$.

10. The acute angles of a right triangle measure $(x - y)°$ and $(x + y)°$.

11. Each leg of a 45-45-90 triangle is $2x$ and the hypotenuse is $12\sqrt{2}$.

12. An angle with measure $(3x + 6)°$ is bisected, forming two angles each with measure $(2x)°$.

13. A rhombus has opposite angles with measures $(6x - 17)°$ and $(2x + 35)°$.

14. Three angles of an isosceles trapezoid are $(6x - 13)°$, $49°$, and $49°$.

15. $\dfrac{x}{28} = \dfrac{3}{7}$

16. $\dfrac{4}{7} = \dfrac{6}{x}$

17. $\dfrac{2}{5} = \dfrac{2x - 4}{15}$

18. $\dfrac{7}{3} = \dfrac{x + 4}{x}$

19. $\dfrac{x - 2}{x + 2} = \dfrac{4}{5}$

20. $\dfrac{x}{x - 1} = \dfrac{x}{x + 2}$

B 21. The supplement of an angle is six times the complement of the angle. What is the measure of the angle?

22. A segment 54 cm long is cut into two pieces with lengths in the ratio $5:4$. What is the length of the longer piece?

23. The acute angles of a right triangle are in the ratio $5:13$. What is the measure of the larger angle?

24. The angles of a triangle are in the ratio $1:3:8$. What is the measure of each?

25. In a regular polygon, the ratio of the measure of an interior angle to the measure of an exterior angle is $2:1$. How many sides does the polygon have?

In Exercises 26–28, $\angle A$ is an acute angle of a right triangle.

26. Sin $A = \frac{3}{8}$ and the side opposite $\angle A$ is 15 cm. How long is the hypotenuse?

27. Cos $A = \frac{2}{3}$ and the hypotenuse is 7.5 cm. How long is the leg adjacent to $\angle A$?

28. Tan $A = \frac{4}{9}$ and the shorter leg is 6 cm. What is the length of the longer leg?

Multiple-Choice Exercises

Write the letter which gives the best answer.

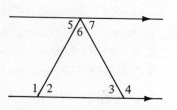

A 1. Which pair of angles do we know are equal?

    a. $\angle 2$ and $\angle 6$ **b.** $\angle 6$ and $\angle 7$

    c. $\angle 3$ and $\angle 5$ **d.** $\angle 3$ and $\angle 7$

    e. none of these

2. If $\angle 2 = \angle 3$, which pair of angles do we know are equal?

    a. $\angle 2$ and $\angle 6$ **b.** $\angle 6$ and $\angle 7$

    c. $\angle 3$ and $\angle 5$ **d.** $\angle 3$ and $\angle 6$ **e.** none of these

3. Which of the following is *not* a method for proving triangles congruent?

    a. ASA **b.** SAS **c.** AAS **d.** SSA **e.** SSS

4. The sides of a triangle can have these measures:

    a. 4, 5, 9 **b.** 16, 22, 5 **c.** 17, 10, 9 **d.** 7, 10, 19 **e.** none of these

5. An exterior angle of a regular polygon *cannot* have the measure:

    a. 120° **b.** 90° **c.** 50° **d.** 40° **e.** 30°

6. Which of the following is *not* in simplest form?

    a. $\sqrt{6}$ **b.** $3\sqrt{15}$ **c.** $9\sqrt{2}$ **d.** $5\sqrt{18}$ **e.** $11\sqrt{11}$

7. The lengths of the two legs and the hypotenuse of a right triangle can be:

    a. 2, 3, 5 **b.** 4, 9, $\sqrt{13}$ **c.** $\dfrac{1}{2}, \dfrac{1}{2}, 1$ **d.** $\sqrt{2}, \sqrt{3}, \sqrt{6}$ **e.** none of these

8. If $\triangle ABC \sim \triangle A'B'C'$, $AB = 6$ and $A'B' = 9$, the lengths of another pair of corresponding sides could be:

    a. 3 and 4 **b.** 42 and 48 **c.** 12 and 18 **d.** 7 and 10 **e.** none of these

9. If $\triangle MJL \cong \triangle AEG$, then it is also true that:

    a. $ML = EG$ **b.** $JM = AG$ **c.** $\angle M = \angle E$ **d.** $\angle M = \angle G$ **e.** none of these

B 10. The altitude to the hypotenuse of a 30-60-90 triangle is 6 cm. The hypotenuse is:

    a. 12 cm **b.** $12\sqrt{3}$ cm **c.** $3\sqrt{3}$ cm **d.** $8\sqrt{3}$ cm **e.** none of these

11. The legs of a right triangle are 6 and 8. The length of the median to the hypotenuse is:

    a. 10 **b.** 5 **c.** $\dfrac{20}{3}$ **d.** $\dfrac{5}{2}$ **e.** none of these

12. If l, j, k, and n are coplanar lines such that $l \parallel j$, $k \parallel n$, and $j \perp n$, then it is *not* true that:

    a. $l \perp n$ **b.** $k \perp j$ **c.** $j \parallel n$ **d.** $n \parallel k$ **e.** $l \perp k$

13. In $\triangle ABC$, D is the midpoint of $\overline{BC}$ and E is the midpoint of $\overline{AB}$. $\overline{AD}$ and $\overline{CE}$ intersect at point F. Then:

    a. F is equidistant from the three sides of the triangle.

    b. $\overrightarrow{CE}$ is the bisector of $\angle ACB$.

    c. $\overleftrightarrow{AD}$ is the perpendicular bisector of $\overline{BC}$.

    d. $AF = 2 \cdot DF$

    e. none of these

14. Which proportion is *not* equivalent to $\dfrac{a}{b} = \dfrac{c}{d}$?

    a. $\dfrac{a}{c} = \dfrac{b}{d}$ **b.** $\dfrac{b}{a + b} = \dfrac{d}{c + d}$ **c.** $\dfrac{a}{d} = \dfrac{c}{b}$ **d.** $\dfrac{b}{a} = \dfrac{d}{c}$ **e.** $\dfrac{c}{a} = \dfrac{d}{b}$

15. The three sides of an isosceles right triangle are in the ratio:

    a. $1 : \sqrt{3} : 2$ **b.** $1 : 1 : 2$ **c.** $3 : 4 : 5$ **d.** $1 : 1 : \sqrt{2}$ **e.** none of these

Completion Exercises

Complete the statements in the best way.

A 1. If both pairs of opposite sides of a quadrilateral are equal, then the quadrilateral must be a __?__.

2. If a trapezoid has equal legs, then it is an __?__ trapezoid.

3. The equation $\frac{a}{b} = \frac{c}{d}$ is called a __?__.   4. When $\frac{a}{b} = \frac{c}{d}$, then $\frac{a}{c} = \frac{?}{?}$.

5. If two angles of one triangle are equal to two angles of another triangle, then the triangles must be __?__.

6. When two planes intersect, they intersect in a __?__.

7. The endpoint of $\overrightarrow{HK}$ is point __?__.

8. If the legs of one right triangle equal the legs of another right triangle, then the triangles are __?__.

9. If the vertex angle of an isosceles triangle measures 110°, then each base angle measures __?__°.

10. If the measures of two angles have a sum of 180°, then the angles are __?__.

11. If the measure of each exterior angle of a regular polygon is 90°, then the polygon is a __?__.

12. If $\overrightarrow{AD}$ bisects $\angle BAC$, and $\angle DAB = 48°$, then $\angle BAC = $ __?__°.

13. If $\angle A$ is an acute angle of a right triangle then $\sin A = \dfrac{?}{\text{hypotenuse}}$.

14. In $\triangle TEX$, J lies on $\overline{ET}$ and K lies on $\overline{EX}$. If $\overline{JK} \parallel \overline{TX}$, then $\dfrac{EJ}{ET} = \dfrac{EK}{?}$.

15. If two points on a number line have coordinates p and q, then the coordinate of the midpoint is __?__.

16. If a, b, c, and d are nonzero numbers such that $ab = cd$, then $\dfrac{b}{d} = \dfrac{?}{?}$.

17. If $\triangle HKM \cong \triangle ADF$, then $\angle HMK = \angle$ __?__.

18. If two triangles are similar, then corresponding sides are __?__ and corresponding angles are __?__.

19. If two parallel lines are cut by a transversal, then interior angles on the same side of the transversal are __?__.

B 20. If the diagonals of a parallelogram are perpendicular to each other and equal in length, then the parallelogram must be a __?__.

21. If the hypotenuse of $\triangle ABC$ is 10 cm and $\sin B = \frac{2}{5}$, then the leg opposite $\angle B$ is __?__ cm.

22. If a building is 80 m high and the distance from point P on the ground to the top of the building is 160 m, then the angle of elevation of the top of the building from point P is __?__°.

23. The contrapositive of the statement "If $x = 3$, then $x^2 = 9$" is __?__.

Miscellaneous Exercises

In the diagram, $\overline{LM} \| \overline{HJ}$.

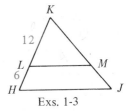

Exs. 1-3

A 1. Name two similar triangles.

2. If $KM = 16$, then $JM = \underline{\ ?\ }$.

3. If $LM = 14$, then $HJ = \underline{\ ?\ }$.

4. The equation $a^2 + b^2 = c^2$ states the $\underline{\ ?\ }$ Theorem.

5. If $a = 12$, $b = 9$, and $c = 15$, then $\sin A = \underline{\ ?\ }$ and $\tan B = \underline{\ ?\ }$.

6. If $\angle A = 53°$, then $\angle B = \underline{\ ?\ }°$.

7. If $\angle A = 53°$ and $c = 10$, then $a \approx \underline{\ ?\ }$. (Use the Table of Trigonometric Ratios on page 225.)

8. If $\angle B = 30°$ and $b = 8$, then $a = \underline{\ ?\ }$ and $c = \underline{\ ?\ }$.

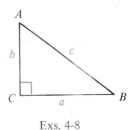

Exs. 4-8

The diagram shows trapezoid $RSTU$ with median $\overline{MN}$.

9. If $RS = 15$ and $UT = 11$, $MN = \underline{\ ?\ }$.

10. If $RS = 15$ and $UT = 11$, a diagonal of trap. $RSTU$ divides $\overline{MN}$ into two segments with lengths $\underline{\ ?\ }$ and $\underline{\ ?\ }$.

11. If $\angle NMR = 105°$, then $\angle URS = \underline{\ ?\ }°$.

12. Draw trap. $RSTU$ on your paper. Mark point G on $\overline{RS}$ so that $\overline{GT} \| \overline{RU}$. Draw $\overline{GT}$, intersecting $\overline{MN}$ at point H. Name each angle that is equal to $\angle R$.

Exs. 9-13

13. How many trapezoids are there in your drawing for Exercise 12?

14. Points A and B on a number line have coordinates -7.5 and 4.5. What is the coordinate of the midpoint of $\overline{AB}$?

15. A square has sides of length 6. What is the length of the diagonal?

16. An equilateral triangle has sides of length 6. What is the length of an altitude?

17. The ratio of one acute angle of a right triangle to the other acute angle is $1:2$. What is the measure of each acute angle?

18. Each interior angle of a regular polygon is $171°$. How many sides does the polygon have?

19. The perimeter of a rhombus is 52 and one diagonal is 10. What is the length of the other diagonal?

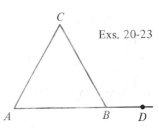

Exs. 20-23

20. If $\angle CBD = 121°$ and $\angle A = 59°$, then $\angle C = \underline{\ ?\ }°$.

21. If $\angle C = 48°$ and $AC = BC$, then $\angle CBD = \underline{\ ?\ }°$.

22. If $AB = AC$ and $\angle CBD = 126°$, then $\angle A = \underline{\ ?\ }°$.

23. If $\overrightarrow{BE}$ bisects $\angle CBD$, $AC = BC$, and $\angle EBC = 67°$, then $\angle C = \underline{\ ?\ }°$.

B 24. In $\triangle RST$, $\angle T = 90°$, $ST = 10$, and $\sin R = \frac{5}{13}$. $RT = \underline{\ ?\ }$.

25. The height of an equilateral triangle is $5\sqrt{3}$. What is the length of a side?

26. A square has diagonals of length 12. What is the length of each side?

27. The vertex angle of an isosceles triangle is 120°. The base is 24 cm. How long is the altitude to the base?

28. The median of a trapezoid is 22.5 and the longer base is 27. How long is the other base?

29. Each side of a regular hexagon is 5. Find the length of one of the shorter diagonals.

In the diagram, $\overline{DE}\|\overline{AB}$.

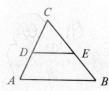

30. If $AC = 20$, $DC = 12$, and $CE = 14$, then $BC = $? .

31. If $DE = 10$, $AB = 15$, and $CD = 8$, then $AD = $? .

32. If $CE = 18$, $DC = 16$, and $EB = 8$, then $AC = $? .

Exs. 30-32

33. The endpoints of a segment on a number line have coordinates $2x - 5$ and $10x + 4$. The coordinate of the midpoint is $5x + 1$. Find the value of x.

34. The bases of an isosceles trapezoid are 33 and 43. The distance between the bases is 12. How long is a leg?

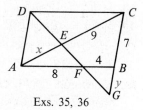

In the diagram, $\overline{AB}\|\overline{DC}$ and $\overline{AD}\|\overline{GC}$.

35. Find the value of x.   **36.** Find the value of y.

Exs. 35, 36

Proof Exercises

A **1.** Given: $l\|m$; $\angle 3 = \angle 5$
Prove: $l\|n$

2. Given: $l\|m$; $m\|n$
Prove: $\angle 12 = \angle 6$

3. Given: $l\|m$; $m\|n$
Prove: $\angle 1$ and $\angle 6$ are supplementary.

4. Given: $\angle 1 = \angle 3$; $\angle 10 = \angle 8$
Prove: $l\|n$

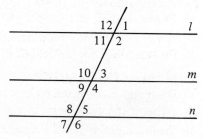

Exs. 1-4

5. Given: $\overline{CD}\|\overline{AB}$; $CD = AB$
Prove: $\triangle CDA \cong \triangle ABC$

6. Given: $AB = CD$; $CB = AD$
Prove: $\triangle ABC \cong \triangle CDA$

7. Given: $\angle ACB = \angle CAD$; $CB = AD$
Prove: $\angle CAB = \angle ACD$

8. Given: $\angle B = \angle D$; $\angle BCA = \angle DAC$
Prove: $AB = CD$

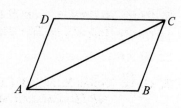

Exs. 5-8

9. Given: $\angle 1 = \angle 2$
   Prove: $\angle 3 = \angle 4$

10. Given: $BE = BD$; $\angle 3 = \angle 4$
    Prove: $\angle 2 = \angle 4$

11. Given: $BC = BA$; $\overline{DE} \| \overline{AC}$
    Prove: $\angle 3 = \angle 2$

12. Given: $\angle P = \angle S$
    Prove: $\angle M = \angle R$

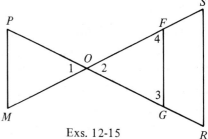

Exs. 9-11

B 13. Given: $\triangle MPO \sim \triangle SRO$; $\angle S = \angle 4$
    Prove: $\triangle MPO \sim \triangle FGO$

14. Given: $\overline{FG} \| \overline{SR}$; $\triangle MPO \sim \triangle SRO$; $PO = GO$
    Prove: $\triangle MPO \cong \triangle FGO$

15. Given: $\angle P = \angle S$; $\triangle MPO \cong \triangle GFO$
    Prove: $\overline{FG} \| \overline{SR}$

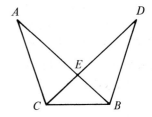

Exs. 12-15

16. Given: $\triangle AEC \cong \triangle DEB$
    Prove: $\angle ECB = \angle EBC$

17. Given: $AC = DB$; $\angle ACB = \angle DBC$
    Prove: $AB = DC$

18. Given: $CE = BE$; $CD = BA$
    Prove: $\angle A = \angle D$

19. Given: $\triangle ACB \cong \triangle DBC$
    Prove: $\triangle CEB$ is isosceles.

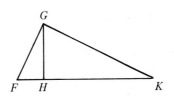

Exs. 16-19

20. Given: $\triangle FHG \sim \triangle GHK$; $\overline{GH} \perp \overline{FK}$
    Prove: $\angle FGK$ is a right angle.

21. Given: $\angle FGK$ is a right angle; $\overline{GH} \perp \overline{FK}$
    Prove: $(FK)^2 = (FH)^2 + 2(HG)^2 + (HK)^2$

22. Given: $\angle TAD = \angle TCB$
    Prove: $\triangle ANB \sim \triangle CND$

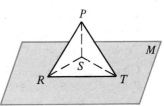

Exs. 20, 21

23. Given: $\triangle TBC \sim \triangle TDA$
    Prove: $\triangle BNA \sim \triangle DNC$

24. Given: $\angle B = \angle D$
    Prove: $\dfrac{TB}{TD} = \dfrac{TC}{TA}$

25. Given: $\angle BAN = \angle DCN$
    Prove: $TC \cdot DA = TA \cdot BC$

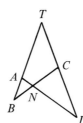

Exs. 22-25

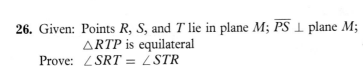

26. Given: Points R, S, and T lie in plane M; $\overline{PS} \perp$ plane M;
    $\triangle RTP$ is equilateral
    Prove: $\angle SRT = \angle STR$

Ex. 26

7

Circles

Tangents, Arcs, and Chords

Objectives

1. Define a circle, a sphere, and terms related to them.
2. Recognize circumscribed and inscribed polygons and circles.
3. Apply the theorems that relate tangents and radii.
4. Apply the properties involving arcs and central angles.
5. Apply the theorems about the chords of a circle.

7-1 *Basic Terms*

A **circle** ($\odot$) is the set of all points in a plane that are a given distance from a given point in the plane. The given point is the **center,** and the given distance is the **radius.** The circle shown is called circle O ($\odot O$) since its center is point O.

The radius of circle O is 6.
A lies inside the circle.　$OA < 6$
B lies on the circle.　$OB = 6$
C lies outside the circle.　$OC > 6$

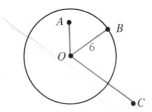

A **secant** is a line that intersects a circle in two points.
A **chord** is a segment that joins two points on a circle.
A **diameter** is a chord that passes through the center.

Secants: $\overleftrightarrow{DE}$, $\overleftrightarrow{FG}$
Chords: $\overline{DE}$, $\overline{FG}$
Diameter: $\overline{FG}$

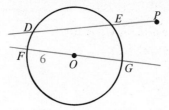

A segment such as $\overline{PD}$ above is referred to as a *secant drawn to a circle from an outside point.*

The words *diameter* and *radius* can refer to lengths of segments or to the segments themselves. Thus the radius of $\odot O$ above is 6 and $\overline{OF}$ is a radius. $\overline{OF}$ and $\overline{OG}$ are *radii* (plural of *radius*). From the definitions it follows that:

> A diameter equals two times a radius.
> All radii of a circle are equal.

Equal circles are circles with equal radii. You may state this in another way: Radii of equal circles are equal.

Concentric circles are circles that lie in the same plane and have the same center. The rings of the target illustrate concentric circles.

A polygon is **inscribed in a circle** and the circle is **circumscribed about the polygon** when each vertex of the polygon lies on the circle.

Inscribed polygons

Circumscribed circles

Remove the phrase *in a plane* from the definition of a circle and you have the definition of a sphere. A **sphere** is the set of all points that are a given distance from a given point. Many of the terms used for circles are also used for spheres.

Classroom Exercises

1. Using ⊙*O*, name two radii, a diameter, three chords, and two secants.

2. Repeat Exercise 1, using sphere *P*.

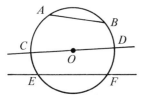

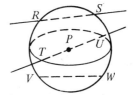

3. What is the diameter of a circle whose radius is 5? 6½? 9.3? *n*?

4. What is the radius of a sphere whose diameter is 18? 15? 3.2? *j*?

5. Consider a circle with several parallel chords. What do you guess about the midpoints of all such chords?

6. Point *A* lies on circle *O*. How many chords can you draw that contain *A*? How many diameters?

7. Point *B* lies inside sphere *S*. How many chords can you draw that contain *B*? How many radii?

8. Is △*ABC* an inscribed or a circumscribed triangle? (See diagram below.)

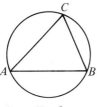

Ex. 8

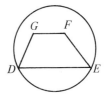

Ex. 9

9. Explain why it is not correct to call *DEFG* an inscribed quadrilateral. (See diagram above.)

10. Suppose ⊙*O* and ⊙*P* are equal and the radius of ⊙*O* is 5. What is the diameter of ⊙*P*?

11. Suppose ⊙*X* and ⊙*Y* are equal and the diameter of ⊙*X* is 3. What is the radius of ⊙*Y*?

12. Plane *P* passes through the center of sphere *O*.

   a. Explain why *OA* = *OB* = *OC*.

   b. Explain why the intersection of the plane and the sphere is a circle. (This circle is called a *great circle* of the sphere.)

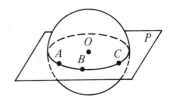

Written Exercises

Point *A* lies inside circle *O* and point *B* lies outside. In how many points does each figure named intersect the circle?

A   **1.** $\overline{AB}$ **2.** $\overrightarrow{AB}$ **3.** $\overrightarrow{BA}$ **4.** $\overleftrightarrow{AO}$

Point *C* lies inside sphere *P* and point *D* lies outside. In how many points does each figure named intersect the sphere?

   5. $\overline{CD}$ **6.** $\overrightarrow{CD}$ **7.** $\overrightarrow{DC}$ **8.** $\overline{CP}$

9. Sphere *Z* is cut by a plane that passes through center *Z*. Describe, as completely as you can, the figure in which the plane intersects the sphere.

10. Repeat Exercise 9, but this time the plane does *not* pass through *Z*.

For each exercise, draw a diagram showing the figure inscribed in a circle. If a drawing cannot be made, write *impossible*.

11. A square

12. An isosceles triangle

13. An obtuse triangle

14. A nonregular pentagon

15. A parallelogram without any right angle.

16. A quadrilateral *ABCD*, with $\overline{AC}$ a diameter.

Circle *O* has radius 8. In each exercise, radii $\overline{OX}$ and $\overline{OY}$ are drawn so that ∠*XOY* has the measure given. Find the length of $\overline{XY}$.

B   **17.** ∠*XOY* = 60°   **18.** ∠*XOY* = 90°   **19.** ∠*XOY* = 120°   **20.** ∠*XOY* = 180°

21. Write a definition of *radius of a sphere*.

22. Write a definition of *chord of a sphere*.

23. The radius of a sphere is 13. The distance from the center of the sphere to a certain plane is 5. What is the radius of the circle in which the plane intersects the sphere?

24. The radius of a sphere is 25. A plane intersects the sphere in a circle whose radius is 24. What is the distance between the center of the sphere and the plane?

C   **25.** The distance between point *Q* and a certain plane is 5. Sphere *Q* intersects the plane in a circle whose area is 75π. Find the radius of sphere *Q*.

26. The radius of ⊙*O* is 5; point *X* lies inside the circle; line *l* lies in the plane of ⊙*O* and passes through *X*. You can prove something that seems obvious: Line *l* cuts ⊙*O* in two points. Write the key steps of a proof. (*Hint:* Draw a perpendicular from *O* to *l* and let *Z* be the point of intersection. On *l* take points *A* and *B* at distance $\sqrt{25 - (OZ)^2}$ from *Z*. Show that both *A* and *B* lie on ⊙*O*.)

27. Write an indirect proof of the statement: A straight line intersects a circle in at most two points.

7-2 *Tangents*

A **tangent** to a circle is a line that lies in the plane of the circle and meets the circle in exactly one point, called the **point of tangency.** A segment that is part of the tangent line and that contains the point of tangency is also called a tangent.

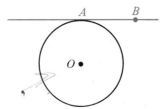

$\overleftrightarrow{AB}$ is tangent to $\odot O$.
$\odot O$ is tangent to $\overleftrightarrow{AB}$.
A is the point of tangency.
$\overline{AB}$ is tangent to $\odot O$.

Theorem 7-1   If a line is tangent to a circle, then the line is perpendicular to the radius drawn to the point of tangency.

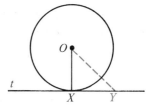

Given: Line t is tangent to $\odot O$ at point X.

Prove: $\overline{OX} \perp t$

Proof (indirect):
Suppose $\overline{OX}$ is not perpendicular to t. Then there must be some other segment, $\overline{OY}$, through O that is perpendicular to t. Then $OY < OX$. (The perpendicular segment from a point to a line is the shortest segment from the point to the line.) But this contradicts the fact that $OY > OX$ (because Y lies outside $\odot O$, while X lies on $\odot O$). What we supposed to be true, that $\overline{OX}$ is not perpendicular to t, must be false. It follows that $\overline{OX} \perp t$.

Corollary   Tangents to a circle from a point are equal.

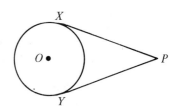

   The corollary tells us that if $\overline{PX}$ and $\overline{PY}$ are tangent to $\odot O$ at X and Y, then $PX = PY$.
   For proof of the corollary, see Classroom Exercise 7.

Theorem 7-2   If a line in the plane of a circle is perpendicular to a radius at its outer endpoint, then the line is tangent to the circle.

Given: Line *m* lies in the plane of $\odot P$;
     $m \perp \overline{PZ}$ at *Z*.

Prove: *m* is tangent to $\odot P$.

The proof is left as Exercise 17.

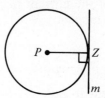

A line that is tangent to each of two coplanar circles is called a **common tangent.**

Common *internal* tangents intersect the segment joining the centers.

Common *external* tangents do *not* intersect the segment joining the centers.

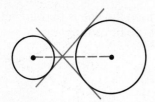

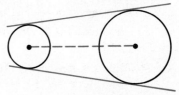

Two *circles are tangent* to each other when they are coplanar and are tangent to the same line at the same point.

$\odot W$ and $\odot X$ are *externally* tangent.

$\odot Y$ and $\odot Z$ are *internally* tangent.

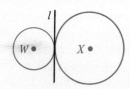

When each side of a polygon is tangent to a circle, the polygon is said to be **circumscribed about the circle.** The circle is **inscribed in the polygon.**

Circumscribed polygons

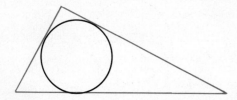

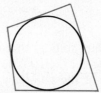

Inscribed circles

Classroom Exercises

How many common external tangents and how many common internal tangents can be drawn to the two circles?

1.

2.

3.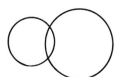

4.

5.

6.

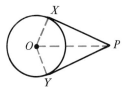

7. State the key steps of a proof of the Corollary to Theorem 7-1. That is, show that if $\overline{PX}$ and $\overline{PY}$ are tangent to $\odot O$, then $PX = PY$.

8. If $\overline{PX}$ and $\overline{PY}$ are tangent to $\odot O$, what can you prove about $\angle P$ and $\angle O$? What are the key steps of a proof?

Written Exercises

Draw two circles, with all their common tangents, in such a way that the number of tangents is the number stated.

A   1. One   **2.** Two   **3.** Three   **4.** Four

In Exercises 5 and 6, O and P are centers of the circles shown. Find the lengths indicated.

5.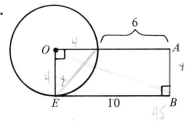

$AB = \underline{\ ?\ }$

6.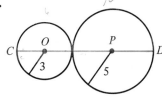

$CD = \underline{\ ?\ }$

In Exercises 7–10, O and P are centers of the circle shown. Find the lengths indicated.

7.

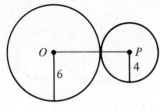

$OP = \underline{\ ?\ }$

$= 50$

8.

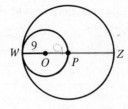

$OZ = \underline{\ ?\ }$

B 9.

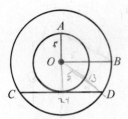

$OA = 5$; $CD = 24$; $OB = \underline{\ ?\ }$

10.

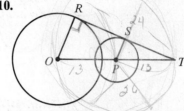

$RT = 24$; $OT = 26$; $OP = 13$;
$SP = \underline{\ ?\ }$

11. At a point on a sphere, how many lines can be drawn tangent to the sphere?

12. Discover and prove a theorem about the two lines drawn tangent to a circle at the ends of a diameter.

13. Two tangents are drawn to a circle from a point outside the circle. A ray is drawn from the point through the center of the circle. What could you prove about this ray? You don't need to write a proof.

For Exercises 14–16, refer to the diagrams below.

14. The circle is inscribed in the triangle. Find the value of x.

15. Each of the circles with centers shown is tangent to the other two circles. $AB = 8$, $BC = 10$, and $AC = 12$. Find the radius of each circle.

16. Three equal circles are shown. How many circles tangent to all three of the given circles can be drawn?

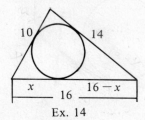

Ex. 14

Ex. 15

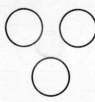

Ex. 16

C 17. Prove Theorem 7-2, page 250. (*Hint:* Write an indirect proof. Suppose that *m* is not tangent to ⊙*P.* Then there is a second point, call it *W*, in which *m* intersects ⊙*P.* Compare *PW* and *PZ.*)

18. Circles *O* and *P* are externally tangent at *X.*

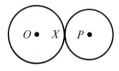

    a. Explain why $\overleftrightarrow{OP}$ must pass through *X.*

    b. What if the circles were internally tangent at *X*? Would your explanation still be good?

19. A circle inscribed in right △*ABC* is tangent to hypotenuse $\overline{AB}$ at *T.* If *AT* = 5 and *BT* = 12, find the perimeter of the triangle.

20. The radius of the circle inscribed in a particular right triangle is 1 while the radius of the circumscribed circle is 10. Find the perimeter of the triangle.

7-3 *Arcs and Central Angles*

An **arc** is part of a circle. There are three kinds of arcs.

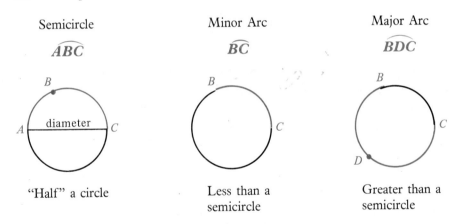

Semicircle	Minor Arc	Major Arc
$\widehat{ABC}$	$\widehat{BC}$	$\widehat{BDC}$
"Half" a circle	Less than a semicircle	Greater than a semicircle

$\widehat{ABC}$ is read "arc *ABC.*" To name a semicircle or a major arc, three letters must be used.

A **central angle** of a circle is an angle whose vertex is the center of the circle. For every minor arc there is a central angle.

The **measure of a minor arc** is defined to be the measure of its central angle. In the diagram, ∠*TOK* is the central angle of $\widehat{TK}.$ To indicate that the measure of ∠*TOK* equals the measure of $\widehat{TK},$ we shall write ∠*TOK* = $\widehat{TK}.$ (In some texts you will find this written as *m* ∠*TOK* = *m* $\widehat{TK}.$) If ∠*TOK* = 60°, then $\widehat{TK}$ = 60°.

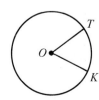

The **measure of a semicircle** is 180°. The **measure of a major arc** is found as shown below.

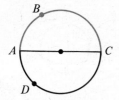

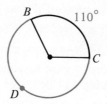

Semicircle

$\overset{\frown}{ABC} = \overset{\frown}{ADC} = 180°$

Major Arc

$\overset{\frown}{BDC} = 360° - \overset{\frown}{BC}$
$= 360° - 110° = 250°$

The diagram at the right above can be used to illustrate the addition of arc measures. We write $\overset{\frown}{BD} + \overset{\frown}{DC} = \overset{\frown}{BDC}$ as an abbreviation for "the *measure* of $\overset{\frown}{BD}$ plus the *measure* of $\overset{\frown}{DC}$ equals the *measure* of $\overset{\frown}{BDC}$."

In the same circle or in equal circles, arcs that have equal measures are called **equal arcs.** Circle O and circle P are equal circles since their radii are equal. $\overset{\frown}{DE}$, $\overset{\frown}{GF}$, and $\overset{\frown}{JK}$ are equal arcs since each has measure 62°.

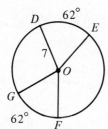

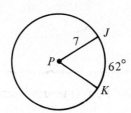

You can see that each of these equal arcs has a 62° central angle and that the central angles are equal. In Exercise 16 you will prove in general that equal minor arcs have equal central angles.

Classroom Exercises

In Exercises 1–9, find the measure of each arc.

1. $\overset{\frown}{HJ}$ 2. $\overset{\frown}{JK}$ 3. $\overset{\frown}{HK}$

4. $\overset{\frown}{HNK}$ 5. $\overset{\frown}{HNJ}$ 6. $\overset{\frown}{NHJ}$

7. $\overset{\frown}{NKJ}$ 8. $\overset{\frown}{JKH}$ 9. $\overset{\frown}{NK}$

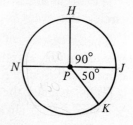

Using the letters shown in the diagram, name:

10. Two equal central angles

11. Two equal minor arcs

12. Any two major arcs

Exs. 1–12

Find the measures indicated.

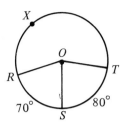

13. $\angle ROS = \underline{\;?\;}°$

15. $\angle ROT = \underline{\;?\;}°$

17. $\widehat{RXT} = \underline{\;?\;}°$

14. $\angle TOS = \underline{\;?\;}°$

16. $\widehat{RST} = \underline{\;?\;}°$

18. $\widehat{TXS} = \underline{\;?\;}°$

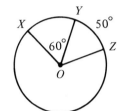

19. $\widehat{XY} = \underline{\;?\;}°$

21. $\angle XOZ = \underline{\;?\;}°$

20. $\angle YOZ = \underline{\;?\;}°$

22. $\widehat{XZ} = \underline{\;?\;}°$

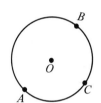

23. a. Explain why it is not correct to say $\widehat{AB} + \widehat{BC} = \widehat{AC}$.

b. Complete the statement correctly: $\widehat{AB} + \widehat{BC} = \underline{\;?\;}$.

Written Exercises

$\overrightarrow{OW}$ bisects central $\angle ZOX$. Name the arcs indicated.

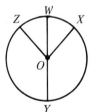

A 1. Two semicircles

2. Five minor arcs

3. Five major arcs

4. Two pairs of equal minor arcs

Find the measure of $\angle 1$.

5.

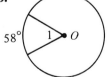

6.

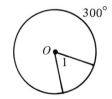

7.

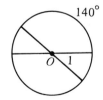

8.

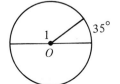

9.

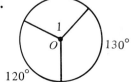

10.

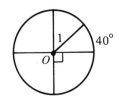

11. At one o'clock the hands of a clock form an angle of __?__ °.

12. At four o'clock the hands of a clock form an angle of __?__ °.

13. At three o'clock the hands form a 90° angle, but at five minutes after three they do not form a 60° angle. Explain.

14. Place points X, Y, and Z on a circle in such positions that the statement $\widehat{XY} + \widehat{YZ} = \widehat{XZ}$ is not correct.

15. In this exercise you will prove that in equal circles, equal central angles have equal arcs.

Given: Equal circles O and P; $\angle 1 = \angle 2$
Prove: $\widehat{WX} = \widehat{YZ}$

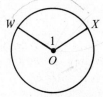
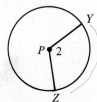

16. Draw a diagram and prove: In a circle, equal minor arcs have equal central angles.

B 17. $\overline{TS}$ is a diameter of $\odot O$.
    a. If $\angle 1 = 32°$, then $\widehat{RS} = $ __?__ °.
    b. If $\widehat{RS} = 70°$, then $\angle 1 = $ __?__ °.

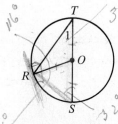

18. $\overline{AB}$ is a diameter of $\odot P$.
    a. If $\widehat{BC} = 132°$, then $\angle C = $ __?__ °.
    b. If $\widehat{BC} = n°$, then $\angle C = $ __?__ °.

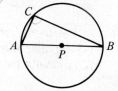

19. Given: $\overline{XW}$ is a diameter of $\odot O$;
    $\overline{XY} \parallel \overline{OZ}$
What can you prove about $\widehat{YZ}$?
Write a complete proof.

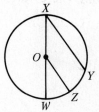

20. Circle O and circle P lie on sphere O and in parallel planes. $\overline{OT}$ is perpendicular to both of those planes. $\widehat{JK} = 30°$ and $PO = 20$. Find the length of $\overline{PT}$.

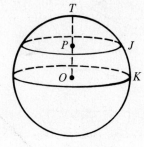

In the diagram shown, sphere O represents Earth (E), and point S represents a satellite. Segments from S tangent to E meet E at points on $\odot P$. Any two points above $\odot P$ on E can communicate with each other via S. Points A and B are extreme communication points. The distance on E between A and B equals the length of $\overset{\frown}{ACB}$: $2 \times \dfrac{n}{360} \times$ circumference of E. Radius of $E \approx 6400$ km. Circumference of $E \approx 40200$ km.

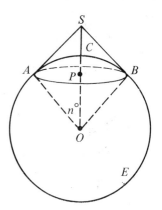

C 21. Suppose S is 200 km above E, that is, $SC = 200$. Find the distance, rounded to the nearest 100 km, on E between A and B. (*Hint:* Evaluate cos n, then n.)

22. Suppose S were 400 km from E, twice as far as in Exercise 21. Would the distance between A and B be twice as great as in Exercise 21?

7-4 *Arcs and Chords*

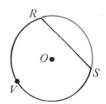

In circle O, $\overline{RS}$ cuts off two arcs, $\overset{\frown}{RS}$ and $\overset{\frown}{RVS}$. We speak of $\overset{\frown}{RS}$, the minor arc, as the *arc of chord* $\overline{RS}$.

Theorem 7-3 In the same circle or in equal circles,
(1) Equal chords have equal arcs;
(2) Equal arcs have equal chords.

We prove (1) for equal circles here.

Given: Equal circles O and P; $AB = CD$

Prove: $\overset{\frown}{AB} = \overset{\frown}{CD}$

Proof:

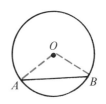

Statements	Reasons
1. Draw $\overline{OA}$, $\overline{OB}$, $\overline{PC}$, $\overline{PD}$.	1. Through any two points ___?___.
2. $OA = PC$; $OB = PD$	2. ___?___
3. $AB = CD$	3. Given
4. $\triangle AOB \cong \triangle CPD$	4. ___?___
5. $\angle O = \angle P$	5. ___?___
6. $\overset{\frown}{AB} = \angle O$; $\overset{\frown}{CD} = \angle P$	6. The measure of a minor arc is the measure of its central angle.
7. $\overset{\frown}{AB} = \overset{\frown}{CD}$	7. Substitution Property

The next theorem involves the idea of bisecting an arc. As you would expect, M is called the *midpoint* of $\widehat{JMK}$ if $\widehat{JM} = \widehat{MK}$. Any line, segment, or ray that contains M bisects $\widehat{JMK}$.

Theorem 7-4   A diameter that is perpendicular to a chord bisects the chord and its arc.

Given: $\odot O$; $\overline{CD} \perp \overline{AB}$

Prove: $AZ = BZ$; $\widehat{AD} = \widehat{BD}$

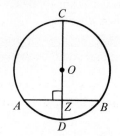

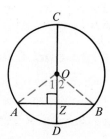

Proof:

Statements	Reasons
1. Draw $\overline{OA}$ and $\overline{OB}$.	1. Through any two points ___?___.
2. $OA = OB$	2. ___?___
3. $OZ = OZ$	3. ___?___
4. $\overline{CD} \perp \overline{AB}$	4. ___?___
5. Rt. $\triangle OZA \cong$ rt. $\triangle OZB$	5. *HL* Theorem
6. $AZ = BZ$; $\angle 1 = \angle 2$	6. ___?___
7. $\widehat{AD} = \angle 1$; $\widehat{BD} = \angle 2$	7. The measure of an arc is ___?___.
8. $\widehat{AD} = \widehat{BD}$	8. ___?___

You will prove part (1) of the next theorem as Classroom Exercise 5.

Theorem 7-5   In the same circle or in equal circles,
   (1) Equal chords are equally distant from the center;
   (2) Chords equally distant from the center are equal.

Classroom Exercises

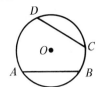

1. Tell how you would prove the first part of Theorem 7-3 for one circle.
   Given: $\odot O$ with $AB = CD$
   Prove: $\overparen{AB} = \overparen{CD}$

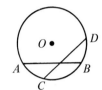

2. Would your proof in Exercise 1 be any different if the chords intersected inside the circle?

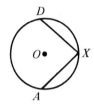

3. Would your proof in Exercise 1 be any different if the chords intersected on the circle? That is, if B and C were the same point X?

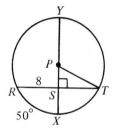

4. $\overline{XY}$ is a diameter of $\odot P$. Find the values indicated.
   a. $ST = \underline{\ ?\ }$ b. $RT = \underline{\ ?\ }$ c. $\overparen{XT} = \underline{\ ?\ }°$
   d. $\overparen{YT} = \underline{\ ?\ }°$ e. $\overparen{RY} = \underline{\ ?\ }°$ f. $\overparen{RXT} = \underline{\ ?\ }°$
   g. If $PT = 10$, then $PS = \underline{\ ?\ }$.

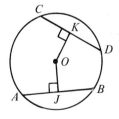

5. Explain how you would prove Theorem 7-5 (1): In the same circle or in equal circles, equal chords are equally distant from the center.
   Given: $\odot O$ with $AB = CD$; $\overline{OJ} \perp \overline{AB}$; $\overline{OK} \perp \overline{CD}$
   Prove: $OJ = OK$

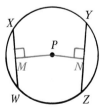

6. The radius of $\odot P$ is 7; $YZ = 10$; $XW = 8$
   a. $NY = \underline{\ ?\ }$; $MX = \underline{\ ?\ }$
   b. Think of rt. $\triangle PMX$ and rt. $\triangle PNY$.
      $PN = \underline{\ ?\ }$; $PM = \underline{\ ?\ }$

7. Exercise 6 suggests that if a circle has two unequal chords, the longer one is nearer the center. Do you think this is true for all circles?

Written Exercises

In these exercises when a point is labeled *O*, that point is the center of the circle.
Find the lengths indicated.

A 1.

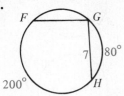

FG = _?_

2.

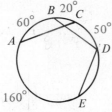

AC = 8; BD = 7;
ED = _?_

3.

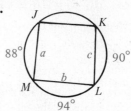

JK = _?_

4.

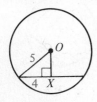

OX = _?_

5.

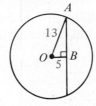

AB = _?_

6.

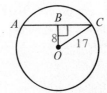

AC = _?_

7.

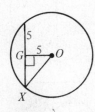

OX = _?_

8.

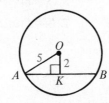

AB = _?_

9.

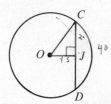

CD = 40; OJ = 15;
OC = _?_

10. Given: $\angle R = \angle S$
Explain why $\overset{\frown}{RT}$ must equal $\overset{\frown}{TS}$.

B 11. Given: $AB = CD$
Prove: $AC = BD$

12. Given: $\overset{\frown}{AB} = \overset{\frown}{CD}$; $\angle A = \angle D$
Prove: $BX = CX$

13. **a.** Draw three circles and inscribe a differently shaped quadrilateral *ABCD* in each.

 b. Use a protractor to measure the four angles of each quadrilateral.

 c. Compare ∠*A* and ∠*C*, and also ∠*B* and ∠*D*.

 d. Use inductive reasoning to predict what may be true about any quadrilateral inscribed in a circle.

14. Suppose you wish to *disprove* the statement:

 In a circle, if $\overset{\frown}{AB} = 2 \cdot \overset{\frown}{CD}$, then $AB = 2 \cdot CD$.

 Show that you have a counterexample when you let $\overset{\frown}{CD}$ be a 60° arc.

Find the lengths indicated.

15.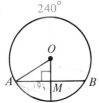

 $AB = 20;\ OA = \underline{\ ?\ }$

16.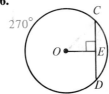

 $CD = 20;\ OE = \underline{\ ?\ }$

17.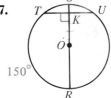

 $OK = 8\sqrt{3};\ SR = \underline{\ ?\ }$

18. In a sphere with radius *r*, the length of a chord is *c*. How far from the center of the sphere is that chord? Write your answer in terms of *r* and *c*.

C 19. Given: ⊙*O* with $\overset{\frown}{AE} = \overset{\frown}{CD} = j°$;
 $\overset{\frown}{BC} = 2j°;\ OT = 2x;\ AC = 10x + 2;$
 $ED = 10x - 2$
 Find the length of $\overline{BC}$ in terms of *j*.

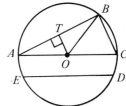

20. Given: ⊙*P* with $FG = HJ;\ \overline{FG}\|\overline{HJ}$
 Prove: $WX = ZY$
 (*Note:* You cannot assume that $\overline{FJ}$ contains *P*.)

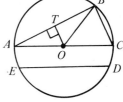

State and prove a theorem suggested by each figure.

21.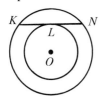

22.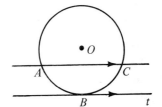

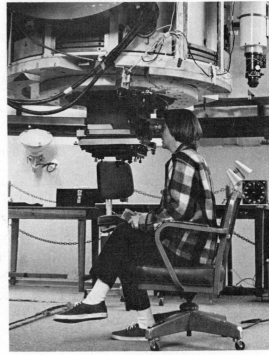

Career Notes

Astronomer

What is the chemical composition of a star in a far-away galaxy? Is it similar to that of our own sun? Astronomers deal with such questions about the universe.

A modern astronomer uses a spectroscope to analyze light from stars, thereby determining their chemical composition. Other electronic means are used to analyze x rays and radio waves emitted by stellar bodies. Although we often picture astronomers viewing the sky through a telescope for long hours, more of their time is spent analyzing astronomical data, suggesting and testing theories about outer space, often with a computer.

Astronomy has interested many of history's geometers. Among these are Isaac Newton, Carl Friedrich Gauss, and Ptolemy. These mathematicians spent days doing calculations which today's astronomer could do in seconds with a computer.

Most astronomers have a Ph.D. degree in astronomy, which requires advanced work in mathematics. They may work as researchers in universities or in government space programs.

SELF-TEST

1. In the diagram of $\odot O$ name:
   a. two radii   **b.** a secant
   c. a diameter   **d.** two chords

2. **a.** Sketch a quadrilateral inscribed in a circle.
   b. Sketch a triangle circumscribed about a circle.

3. Suppose a plane intersects a sphere in more than one point. The intersection is a ___?___ .

4. $\odot O$ and $\odot P$ are equal, and the radius of $\odot O$ is 6. The diameter of $\odot P$ is ___?___ .

5. $\odot A$ and $\odot B$ are externally tangent. How many common tangents can be drawn to $\odot A$ and $\odot B$?

Given: Circle Q with $\overline{TX}$ tangent to $\odot Q$ at X.

6. If $\overline{QX}$ is drawn, what is true about $\overline{QX}$ and $\overline{TX}$?

7. Suppose $\overline{TY}$ is drawn tangent to $\odot Q$ at Y. What is true about $\overline{TX}$ and $\overline{TY}$?

A, B, C, and X lie on $\odot O$.

8. $\widehat{AB} = \underline{\ ?\ }°$

9. $\widehat{AXB} = \underline{\ ?\ }°$

10. Suppose $\widehat{AB} = \widehat{BC}$. What theorem supports this statement: $AB = BC$?

The diagram shows $\odot P$, with $\overline{RT} \perp \overline{EF}$.

11. If $EF = 16$, then $ED = \underline{\ ?\ }$.

12. If $\widehat{FTE} = 122°$, then $\widehat{RE} = \underline{\ ?\ }°$.

13. Suppose $PR = 13$ and $EF = 24$. A chord $\overline{GH}$ is to be drawn as long as $\overline{EF}$. How far from P must $\overline{GH}$ be?

Angles and Segments

Objectives

1. **Solve problems and prove statements involving inscribed angles.**
2. **Solve problems and prove statements involving angles formed by secants and tangents.**
3. **Solve problems involving lengths of chords, secant segments, and tangent segments.**

7-5 *Inscribed Angles*

In the diagrams below, the angles are said to **intercept** the arcs shown in color. Note that an angle can intercept two arcs.

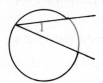 

An **inscribed angle** is an angle whose vertex is on a circle and whose sides contain chords of the circle. In the diagrams above, only $\angle 1$ and $\angle 2$ are inscribed angles. $\angle 1$ intercepts a minor arc, and $\angle 2$ intercepts a major arc. An inscribed angle may also intercept a semicircle.

The next theorem tells how the measure of an inscribed angle compares with the measure of its intercepted arc.

Theorem 7-6 An inscribed angle is equal to half its intercepted arc.

Given: $\angle ABC$ inscribed in $\odot O$

Prove: $\angle ABC = \frac{1}{2}\widehat{AC}$

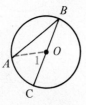

| I. Point O lies on $\angle ABC$. | II. Point O lies inside $\angle ABC$. | III. Point O lies outside $\angle ABC$. |

Proof of Case I:

Statements	Reasons
1. Draw $\overline{OA}$.	1. Through any two points __?__ .
2. $OB = OA$	2. Radii of a $\odot$ are equal.
3. $\angle A = \angle B$	3. If two sides of a $\triangle$ are =, __?__ .
4. $\angle A + \angle B = \angle 1$	4. __?__
5. $\angle B + \angle B = \angle 1$	5. Substitution Property
6. $\angle B = \frac{1}{2} \angle 1$	6. Division Postulate
7. $\widehat{AC} = \angle 1$	7. The measure of a minor arc is __?__ .
8. $\angle B = \frac{1}{2}\widehat{AC}$	8. Substitution Property

The proofs of Cases II and III are left as Classroom Exercises 8 and 9. Proofs of the following corollaries will be considered in Classroom Exercises 1–3.

Corollary 1 If two inscribed angles intercept the same arc, then the angles are equal.

Corollary 2 If a quadrilateral is inscribed in a circle, then opposite angles are supplementary.

Corollary 3 An angle inscribed in a semicircle is a right angle.

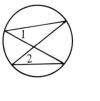

Corollary 1	Corollary 2	Corollary 3
$\angle 1 = \angle 2$	$\angle D$ is supp. to $\angle F$	If $\overset{\frown}{RUN}$ is a semicircle,
	$\angle E$ is supp. to $\angle G$	$\angle U$ is a right angle.

Study the diagrams from left to right below. Point C moves along the circle closer and closer to point A. Finally, in (4), point C has merged with A, and one side of $\angle A$ has become a tangent.

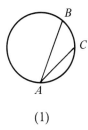

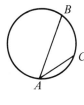

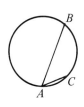

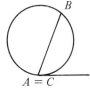

| (1) | (2) | (3) | (4) |

If you apply Theorem 7-6 to diagrams (1), (2), and (3), you find $\angle A = \frac{1}{2}\overset{\frown}{BC}$. This equation applies to diagram (4), too. See Theorem 7-7, which you will prove in Exercises 8–10.

Theorem 7-7 An angle formed by a chord and a tangent is equal to half the intercepted arc.

If $\overrightarrow{AT}$ is tangent to the circle and $\overset{\frown}{BA} = 120°$, then $\angle BAT = 60°$.

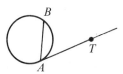

Classroom Exercises

1. Explain why Corollary 1 is true. That is, explain why $\angle 1 = \angle 2$.

Ex. 1

Ex. 2

Ex. 3

2. Explain why Corollary 2 is true. That is, explain why $\angle D$ and $\angle F$ are supplementary. (*Hint:* Let $\widehat{EFG} = n°$. Evaluate $\angle D$ and $\angle F$ in terms of n.)

3. Explain why Corollary 3 is true. That is, explain why if $\widehat{RUN}$ is a semicircle, $\angle U$ is a right angle.

4. In a quadrilateral $ABCD$, $\angle A = 70°$, $\angle B = 80°$, $\angle C = 90°$, and $\angle D = 120°$. Can quad. $ABCD$ be inscribed in a circle? Explain.

5. If a parallelogram is inscribed in a circle, then the parallelogram must be a ___?___ .

6. Suppose two angles inscribed in a circle are equal. What can you say about the intercepted arcs? Explain.

7. $\overleftrightarrow{XY}$ is tangent to the circle at T. State the degree measures of as many of the angles shown as possible.

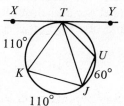

8. State the key steps of a proof of Case II of Theorem 7-6. Use the diagram on page 264. (*Hint:* Draw diameter $\overline{BX}$ and apply Case I.)

9. Repeat Exercise 8 for Case III of Theorem 7-6.

Written Exercises

Find the values of x, y, and z. O stands for the center of the circle in these exercises.

A 1.

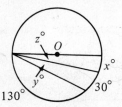

2.

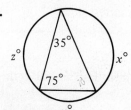

3.

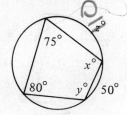

4.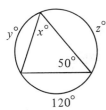

5.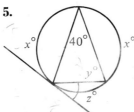

6.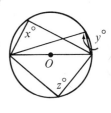

7. In this exercise you will prove that if two chords of a circle are parallel, then they cut off equal arcs.

Given: $\overline{AB} \parallel \overline{CD}$

Prove: $\widehat{AC} = \widehat{BD}$

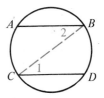

The diagrams for Exercises 8–10 show three cases of Theorem 7-7. For each, you are given $\odot O$ with $\angle BAT$ formed by chord $\overline{BA}$ and tangent $\overrightarrow{AT}$.

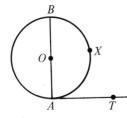

I. O on $\angle BAT$

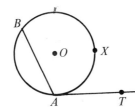

II. O inside $\angle BAT$

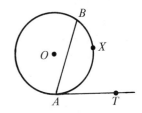

III. O outside $\angle BAT$

8. Key steps of a proof that $\angle BAT = \frac{1}{2}\widehat{BXA}$ are given for Case I. Write a reason for each step.

  1. $\overline{OA} \perp \overrightarrow{AT}$, or $\angle BAT = 90°$
  2. $\widehat{BXA}$ is a semicircle, so $\widehat{BXA} = 180°$.
  3. $\frac{1}{2}\widehat{BXA} = 90°$
  4. $\angle BAT = \frac{1}{2}\widehat{BXA}$

B **9.** Write the key steps of a proof that $\angle BAT = \frac{1}{2}\widehat{BXA}$ in Case II. (*Hint:* Draw diameter $\overline{AY}$. Use Theorem 7-6, Case I of Theorem 7-7, and the Addition Property.)

10. Write the key steps of a proof that $\angle BAT = \frac{1}{2}\widehat{BXA}$ in Case III.

11. Given: $\overleftrightarrow{EG}$ is tangent to $\odot O$ at E; $ED = EF$
What can you prove? Write a complete proof.

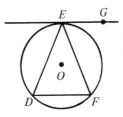

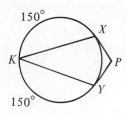

12. $\overline{PX}$ and $\overline{PY}$ are tangent to the circle at X and Y. $\angle P = \underline{\ ?\ }°$

In Exercises 13–14, quad. *ABCD* is inscribed in a circle. Give numerical answers.

13. If $\angle A = x°$, $\angle B = (x + 5)°$, and $\angle C = (x + 10)°$, then $\angle D = \underline{\ ?\ }°$.

14. If $\angle B = (x^2)°$, $\angle C = (7x + 5)°$, and $\angle D = (8x)°$, then $\angle A = \underline{\ ?\ }°$.

C 15. Given: Quadrilateral *ABCD* with *A*, *B*, and *C* on $\odot O$;
$\qquad\qquad \angle A + \angle C = 180°$
Prove: *D* is on $\odot O$.
(*Hint:* Use an indirect proof. Sketch two alternatives: (1) *D* is outside $\odot O$ and (2) *D* is inside $\odot O$. In each case, suppose $\overleftrightarrow{AD}$ intersects the circle at *X*, draw $\overline{CX}$, and show that (1) and (2) lead to contradictions.)

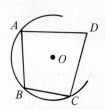

16. Prove: If *R* and *S* are points on a circle and if *T* is a point inside the circle, then $\angle RTS > \frac{1}{2}\widehat{RS}$.

17. Given: $\overline{BE} \| \overline{CD}$; $\overline{AB}$ is tangent to the circle at *B*. Find two triangles that are similar. Then prove the similarity.

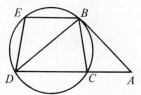

18. Circle *P* is inscribed in $\triangle ABC$, and $\odot Q$ is circumscribed about $\triangle ABC$. $\overrightarrow{AP}$ intersects $\odot Q$ in a point *Z*. Discover and prove a relationship between $\overline{ZB}$, $\overline{ZP}$, and $\overline{ZC}$.

7-6 *Other Angles*

In the diagram, $\angle 1$ is formed by two chords. The dashed line has been drawn to form a triangle.

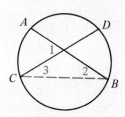

$\angle 1 = \angle 2 + \angle 3$ (Why?)
$\angle 2 = \frac{1}{2}\widehat{AC}$ and $\angle 3 = \frac{1}{2}\widehat{BD}$ (Why?)

Substituting, we get

$\angle 1 = \frac{1}{2}\widehat{AC} + \frac{1}{2}\widehat{BD} = \frac{1}{2}(\widehat{AC} + \widehat{BD})$.

We have just proved the theorem below.

> **Theorem 7-8**   An angle formed by two chords intersecting inside a circle is equal to half the sum of the intercepted arcs.

The next theorem will be proved in Classroom Exercise 11 for the case of two secants. See Exercises 25 and 26 for the other cases.

> **Theorem 7-9**   An angle formed by two secants, two tangents, or by a secant and a tangent drawn from a point outside a circle is equal to half the difference of the intercepted arcs.

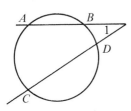
$$\angle 1 = \tfrac{1}{2}(\widehat{AC} - \widehat{BD})$$

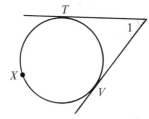
$$\angle 1 = \tfrac{1}{2}(\widehat{TXV} - \widehat{TV})$$

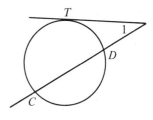
$$\angle 1 = \tfrac{1}{2}(\widehat{TC} - \widehat{TD})$$

Classroom Exercises

Chords, secants, and tangents of circles are shown. State the measure of each numbered angle.

1.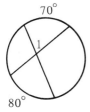

2.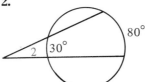

3.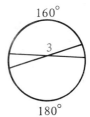

4.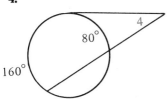

5.

6.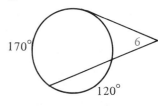

State an equation you can use to find *x*. Then state the value of *x*.

7.
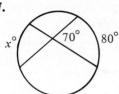

8.
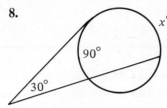

9.
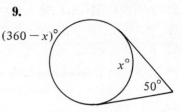

10. Because ∠1 is a central angle, you know that $\angle 1 = \overset{\frown}{AC}$. On the other hand, Theorem 7-8 tells you that ∠1, an angle formed by two chords, equals $\frac{1}{2}(\overset{\frown}{AC} + \overset{\frown}{BD})$. Is there any contradiction? Explain.

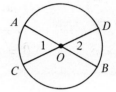

11. Key steps are shown for a proof of Theorem 7-9 for the case of two secants. State a reason for each step.

Given: Secants $\overleftrightarrow{PA}$ and $\overleftrightarrow{PC}$

Prove: $\angle 1 = \frac{1}{2}(\overset{\frown}{AC} - \overset{\frown}{BD})$

Key Steps of proof:

1. Draw $\overline{AD}$.
2. $\angle 1 + \angle 2 = \angle 3$
3. $\angle 1 = \angle 3 - \angle 2$
4. $\angle 3 = \frac{1}{2}\overset{\frown}{AC}$; $\angle 2 = \frac{1}{2}\overset{\frown}{BD}$
5. $\angle 1 = \frac{1}{2}\overset{\frown}{AC} - \frac{1}{2}\overset{\frown}{BD}$
6. $\angle 1 = \frac{1}{2}(\overset{\frown}{AC} - \overset{\frown}{BD})$

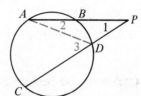

Written Exercises

$\overleftrightarrow{EF}$ is tangent to $\odot O$; $\overline{AD}$ is a diameter; $\overset{\frown}{AB} = 30°$; $\overset{\frown}{CD} = 40°$; $\overset{\frown}{DE} = 50°$.

A 1–10. Copy the diagram so that you can write arc measures alongside the arcs. Find the measure of each numbered angle.

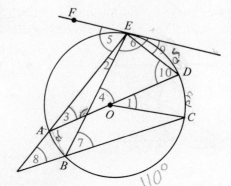

Complete.

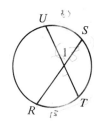

11. If $\overset{\frown}{US} = 60°$ and $\overset{\frown}{RT} = 80°$, $\angle 1 = \underline{\ ?\ }°$.

12. If $\overset{\frown}{UR} = 130°$ and $\overset{\frown}{ST} = 100°$, $\angle 1 = \underline{\ ?\ }°$.

13. If $\angle 1 = 60°$ and $\overset{\frown}{US} = 50°$, $\overset{\frown}{RT} = \underline{\ ?\ }°$.

14. If $\overset{\frown}{RT} = 2 \cdot \overset{\frown}{US}$ and $\angle 1 = 63°$, $\overset{\frown}{RUT} = \underline{\ ?\ }°$.

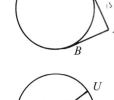

$\overline{NA}$ and $\overline{NB}$ are tangents.

15. If $\overset{\frown}{ACB} = 210°$, $\angle N = \underline{\ ?\ }°$.

16. If $\overset{\frown}{AB} = 100°$, $\angle N = \underline{\ ?\ }°$.

17. If $\angle N = 90°$, $\overset{\frown}{AB} = \underline{\ ?\ }°$.

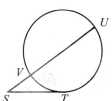

$\overline{ST}$ is a tangent.

18. If $\overset{\frown}{UT} = 110°$ and $\overset{\frown}{VT} = 30°$, $\angle S = \underline{\ ?\ }°$.

19. If $\angle S = 40°$ and $\overset{\frown}{VT} = 40°$, $\overset{\frown}{UT} = \underline{\ ?\ }°$.

20. If $\angle S = 38°$ and $\overset{\frown}{UT} = 120°$, $\overset{\frown}{VT} = \underline{\ ?\ }°$.

For Exercises 21 and 22, sketch a regular octagon *ABCDEFGH* inscribed in a circle.

B 21. Diagonals $\overline{AD}$ and $\overline{CG}$ intersect to form an acute angle of $\underline{\ ?\ }°$.

22. If $\overrightarrow{BA}$ and $\overrightarrow{FG}$ intersect at *P*, $\angle APG = \underline{\ ?\ }°$.

23. A quadrilateral circumscribed about a circle has angles of $80°, 90°, 94°,$ and $96°$. Find the measures of the four non-overlapping arcs determined by the points of tangency.

24. In the diagram, $\triangle ABC$ is circumscribed about the circle. Find the measures of the angles of $\triangle PRS$.

25. Prove Theorem 7-9 for the case of two tangents. (*Hint:* Draw $\overline{TV}$ in the diagram on page 269.)

26. Prove Theorem 7-9 for the case of a secant and a tangent. (*Hint:* Draw $\overline{TD}$ in the diagram on page 269.)

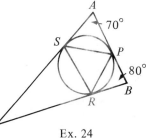

Ex. 24

27.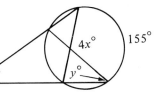

$x = \underline{\ ?\ }; y = \underline{\ ?\ }$

28.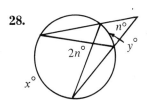

$x{:}y = \underline{\ ?\ }$

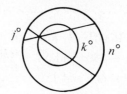

C 29. Write an equation involving *j*, *k*, and *n*.

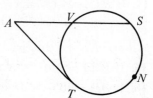

In Exercises 30–31, $\overline{AT}$ is tangent to the circle. In each exercise, state as much as you can about the measure of ∠*A*.

30. Given: $\widehat{VS} = 90°$; $\widehat{VT} > 80°$; $\widehat{SNT} > 160°$

31. Given: $110° > \widehat{VT} > \widehat{VS} > 100°$

32. Given that $\overline{OA}$, $\overline{OB}$, $\overline{PC}$, and $\overline{PD}$ are tangents, write the key steps of a proof that $GH = JK$. (*Hint:* You need auxiliary segments. Draw a big diagram.)

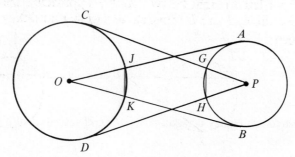

7-7 *Circles and Lengths of Segments*

Theorem 7-10 When two chords intersect inside a circle, the product of the segments of one chord equals the product of the segments of the other.

Given: $\overline{RS}$ and $\overline{TU}$ intersect in *K*.

Prove: $a \cdot b = c \cdot d$

Key steps of proof:

1. Draw $\overline{RT}$ and $\overline{US}$. (Through any two points __?__.)
2. ∠*R* = ∠*U*; ∠*T* = ∠*S* (If two inscribed angles intercept the same arc, __?__.)
3. △*RKT* ~ △*UKS* (AA Similarity Postulate)
4. $\dfrac{a}{d} = \dfrac{c}{b}$ (Why?)
5. $ab = cd$ (A property of proportions)

Theorem 7-11   When two secants are drawn to a circle from an outside point, the product of one secant and its external segment equals the product of the other secant and its external segment.

Given: Secants $\overline{KR}$ and $\overline{KT}$ drawn to the circle from point K.

Prove: $KR \cdot KS = KT \cdot KU$

For a proof, see Classroom Exercise 7.

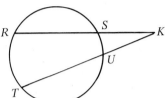

Study the diagrams from left to right below. As $\overline{KT}$ approaches the position of a tangent to the circle, T and U move closer together until they merge and $\overline{KT}$ is a tangent.

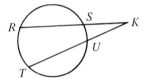 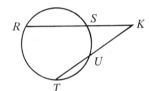 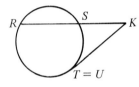

From Theorem 7-11 we know that $KT \cdot KU = KR \cdot KS$. When $T = U$, the equation becomes $(KT)^2 = KR \cdot KS$. The result is stated below.

Theorem 7-12   When a tangent and a secant are drawn to a circle from an outside point, the square of the tangent is equal to the product of the secant and its external segment.

Classroom Exercises

Chords, secants, and tangents are shown. State an equation you could use to find the value of x. You do not have to solve the equation.

1.

2.

3.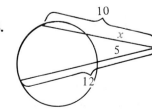

4.

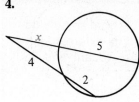

5.

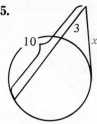

6.

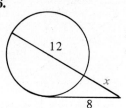

7. Supply reasons to complete the proof of Theorem 7-11.

Given: Secants $\overline{KR}$ and $\overline{KT}$ drawn to the circle from point K.

Prove: $KR \cdot KS = KT \cdot KU$

Proof:

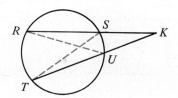

Statements	Reasons
1. Draw $\overline{RU}$ and $\overline{TS}$.	1. ___?___
2. $\angle R = \angle T$	2. ___?___
3. $\angle K = \angle K$	3. ___?___
4. $\triangle RUK \sim \triangle TSK$	4. ___?___
5. $\dfrac{KR}{KT} = \dfrac{KU}{KS}$	5. ___?___
6. $KR \cdot KS = KT \cdot KU$	6. ___?___

Written Exercises

Chords, secants, and tangents are shown. Find the value of x.

A **1.**

2.

3.

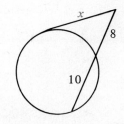

4.

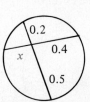

5.

6.

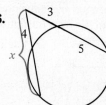

7.
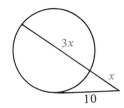

8.
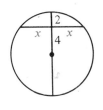

9.
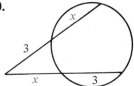

10.
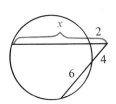

11.
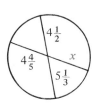

12.
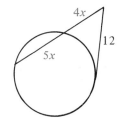

B 13. Given: $\odot O$ and $\odot P$ are externally tangent at T.
Prove: $AC \cdot AB = AE \cdot AD$

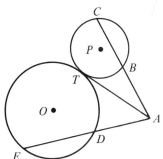

14. Given: $\overline{KU}$ is tangent to $\odot Q$;
$\overline{KZ}$ is tangent to $\odot S$.
Prove: $KU = KZ$

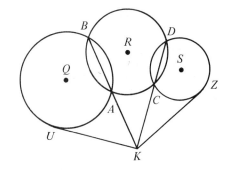

Chords $\overline{FG}$ and $\overline{HJ}$ intersect at I. Find the indicated length.

Example: $FI = 6;\ GI = 8;\ HJ = 16;\ HI = \underline{\ ?\ }$

Solution: Let $HI = x$. Then $JI = 16 - x$
$$x(16 - x) = 6 \cdot 8$$
$$16x - x^2 = 48$$
$$x^2 - 16x + 48 = 0$$
$$(x - 4)(x - 12) = 0$$
$$x = 4 \text{ or } x = 12$$
$$HI = 4 \text{ or } HI = 12$$

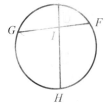

15. $FI = 4;\ GI = 5;\ HJ = 12;\ HI = \underline{\ ?\ }$
16. $HJ = 10;\ HI = 4;\ FG = 11;\ FI = \underline{\ ?\ }$
17. $FI = 4;\ GI = 15;\ HI = 6;\ HJ = \underline{\ ?\ }$
18. $JI = 6;\ HI = 8;\ GI = 3 \cdot FI;\ FI = \underline{\ ?\ }$

$\overline{PD}$ is a tangent. Complete.

19. $PD = 4$; $PA = 2$; $AB = $?

20. $PA = 4$; $AB = 8$; $NC = 13$; $PN = $?

21. $PA = AB$; $PD = 5$; $PA = $?

22. $PN = NC = 8$; $PD = $?

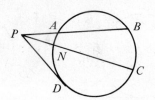

C 23. $DE = 24$; $EG = 14$; $GH = 32$;
$EF = $?

24. $\overline{AB}$ and $\overline{AC}$ are tangent to $\odot O$;
x is known to be an integer.
Find x.

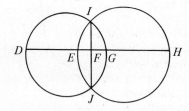

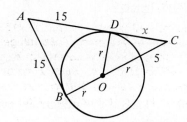

SELF-TEST

1. If $\overset{\frown}{RT} = 44°$, then $\angle 1 = $? °.

2. If $\overset{\frown}{RT} = 48°$ and $\overset{\frown}{US} = 62°$, then $\angle 2 = $? °.

3. If $\overleftrightarrow{UV}$ is a tangent and $\angle RUV = 80°$, then $\overset{\frown}{RU} = $? °.

4. If $UN = 16$, $TN = 6$, and $SN = 12$, then $RN = $?.

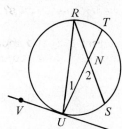

5. If $\angle 1 = 25°$ and $\overset{\frown}{BD} = 40°$, then $\overset{\frown}{AC} = $? °.

6. If $\overset{\frown}{AC} = 100°$, $\overset{\frown}{CD} = 110°$, and $\overset{\frown}{AB} = 90°$, then
$\angle 1 = $? °.

7. If $PA = 8$, $PB = 5$, and $PC = 10$, then $PD = $?.

8. Given: Points $W, X, Y,$ and Z lie on a circle; $TZ = TY$

Prove: $\angle 2 = \angle 4$

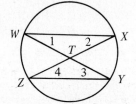

CALCULATOR KEY-IN

The ratio of the circumference of a circle to the diameter is a constant for all circles. The ratio is denoted by the Greek letter π (*pi*).

Circumference C

diameter d

$$\pi = \frac{C}{d}$$

If you were to wrap a string around a circle to measure the circumference, C, and then measure the diameter, d, you would find that $\frac{C}{d} \approx 3$.

π is an irrational number (about 3.14159). This means that it cannot be expressed exactly as the ratio of two integers. Decimal approximations of π have been computed to thousands of decimal places. We can easily look up values in reference books. Such was not always the case. In the past, mathematicians had to rely on their cleverness to compute an approximate value of π. One of the earliest approximations was that of Archimedes, who found that $3\frac{1}{7} > \pi > 3\frac{10}{71}$.

Exercises

1. Find decimal approximations of $3\frac{1}{7}$ and $3\frac{10}{71}$. Did Archimedes approximate π correct to hundredths?

In Exercises 2–5, find approximations for π. The more terms or factors you use, the better your approximations will be. But you can't use them all!

2. $\pi \approx 2\sqrt{3}(1 - \dfrac{1}{3 \cdot 3} + \dfrac{1}{3^2 \cdot 5} - \dfrac{1}{3^3 \cdot 7} + \dfrac{1}{3^4 \cdot 9} - \dfrac{1}{3^5 \cdot 11} + \cdots)$ (Sharpe, 18th century)

3. $\pi \approx 2 \cdot \dfrac{2}{1} \cdot \dfrac{2}{3} \cdot \dfrac{4}{3} \cdot \dfrac{4}{5} \cdot \dfrac{6}{5} \cdot \dfrac{6}{7} \cdot \dfrac{8}{7} \cdot \dfrac{8}{9} \cdot \cdots$ (Wallis, 17th century)

4. This exercise is for calculators that have a square root function and a memory.

$$\pi \approx 2 \div \left(\sqrt{0.5} \cdot \sqrt{0.5 + 0.5\sqrt{0.5}} \cdot \sqrt{0.5 + 0.5\sqrt{0.5 + 0.5\sqrt{0.5}}} \cdot \cdots \right)$$

(Vieta, 16th century)

5. $\pi \approx 4 \cdot (1 - \dfrac{1}{3} + \dfrac{1}{5} - \dfrac{1}{7} + \dfrac{1}{9} - \dfrac{1}{11} + \cdots)$ (Leibniz, 17th century)

Note: Although the expression here is simple in appearance, your approximations will move *very* slowly toward π. If you used one hundred terms within the parentheses, your approximation would not be correct to more than one decimal place.

Chapter Summary

1. If a line is tangent to a circle, then the line is perpendicular to the radius drawn to the point of tangency.

2. If a line in the plane of a circle is perpendicular to a radius at its outer endpoint, then the line is tangent to the circle.

3. In the same circle or in equal circles:
   a. Equal minor arcs have equal central angles.
   b. Equal chords have equal arcs.
   c. Equal arcs have equal chords.
   d. Equal chords are equally distant from the center.
   e. Chords equally distant from the center are equal.

4. A diameter that is perpendicular to a chord bisects the chord and its arc.

5. If two inscribed angles intercept the same arc, then the angles are equal.

6. If a quadrilateral is inscribed in a circle, then opposite angles are supplementary.

7. An angle inscribed in a semicircle is a right angle.

8. Relationships expressed by formulas:

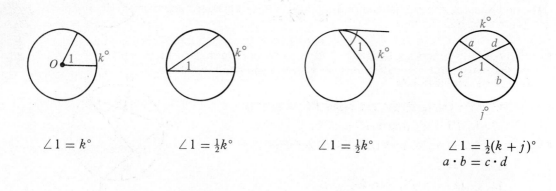

$$\angle 1 = k° \qquad \angle 1 = \tfrac{1}{2}k° \qquad \angle 1 = \tfrac{1}{2}k° \qquad \begin{array}{c}\angle 1 = \tfrac{1}{2}(k + j)° \\ a \cdot b = c \cdot d\end{array}$$

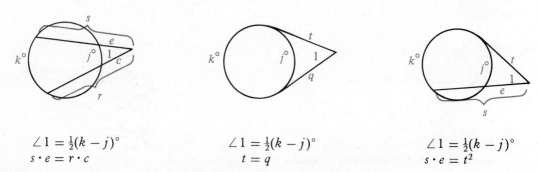

$$\begin{array}{c}\angle 1 = \tfrac{1}{2}(k - j)° \\ s \cdot e = r \cdot c\end{array} \qquad \begin{array}{c}\angle 1 = \tfrac{1}{2}(k - j)° \\ t = q\end{array} \qquad \begin{array}{c}\angle 1 = \tfrac{1}{2}(k - j)° \\ s \cdot e = t^2\end{array}$$

Chapter Review

7-1 Basic Terms

In the diagram of ⊙P name:

1. two radii
2. a diameter
3. three chords
4. an inscribed polygon

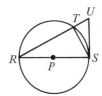

7-2 Tangents

$\overline{EF}$, $\overline{CF}$, and ⊙O are tangent to ⊙A; and $EF = 16$ cm.

5. $OB = \underline{\ ?\ }$ cm
6. $AC = \underline{\ ?\ }$ cm
7. $\angle ACF = \underline{\ ?\ }°$
8. $FC = \underline{\ ?\ }$ cm

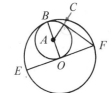

7-3 Arcs and Central Angles

In ⊙P, $\angle WPY = 91°$ and $\widehat{WX} = 48°$.

9. $\widehat{XY} = \underline{\ ?\ }°$
10. $\widehat{WZY} = \underline{\ ?\ }°$
11. $\angle ZPY = \underline{\ ?\ }°$
12. $\widehat{WZ} = \underline{\ ?\ }°$

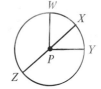

7-4 Arcs and Chords

O is the center of the circle and $\widehat{BC} = 120°$.

13. If $AB = BC$, then $\widehat{AB} = \underline{\ ?\ }°$.
14. If $AB = BC$ and $OE = 3$, then $OF = \underline{\ ?\ }$.
15. $\widehat{BD} = \underline{\ ?\ }°$
16. If $BC = 6\sqrt{3}$, then $BE = \underline{\ ?\ }$.

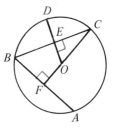

7-5 Inscribed Angles

$\overleftrightarrow{MN}$ is tangent to the circle.

17. $\angle LMK = \underline{\ ?\ }°$
18. $\angle LMN = \underline{\ ?\ }°$
19. $\angle JMK = \angle \underline{\ ?\ }$
20. $\angle JML$ is supplementary to $\angle \underline{\ ?\ }$.

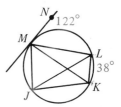

7-6 *Other Angles*

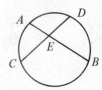

21. If $\overgroup{AC} = 66°$ and $\overgroup{BD} = 52°$, then $\angle AEC = \underline{}°$.

22. If $\angle DEB = 62°$ and $\overgroup{DB} = 47°$, then $\overgroup{AC} = \underline{}°$.

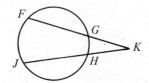

23. If $\overgroup{FJ} = 86°$ and $\overgroup{GH} = 29°$, then $\angle K = \underline{}°$.

24. If $\angle K = 25°$ and $\overgroup{GH} = 25°$, then $\overgroup{FJ} = \underline{}°$.

7-7 *Circles and Lengths of Segments*

Find each value of *x*.

25.

26.
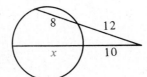

27.
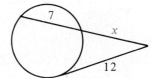

Chapter Test _____

Complete each statement.

1. If each vertex of a polygon lies on a circle, then the polygon is ___?___ in the circle.

2. A segment which joins two points on a circle is called a ___?___.

3. Circles in the same plane and having the same center are called ___?___ circles.

4. A sphere is the set of all ___?___ that are a given distance from a given point.

5. The measure of a minor arc is equal to the measure of ___?___.

6. If a quadrilateral is inscribed in a circle, then the opposite angles are ___?___.

7. An angle inscribed in a semicircle is a ___?___ angle.

8. If a rectangle is inscribed in a circle, then each diagonal of the rectangle is a ___?___ of the circle.

$\overleftrightarrow{CD}$ and $\overleftrightarrow{AD}$ are tangent to $\odot O$.

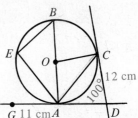

9. $AD = \underline{}$ cm

10. $\angle BEA = \underline{}°$

11. $\angle CAD = \underline{}°$

12. $\angle BAC = \underline{}°$

13. $\angle COA = \underline{}°$

14. $\angle D = \underline{}°$

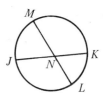

15. If $\widehat{MJ} = 73°$ and $\angle MNJ = 61°$, then $\widehat{KL} = \underline{\ ?\ }°$.

16. If $MN = 12$, $JN = 10$, and $NK = 9$, then $NL = \underline{\ ?\ }$.

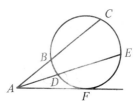

17. If $AB = 10$, $AC = 32$, and $AD = 8$, then $AE = \underline{\ ?\ }$.

18. If $AF = 6$ and $AB = 3$, then $AC = \underline{\ ?\ }$.

19. If $\widehat{DF} = 50°$ and $\widehat{FE} = 90°$, then $\angle FAE = \underline{\ ?\ }$.

20. If $\widehat{CE} = 75°$ and $\widehat{BD} = 15°$, then $\angle CAE = \underline{\ ?\ }°$.

21. Given: $\odot O$ and $\odot P$ with equal radii;
$\widehat{AB} = \widehat{EF}$; $\widehat{BC} = \widehat{FG}$
Write the key steps of a proof that
$\triangle ABC \cong \triangle EFG$.

22. Given: $\overleftrightarrow{BC}$ is tangent to $\odot O$;
$\overline{OE}$ bisects $\overline{AD}$.
Prove: $\triangle AEO \sim \triangle ABC$

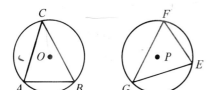

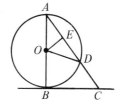

Algebra Review

Solve each equation for the variable shown in color.

Example: $ax + b = c$

Solution: $ax = c - b$

$$x = \frac{c - b}{a}$$

1. $dx - e = f$

2. $k = bx$

3. $2(x + a) = b$

4. $\frac{1}{2}(x - j) = k$

5. $\frac{r}{5} = t$

6. $\frac{n}{g} = k$

7. $\frac{a}{b} = c$

8. $\frac{a}{2b} = c$

9. $A = bh$

10. $V = \frac{1}{3}Bh$

11. $\frac{C}{d} = \pi$

12. $C = 2\pi r$

13. $S = (n - 2)180$

14. $A = \frac{1}{2}h(b + c)$

15. $a^2 + b^2 = (2a)^2$

16. $a^2 + b^2 = c^2$

8

Constructions
and Loci

Basic Constructions

Objectives
1. Perform the basic constructions.
2. Use the basic constructions in original construction exercises.

8-1 *What Construction Means*

In Chapters 1–7 we have used rulers and protractors to draw segments with certain lengths and angles with certain measures. In this chapter, we shall *construct* geometric figures using only two instruments, a *straightedge* and a *compass*. You may use a ruler as a straightedge, but only if you agree not to use the marks on it.

Using a Straightedge in Constructions

   Given two points A and B, we know from Postulate 2 that there is exactly one line through A and B. We agree that we can use a straightedge to draw $\overleftrightarrow{AB}$ or parts of the line, such as $\overline{AB}$ and $\overrightarrow{AB}$.

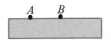

Using a Compass in Constructions

   Given a point O and a length r, we know from the definition of circle that there is exactly one circle with center O and radius r. We agree that we can use a compass to draw the circle or an arc of the circle.

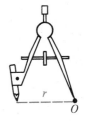

Construction 1 Given a segment, construct a segment equal to the given segment.

Given: $\overline{AB}$

Construct: A segment equal to $\overline{AB}$

Procedure:

1. Use a straightedge to draw a line. Call it *l*.
2. Choose any point on *l* and label it *X*.
3. Set your compass for radius *AB*. Use *X* as center and draw an arc intersecting line *l*. Label the point of intersection *Y*.

$\overline{XY}$ is equal to $\overline{AB}$.

Justification: Since we used *AB* for the radius of $\odot X$, $\overline{XY}$ is equal to $\overline{AB}$.

Construction 2 Given an angle, construct an angle equal to the given angle.

Given: $\angle ABC$

Construct: An angle equal to $\angle ABC$

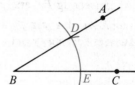

Procedure:

1. Draw a ray. Label it $\overrightarrow{RY}$.
2. Using *B* as center and any convenient radius, draw an arc intersecting $\overrightarrow{BA}$ and $\overrightarrow{BC}$. Label the points of intersection *D* and *E*.
3. Using *R* as center and the same radius as before, draw an arc intersecting $\overrightarrow{RY}$ and label it $\overset{\frown}{XS}$, with *S* at the point of intersection.
4. Using *S* as center and a radius equal to *DE*, draw an arc that intersects $\overset{\frown}{XS}$ at a point *Q*.
5. Draw $\overrightarrow{RQ}$.

$\angle R$ is equal to $\angle B$.

Justification: If $\overline{DE}$ and $\overline{QS}$ are drawn, $\triangle DBE \cong \triangle QRS$ (SSS Postulate). Then $\angle R = \angle B$.

Example: Given ∠ 1 and ∠ 2, construct an angle equal to ∠ 1 + ∠ 2.

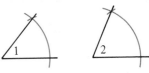

Solution: First use Construction 2 to make ∠ *DEF* equal to ∠ 1. Then use Construction 2 to make ∠ *HED* equal to ∠ 2, as shown, so that ∠ *HEF* = ∠ 1 + ∠ 2.

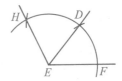

Construction 3   Given an angle, bisect the angle.

Given: ∠ *ABC*

Construct: The ray that bisects ∠ *ABC*

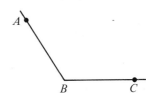

Procedure:

1. Using *B* as center and any convenient radius, draw arcs intersecting $\overrightarrow{BA}$ and $\overrightarrow{BC}$ in points *X* and *Y*.

2. Using *X* and *Y* as centers and any convenient radius, draw arcs that intersect at a point *Z*.

3. Draw $\overrightarrow{BZ}$.

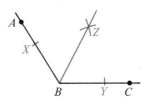

$\overrightarrow{BZ}$ bisects ∠ *ABC*.

Justification: If $\overline{XZ}$ and $\overline{YZ}$ are drawn, △*XBZ* ≅ △*YBZ* (SSS Postulate). Then ∠ *XBZ* = ∠ *YBZ*, and $\overrightarrow{BZ}$ bisects ∠ *ABC*.

In construction exercises, you won't ordinarily have to write out the procedure and the justification. You should be able to supply them when asked to do so, however.

Classroom Exercises

1. Draw some convenient segment $\overline{AB}$.

   a. Construct $\overline{XY}$ so that $XY = AB$.

   b. Using *X* and *Y* as centers, and a radius equal to *AB*, draw arcs that intersect. Label the point of intersection *Z*.

   c. Draw $\overline{XZ}$ and $\overline{YZ}$.   **d.** What kind of triangle is △*XYZ*?

2. Explain how you could construct a 30° angle. (*Hint:* See Exercise 1 and Construction 3.)

3. Explain how to construct a 15° angle.   4. Explain how to construct a 75° angle.

5. Given: △JKM
   Explain how to construct a triangle that is congruent to △JKM.

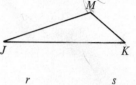

6. Suppose you are given the three lengths shown and are asked to construct a triangle whose sides have lengths r, s, and t. Comment about the task. State the theorem from Chapter 4 that applies.

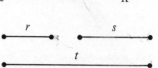

7. ∠1 and ∠2 are given. You see two attempts at constructing an angle equal to ∠1 + ∠2. Are both constructions satisfactory?

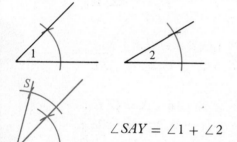

∠SAY = ∠1 + ∠2

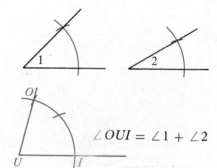

∠OUI = ∠1 + ∠2

Written Exercises

On your paper, draw two segments roughly like those shown. Then for Exercises 1–4 construct a segment having the indicated length.

c ●————●

d ●——————●

A 1. $c + d$ 2. $d - c$ 3. $2c + d$ 4. $3d - 2c$

5. Draw any acute △ACU. Use the SSS method to construct a triangle congruent to △ACU.

6. Draw any obtuse △OBT. Use the SSS method to construct a triangle congruent to △OBT.

7. Repeat Exercise 5, but use the SAS method.

8. Repeat Exercise 6, but use the ASA method.

9. Using any convenient side length, construct an equilateral triangle.

10. Construct a 30° angle.

On your paper, draw two angles roughly like those shown. Then for Exercises 11–16 construct an angle having the indicated measure.

B 11. $(x + y)°$ **12.** $(x - y)°$ **13.** $2y°$

  14. $\frac{1}{2}x°$ **15.** $\frac{3}{4}x°$ **16.** $(3y - x)°$

17. Draw any acute triangle. Bisect the three angles.

18. Draw any obtuse triangle. Bisect the three angles.

Construct angles having the indicated measures.

  19. 120°   **20.** 150°   **21.** 45°   **22.** 165°

23. Draw any triangle ABC.

  a. Construct $\triangle DEF$ so that $\triangle DEF \sim \triangle ABC$ and $DE = 2AB$.

  b. What is the ratio of the perimeter of $\triangle DEF$ to the perimeter of $\triangle ABC$?

24. Draw any triangle GHI.

  a. Construct $\triangle JKM$ so that $JK = GH$, $\angle J = \frac{1}{2}\angle G$, and $\angle K = \frac{1}{2}\angle H$.

  b. In $\triangle JKM$ constructed in (a), $\angle M$ sometimes turns out to be twice as large as $\angle I$ in $\triangle GHI$. This occurs only if you happen to have drawn $\angle I$ with a certain measure. What is that measure?

C 25. On your paper, draw figures roughly like the ones shown. Then construct a triangle whose three angles equal $\angle 1$, $\angle 2$, and $\angle 3$, and whose circumscribed circle has radius n.

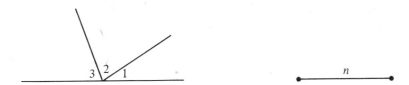

26. On your paper, draw figures roughly like those shown. Construct a parallelogram with an angle equal to $\angle 4$, a longer side equal to s, and a longer diagonal equal to d.

8-2 *Perpendiculars and Parallels*

The perpendicular to a line at a point can be thought of as the bisector of a straight angle whose vertex is that point. Notice how Construction 4 below is related to Construction 3 on page 285.

Construction 4   Given a point on a line, construct the perpendicular to the line at the given point.

Given: Point A on line l

Construct: The perpendicular to l at A

Procedure:

Bisect the straight angle whose vertex is A.

$\overleftrightarrow{AZ}$ is perpendicular to l at A.

In the next construction we do not bisect a straight angle, but we follow a procedure much like the one used in Construction 4.

Construction 5   Given a point outside a line, construct the perpendicular to the line from the point.

Given: Point B outside line l

Construct: The perpendicular to l from B

Procedure:

1. Using B as center and any convenient radius, draw arcs that intersect l in two points X and Y.
2. Using X and Y as centers and any convenient radius, draw arcs that intersect at a point Z.
3. Draw $\overleftrightarrow{BZ}$.

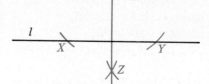

$\overleftrightarrow{BZ}$ is perpendicular to l.

Justification: See Classroom Exercise 1.

Construction 6   Given a segment, construct the perpendicular bisector of the segment.

Given: $\overline{CD}$

Construct: The perpendicular bisector of $\overline{CD}$

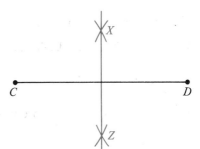

Procedure:

1. Using any convenient radius, construct two arcs having C as center and two arcs having D as center. Call the points of intersection X and Z.
2. Draw $\overline{XZ}$.

$\overline{XZ}$ is the perpendicular bisector of $\overline{CD}$.

Justification: See Classroom Exercise 2.

Construction 7   Given a point outside a line, construct the parallel to the given line through the given point.

Given: Point P outside line l

Construct: The parallel to l, through P

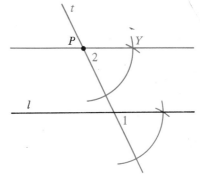

Procedure:

1. Through P draw any line t that intersects l.
2. At P construct $\angle 2$ so that $\angle 2$ and $\angle 1$ are corresponding angles and $\angle 2 = \angle 1$.
3. Draw $\overleftrightarrow{PY}$.

$\overleftrightarrow{PY}$ is parallel to l, through P.

Justification: See Classroom Exercise 3.

Classroom Exercises

1. $\overleftrightarrow{BZ}$ has been constructed as in Construction 5. Supply reasons for these key steps in a justification of the construction.

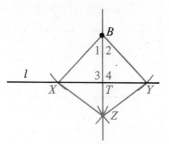

   1. Draw $\overline{BX}$, $\overline{BY}$, $\overline{XZ}$, and $\overline{YZ}$.
   2. $\triangle BXZ \cong \triangle BYZ$
   3. $\angle 1 = \angle 2$
   4. $\triangle BXT \cong \triangle BYT$
   5. $\angle 3 = \angle 4$
   6. $\overleftrightarrow{BZ} \perp l$

2. You have an easy justification of Construction 6 if you note that one radius was used for the four arcs.

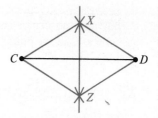

   a. If $\overline{CX}$, $\overline{DX}$, $\overline{CZ}$, and $\overline{DZ}$ are drawn, what special kind of quadrilateral is $CXDZ$?
   b. State a theorem that tells you $\overline{XZ}$ bisects $\overline{CD}$.
   c. State a theorem that tells you $\overline{XZ} \perp \overline{CD}$.

3. Refer to Construction 7, page 289. State the postulate that you have in mind when you use the procedure shown.

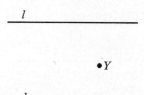

4. Explain how you could use Constructions 5 and 4 to get the line through P that is parallel to l.

5. Suppose you want to construct a circle that is tangent to l at X, and that passes through points X and Y.

   a. Where must the center lie with respect to line l and point X?
   b. Where must the center lie with respect to points X and Y?
   c. Explain how you could carry out the construction.

Written Exercises

Draw a figure roughly like the one shown, but larger. Do the indicated constructions clearly enough so that your method can be understood easily.

A 1. The perpendicular to l at P 2. The perpendicular to l from P

3. The perpendicular bisector of $\overline{JK}$

4. The parallel to l through P

5. The perpendicular to $\overrightarrow{BA}$ at A

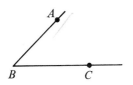

6. The parallel to $\overleftrightarrow{ED}$ through F

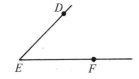

7. The perpendicular to $\overleftrightarrow{HJ}$ from G

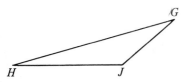

8. A complement of $\angle KMN$

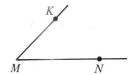

9. Draw an acute triangle. Construct the perpendicular bisectors of the three sides.

10. Draw an acute triangle. Construct the three altitudes.

B **11.** Draw an acute triangle. Construct the three medians. (*Hint:* Use Construction 6 to locate the midpoints.)

12–14. Repeat Exercises 9–11, but begin with obtuse triangles.

15. Draw a segment $\overline{AB}$. Construct a segment $\overline{XY}$ whose length equals $\frac{3}{4}AB$.

Construct an angle with the indicated measure.

16. 45°   **17.** 135°   **18.** $22\frac{1}{2}°$

On your paper draw figures roughly like those shown. Use them in constructing the figures described in Exercises 19–29.

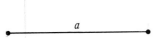   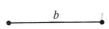

19. A parallelogram with sides a and b and an $n°$ angle

20. The figure described in Exercise 19 but constructed by a different method

21. A rectangle with sides a and b

22. A square with sides $\frac{1}{2}a$

23. A rhombus with diagonals a and b

For Exercises 24–29 see the directions and diagrams on page 291.

24. A nonequilateral parallelogram with diagonals a and b

25. A right triangle with legs a and b

26. A segment with length $\sqrt{a^2 + b^2}$

27. A right triangle with hypotenuse a and one leg b

28. An isosceles triangle with a vertex angle of $n°$ and legs b

C 29. An isosceles triangle with a vertex angle of $n°$ and base a

30. Draw a diagram roughly like the one shown. Without laying your straightedge across any part of the lake, construct more of $\overrightarrow{RS}$.

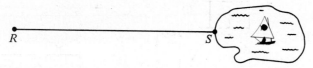

31. Draw three noncollinear points R, S, and T. Construct a triangle whose sides have R, S, and T as midpoints. (*Hint:* How is $\overline{RT}$ related to the side of the triangle that has S as midpoint?)

32. Draw a figure roughly like the one shown. By construction, locate a point X on $\overline{YZ}$ so that $\angle AXY = \angle BXZ$. (*Hint:* First locate a point P so that $\overline{BP} \perp \overline{YZ}$ and $\overline{YZ}$ bisects $\overline{BP}$. Then draw $\overleftrightarrow{AP}$.)

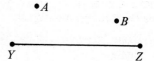

Application _____

Technical Drawing

Using only words, can you easily describe the object shown at the right? Probably not. Unless an object has a very simple geometric shape, a multiview drawing is required to describe it adequately.

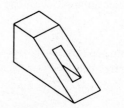

In industry, a clear description of the shape and size of a machine part is needed if the part is to be manufactured as intended by the design engineers. A number of views, systematically arranged, are used. Here is a three-view sketch of the object above. The right side was pictured first. The red lines show how the other views were constructed from the first. Dashed lines indicate edges hidden from view.

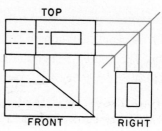

Exercises

In each three-view drawing, one or more lines have been omitted. Sketch the given views and add the missing lines.

1.

2.

3.

4.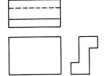

5.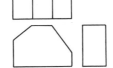

6.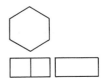

Make a three-view sketch of each object. In Exercise 11, the hole goes completely through the object.

7.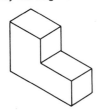

8.

9.

10.

11.

12.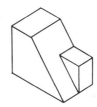

Challenge

Draw a segment and label its length *n*. Then construct a square such that the combined length of a side and a diagonal is *n*.

8-3 *Circles*

Construction 8   Given a point on a circle, construct the tangent to the circle at the point.

Given: Point A on $\odot O$

Construct: The tangent to $\odot O$ at A

Procedure:

1. Draw $\overrightarrow{OA}$.
2. Construct a perpendicular to $\overrightarrow{OA}$ at A. Call it t.

Line t is tangent to $\odot O$ at A.

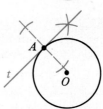

Justification: Because t is perpendicular to radius $\overline{OA}$ at A, t is tangent to $\odot O$.

Construction 9   Given a point outside a circle, construct a tangent to the circle from the point.

Given: Point P outside $\odot O$

Construct: A tangent to $\odot O$ from P

Procedure:

1. Draw $\overline{OP}$.
2. Find the midpoint M of $\overline{OP}$ by constructing the perpendicular bisector of $\overline{OP}$.
3. Using M as center and MP as radius, draw a circle that intersects $\odot O$ in a point X.
4. Draw $\overrightarrow{PX}$.

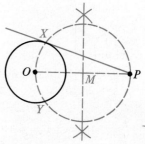

$\overrightarrow{PX}$ is tangent to $\odot O$ from P. $\overrightarrow{PY}$, not drawn, is the other tangent from P.

Justification: Draw $\overline{OX}$. Because $\angle OXP$ is inscribed in a semicircle, $\angle OXP$ is a right angle and $\overrightarrow{PX} \perp \overline{OX}$. Because $\overrightarrow{PX}$ is perpendicular to radius $\overline{OX}$ at its outer endpoint, $\overrightarrow{PX}$ is tangent to $\odot O$.

Construction 10   Given a triangle, circumscribe a circle about the triangle.

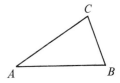

Given: △ABC

Construct: A circle passing through *A*, *B*, and *C*

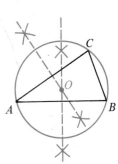

Procedure:

1. Construct the perpendicular bisectors of any two sides of △ABC. Label the point of intersection *O*.
2. Using *O* as center and *OA* as radius, draw a circle.

Circle *O* passes through *A*, *B*, and *C*.

Justification: See Theorem 4-17 on page 143.

Construction 11   Given a triangle, inscribe a circle in the triangle.

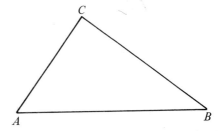

Given: △ABC

Construct: A circle tangent to
$\overline{AB}$, $\overline{BC}$, and $\overline{AC}$

Procedure:

1. Construct the bisectors of ∠*A* and ∠*B*. Label the point of intersection *I*.
2. Construct a perpendicular from *I* to $\overline{AB}$. It intersects $\overline{AB}$ at a point *R*.
3. Using *I* as center and *IR* as radius, draw a circle.

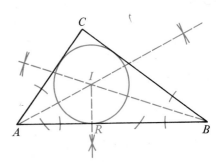

Circle *I* is tangent to $\overline{AB}$, $\overline{BC}$, and $\overline{AC}$.

Justification: See Theorem 4-16 on page 142.

Classroom Exercises

1. Suppose the center of the circle has been lost. Explain how to find it by construction.

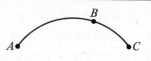

2. Explain how to find the midpoint of $\overset{\frown}{DE}$.

3. Explain how to find a line that is parallel to $\overline{RS}$ and tangent to $\odot P$.

4. Here you see a common method for using just one compass setting for drawing a circle and dividing the circle into six equal arcs. Explain how the method works.

5. Suppose a circle is given. Explain how you can use the method of Exercise 4 to inscribe an equilateral triangle in the circle.

6. Suppose the construction of Exercise 4 has been carried out. Explain how you can then inscribe a regular twelve-sided polygon in the circle.

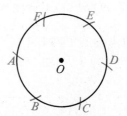

Written Exercises

In Exercises 1 and 2, draw a diagram similar to the one shown, but larger.

A 1. Construct a tangent at *P*.   2. Construct two tangents from *P*.

3. Draw a large, acute triangle. Construct the circumscribed circle.
4. Draw a large, obtuse triangle. Construct the circumscribed circle.
5. Draw a large, acute triangle. Construct the inscribed circle.
6. Draw a large, obtuse triangle. Construct the inscribed circle.
7. Construct a large, right triangle. Construct the circumscribed circle.
8. Construct a large, right triangle. Construct the inscribed circle.

In each exercise, begin with a circle. Do the indicated construction.

B   **9.** A square circumscribed about the circle

   10. A square inscribed in the circle

   11. An equilateral triangle inscribed in the circle

   12. An equilateral triangle circumscribed about the circle

In each of Exercises 13 and 14, begin with a diagram roughly like the one shown, but larger.

   13. Construct a line that is parallel to line *l* and tangent to $\odot O$.

   14. Construct a line that is perpendicular to line *l* and tangent to $\odot O$.

In Exercises 15–17, begin with two circles *P* and *Q* such that $\odot P$ and $\odot Q$ do not intersect and $\odot P$ is larger than $\odot Q$.

C **15.** Construct a circle, with radius equal to *PQ*, that is tangent to $\odot P$ and $\odot Q$.

   16. Construct a common external tangent to $\odot P$ and $\odot Q$. One method is suggested below.

     1. Let the radii of $\odot P$ and $\odot Q$ be *p* and *q*. Draw a circle with center *P* and radius $p - q$.

     2. Construct a tangent to this circle from *Q*, and call the point of tangency *Z*.

     3. Draw $\overrightarrow{PZ}$. $\overrightarrow{PZ}$ intersects $\odot P$ in a point *X*.

     4. With center *X* and radius *ZQ*, draw an arc that intersects $\odot Q$ in a point *Y*.

     5. Draw $\overleftrightarrow{XY}$.

   If you draw $\overline{QY}$, you can show that *XZQY* is a rectangle. The rest of a justification is easy.

   17. Construct a common internal tangent to $\odot P$ and $\odot Q$. (*Hint:* Draw a circle with center *P* and radius $p + q$.)

   18. Begin with $\overleftrightarrow{AB}$ and a circle *O* tangent to $\overleftrightarrow{AB}$ at *A*. Construct a circle that is tangent to $\overleftrightarrow{AB}$ at *B* and tangent to $\odot O$. (*Hint:* Suppose the problem has been solved. You can then consider the common internal tangent.)

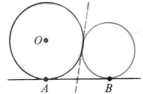

Challenge

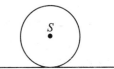

Circle *S* is tangent to line *j*. Construct two circles equal to each other, tangent to each other, above line *j*, with both circles tangent to $\odot S$ and to line *j*.

When an airport is constructed, a civil engineer is certain to be on the job, ensuring that the project is carried out according to plan. The lines and angles of airport runways present a large-scale geometry problem. The civil engineer uses surveyor's tools to lay out boundaries as you might use a compass and straightedge to construct a geometric figure.

Career Notes

Civil Engineer

Civil engineers have traditionally been involved in the practical aspects of designing buildings, highways, tunnels, and bridges. Today they work in an even wider variety of situations. For example, a civil engineer may deal with the problems of traffic flow and public transportation in a large city.

The government agencies, construction companies, and consulting firms which employ civil engineers expect job applicants to have a college degree in engineering, with several courses in mathematics.

8-4 *Special Segments*

Construction 12   Given a segment, divide the segment into a given number of equal parts. (3 shown)

Given: $\overline{AB}$

Construct: Points X and Y on $\overline{AB}$ so that $AX = XY = YB$

Procedure:

1. Draw any ray $\overrightarrow{AZ}$ from A.
2. Using any convenient radius, start with A as center and mark off R, S, and T so that $AR = RS = ST$.
3. Draw $\overline{TB}$.
4. At R and S construct lines parallel to $\overline{TB}$ and intersecting $\overline{AB}$ in X and Y.

$\overline{AX}$, $\overline{XY}$, and $\overline{YB}$ are equal parts of $\overline{AB}$.

Justification: Since the lines constructed parallel cut off equal segments on transversal $\overleftrightarrow{AZ}$, they cut off equal segments on transversal $\overleftrightarrow{AB}$. (It may help you to think of the parallel to $\overline{TB}$ through A.)

Construction 13   Given three segments, construct a fourth segment so that the four segments are in proportion.

Given: Segments with lengths a, b, and c

Construct: A segment of length x such that $\dfrac{a}{b} = \dfrac{c}{x}$

Procedure:

1. Draw an angle HIJ.
2. On $\overrightarrow{IJ}$, mark off $IR = a$ and $RS = b$.
3. On $\overrightarrow{IH}$, mark off $IT = c$.
4. Draw $\overline{RT}$.
5. At S, construct a parallel to $\overline{RT}$, intersecting $\overrightarrow{IH}$ in a point U.

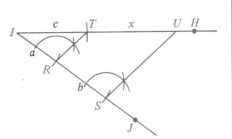

$\overline{TU}$ has length x such that $\dfrac{a}{b} = \dfrac{c}{x}$.

Justification: In $\triangle ISU$, $\overline{RT} \parallel \overline{SU}$. Therefore, $\dfrac{a}{b} = \dfrac{c}{x}$.

Construction 14   Given two segments, construct their geometric mean.

Given: Segments with lengths a and b

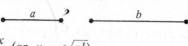

Construct: A segment of length x such that $\dfrac{a}{x} = \dfrac{x}{b}$ (or, $x = \sqrt{ab}$)

Procedure:

1. Draw a line and mark off $RS = a$ and $ST = b$.

2. Locate the midpoint O of $\overline{RT}$ by constructing the perpendicular bisector of $\overline{RT}$.

3. Using O as center draw a semicircle with a radius equal to OR.

4. At S, construct a perpendicular to $\overline{RT}$. The perpendicular intersects the semicircle at a point U.

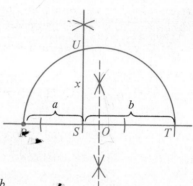

US, or x, is the geometric mean between a and b.

Justification: When $\overline{RU}$ and $\overline{UT}$ are drawn, $\triangle RUT$ is a right triangle. Since $\overline{US}$ is the altitude to the hypotenuse of rt. $\triangle RUT$, $\dfrac{a}{x} = \dfrac{x}{b}$.

Classroom Exercises

This set of exercises will analyze one problem.

Given: Line t; points A and B
Construct: A circle through A and B and tangent to t

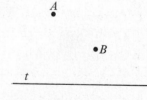

If the problem had been solved, we would have a diagram something like the one shown.

1. Where does the center of the circle lie with respect to $\overline{AB}$?

2. Where does the center of the circle lie with respect to line t and point K?

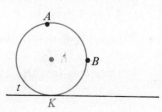

Note that we don't have point K located in the given diagram. Hunting for ideas, we draw $\overleftrightarrow{AB}$. We now have a point J, which we can locate in the given diagram.

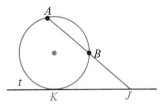

3. State an equation that relates JK to JA and JB.

4. Rewrite your equation in the form $\dfrac{?}{JK} = \dfrac{JK}{?}$.

5. What construction can we use to get the length JK?

In a *separate* diagram we can mark off the lengths JA and JB on some line l and then use Construction 14 to find x such that $\dfrac{JA}{x} = \dfrac{x}{JB}$. Once we have x, which equals JK, we return to the given diagram and draw an arc to locate K.

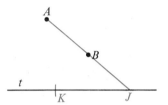

6. Explain how to complete the construction of the circle.

Written Exercises

In each of Exercises 1–4, begin by drawing a segment $\overline{AB}$ that is roughly from 12 to 15 cm long.

A **1.** Divide $\overline{AB}$ into three equal segments.

2. Use Construction 12 to divide $\overline{AB}$ into four equal segments.

3. Use Construction 6, page 289, repeatedly to divide $\overline{AB}$ into four equal segments.

4. Divide $\overline{AB}$ into two segments that have the ratio $2:3$. (*Hint:* Refer to Construction 12. You will need to mark off five equal segments on $\overrightarrow{AZ}$.)

On your paper, draw three segments roughly as long as those shown below. Use your segments in Exercises 5–11. In each exercise, construct a segment that has length x.

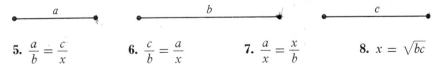

5. $\dfrac{a}{b} = \dfrac{c}{x}$ **6.** $\dfrac{c}{b} = \dfrac{a}{x}$ **7.** $\dfrac{a}{x} = \dfrac{x}{b}$ **8.** $x = \sqrt{bc}$

B **9.** $cx = ab$ (*Hint:* First write a proportion equivalent to the given equation and with x as the last term.)

10. $x = \dfrac{bc}{a}$ **11.** $x = \sqrt{2ab}$

On your paper, draw segments roughly as long as those shown to use in Exercises 12–14.

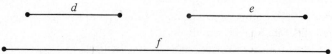

12. Construct an equilateral triangle whose perimeter is equal to *f*.

13. Construct a square whose perimeter is equal to *f*.

14. Construct a segment $\overline{AB}$, with $AB = f$. Divide $\overline{AB}$ into two parts that have the ratio $d:e$.

C **15.** Draw any triangle *HIJ*. Construct a point *X* on $\overline{IJ}$ so that $IX:XJ = HI:HJ$. (*Hint:* For a very easy construction, apply a certain theorem from Chapter 5.)

16. Draw a segment about 16 cm long. Label the endpoints *C* and *D*. Construct a triangle whose perimeter is equal to *CD* and whose sides are in the ratio $2:2:3$.

17. To trisect a general angle *G*, a student tried this procedure:

1. Mark off $\overline{GA}$ equal to $\overline{GB}$.

2. Draw $\overline{AB}$.

3. Divide $\overline{AB}$ into three equal parts using Construction 12.

4. Draw $\overline{GX}$ and $\overline{GY}$.

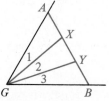

Show that the student did not trisect $\angle G$. That is, show that $\angle 2 \neq \angle 1$. You may use the fact that $GA > GX$. (This fact is proved in Exercise 17, page 155.)

SELF-TEST

For each exercise, draw a diagram similar to, but larger than, the one shown.

1. Construct the bisector of $\angle A$.

2. Construct the altitude from *C* to $\overleftrightarrow{AB}$.

3. Construct the line through *C* parallel to $\overleftrightarrow{AB}$.

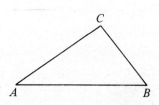

For each exercise, draw a diagram similar to, but larger than, the one shown.

4. Construct a line tangent to $\odot O$ at *X*.

5. Construct a line tangent to $\odot O$ from *Y*.

6. Draw a segment $\overline{DE}$ roughly 10 cm long. Then by construction divide $\overline{DE}$ into three equal parts.

7. Draw a triangle *FGH*. Then construct the circumscribed circle.

Locus

Objectives
1. Describe the locus that satisfies a given condition.
2. Describe the locus that satisfies more than one given condition.
3. Apply the concept of locus in the solution of construction exercises.

8-5 *The Meaning of Locus*

The solutions to some construction problems use the idea of *locus*. The word means location or place. A **locus** is a figure; it is the set of all points, and only those points, that satisfy one or more conditions.

Suppose we have a line *k* in a plane and wish to picture the locus of points in the plane that are 1 cm from *k*. Several points are shown in the first diagram.

All the points satisfying the given conditions are indicated in the next diagram. You see that the required locus is a pair of lines parallel to, and 1 cm from, *k*.

Suppose we wish to picture the locus of points 1 cm from *k* without requiring the points to be *in a plane*. The problem changes. Now you can use a pointer or a pencil to represent line *k*. You should hold the pointer in some position that doesn't tempt you into focusing your attention on any one plane. The required locus is a cylindrical surface with axis *k* and a 1 cm radius, shown below. Of course, the surface does not end where the diagram does.

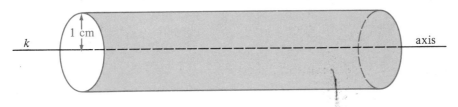

When you are solving a locus problem, always think in terms of three dimensions unless the statement of the problem restricts the locus to a plane.

Classroom Exercises

Exercises 1–3 deal with figures in a plane.

1. Draw a point Z.
   a. Draw several points that are 20 cm from Z.
   b. Indicate all the points that are 20 cm from Z.
   c. What is the locus of points that are 20 cm from a given point Z? (Answer in a sentence.)

2. Draw two parallel lines h and j.
   a. Draw several points that are equidistant from h and j.
   b. Indicate all the points that are equidistant from h and j.
   c. What is the locus of points that are equidistant from two parallel lines h and j? (Answer in a sentence.)

3. Draw an angle.
   a. Draw several points that are equidistant from the sides of the angle. (Each individual point is as far from one side as from the other.)
   b. Indicate all the points equidistant from the sides of the angle.
   c. What is the locus of points equidistant from the sides of a given angle? (Answer in a sentence.)
   d. State two theorems from Chapter 3 related to the locus in part (c).

Exercises 4–7 deal with figures in space.

4. What is the locus of points equidistant from the ceiling and floor of your classroom?

5. What is the locus of points in your classroom that are 1 m from the floor?

6. Choose a point P on the floor of the classroom. What is the locus of points, in the room, that are 1 m from P?

7. What is the locus of points that are equidistant from the ceiling and floor and are also equidistant from the two side walls?

8. Make up a locus problem for which the locus is a sphere.

9. Make up a locus problem for which the locus doesn't contain any points.

Written Exercises

Exercises 1–4 deal with figures in a plane. Draw a diagram showing the locus. Then write a description of the locus.

A **1.** Given two points A and B, what is the locus of points equidistant from A and B?

 2. Given parallel lines j and k, what is the locus of points equidistant from j and k?

 3. Given a point O, what is the locus of points 2 cm from O?

 4. Given a line h, what is the locus of points 2 cm from h?

In Exercises 5–8, begin each exercise with a square $ABCD$ that has sides 4 cm long. Draw a diagram showing the locus of the points inside the square that satisfy the given conditions. Then write a description of the locus.

 5. Equidistant from $\overline{AB}$ and $\overline{CD}$ **6.** Equidistant from points B and D

 7. Equidistant from $\overline{AB}$ and $\overline{BC}$ **8.** Equidistant from all four sides

Exercises 9–12 deal with figures in space.

 9. Given two parallel planes, what is the locus of points equidistant from the two planes?

 10. Given points C and D, what is the locus of points equidistant from C and D?

 11. Given point E, what is the locus of points 3 cm from E?

 12. Given line m, what is the locus of points 3 cm from m?

Exercises 13–17 deal with figures in a plane. (*Note:* If a point in a segment or an arc is not included in the locus, indicate the point by a small circle.)

B **13. a.** Draw an angle *HEX*. Construct the locus of points equidistant from the sides of $\angle HEX$.

 b. Draw two intersecting lines j and k. Construct the locus of points equidistant from j and k.

 14. Draw a segment $\overline{DE}$ and a line n. Construct the locus of points whose distance from n is *DE*.

 15. Draw a segment $\overline{AB}$. Construct the locus of points P such that $\angle APB$ is a right angle.

 16. Draw a segment $\overline{CD}$. Construct the locus of points Q such that $\triangle CQD$ is isosceles with base $\overline{CD}$.

 17. Draw a circle. Construct the locus of the midpoints of all radii of the circle.

Exercises 18–21 deal with figures in space.

 18. Given a square, what is the locus of points equidistant from the sides?

 19. Given a scalene triangle, what is the locus of points equidistant from the vertices?

 20. Given two parallel lines, what is the locus of points equidistant from the lines?

 21. Given a sphere, what is the locus of the midpoints of the radii of the sphere?

C **22.** A goat is tied to a square shed as shown. Using the scale 1 : 100, draw carefully a diagram that shows the region over which the goat can graze.

Ex. 22   Ex. 23

23. A ladder leans against a house. As *A* moves up or down on the wall, *B* moves along the ground. What path is followed by midpoint *M*? (*Hint:* Experiment with a meter stick, a wall, and the floor.)

24. Given circle *O* with radius 5, what is the locus of points *P* such that the two tangents to circle *O* from *P* form a 60° angle? Be precise.

25. Construct a square with a 4 cm side. A 3 cm segment moves so that its endpoints are always on the sides of the square. Construct the locus of the midpoint of the moving 3 cm segment.

26. Given a segment $\overline{CD}$, what is the locus in space of points *P* such that $\angle CPD = 90°$?

27. Given point *A* inside circle *O*, what is the locus of the midpoints of the chords that pass through *A*?

8-6 *Locus Problems*

The plural of locus is *loci*. The following problem involves intersections of loci.

Suppose you are given three noncollinear points *A*, *B*, and *C*. In the plane of *A*, *B*, and *C*, what is the locus of points that are (1) 1 cm from *A* and (2) equidistant from *B* and *C*?

You can take up one part of the problem at a time.

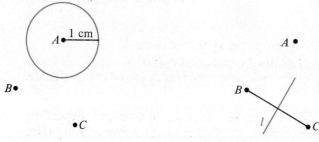

The locus satisfying (1) is circle *A* with radius 1 cm.

The locus satisfying (2) is *l*, the perpendicular bisector of $\overline{BC}$.

The locus of points satisfying *both* conditions must lie on both circle *A* and line *l*. There are three possibilities, depending on the positions of *A*, *B*, and *C*, as shown below.

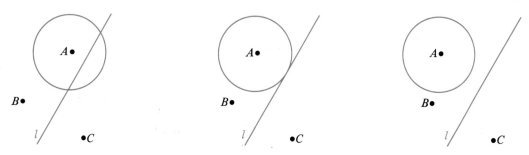

All three can be described in one sentence:

The locus is two points, one point, or no points, depending on the intersection of (1) a circle with center *A* and radius 1 cm and (2) a line which is the perpendicular bisector of $\overline{BC}$.

The example below deals with the corresponding problem in three dimensions.

Example: Given three noncollinear points *A*, *B*, and *C*, what is the locus of points 1 cm from *A* and equidistant from *B* and *C*?

Solution:

The first locus is sphere *A* with radius 1 cm.

The second locus is plane *P*, a perpendicular bisector of $\overline{BC}$.

Possibilities:

The plane might cut the sphere in a circle.
The plane might be tangent to the sphere.
The plane might not have any points in common with the sphere.

Thus, the locus is a circle, one point, or no points, depending on the intersection of (1) a sphere with center *A* and radius 1 cm and (2) a plane which is the perpendicular bisector of $\overline{BC}$.

Classroom Exercises ━━━━━━━━━━━━━━━━━━━━━━━━━━━━━━━━

Exercises 1–4 refer to coplanar figures. Describe the intersections that are possible.

1. A line and a circle

2. Two parallel lines and a third line

3. An angle and a circle

4. Two intersecting lines and a third line

Exercises 5–8 refer to figures in space. Describe the intersections that are possible.

5. A line and a plane

6. A line and a sphere

7. Two spheres

8. A plane and a sphere

9. Let C be the point in the center of your classroom. Describe the locus of points in space that satisfy the given conditions.

a. 3 m from C

b. 3 m from C and equidistant from the ceiling and the floor

c. 3 m from C and 1 m from either the ceiling or the floor

10. Call the four corners on the floor of your classroom A, B, C, and D. Call the center of the ceiling X. Describe the locus of points that satisfy the given conditions.

a. On the floor and equidistant from A, B, C, and D

b. Equidistant from A, B, C, and D

c. Equidistant from A and X

Written Exercises ━━━━━━━━━━━━━━━━━━━━━━━━━━━━━━━━━━

Exercises 1 and 2 refer to plane figures.

A 1. Refer to the problem: Given two points D and E, what is the locus of points 1 cm from D and 2 cm from E?

a. The locus of points 1 cm from D is __?__ .

b. The locus of points 2 cm from E is __?__ .

c. Draw diagrams to show three possibilities with regard to points that satisfy both conditions (a) and (b).

d. Give a one-sentence solution to the problem.

2. Refer to the problem: Given a point A and a line k, what is the locus of points 3 cm from A and 1 cm from k?

a. The locus of points 3 cm from A is __?__ .

b. The locus of points 1 cm from k is __?__ .

c. Draw diagrams to show five possibilities with regard to points that satisfy both conditions (a) and (b).

d. Give a one-sentence solution to the problem.

Exercises 3–6 refer to plane figures. Draw a diagram of the locus. Then write a description of the locus.

3. Points A and B are 3 cm apart. What is the locus of points 2 cm from both A and B?

4. Lines j and k intersect in point P. What is the locus of points equidistant from j and k, and 2 cm from P?

5. Point Q is on line l. What is the locus of points 5 cm from Q and 3 cm from l?

6. Given $\angle A$, what is the locus of points equidistant from the sides of $\angle A$ and 2 cm from vertex A?

In Exercises 7–10, draw diagrams to show the possibilities with regard to points in a plane.

B **7.** Given points C and D, what is the locus of points 2 cm from C and 3 cm from D? (3 possibilities)

8. Given point E and line k, what is the locus of points 3 cm from E and 2 cm from k? (5 possibilities)

9. Given a point A and two parallel lines j and k, what is the locus of points 30 cm from A and equidistant from j and k? (3 possibilities)

10. Given four points P, Q, R, and S, what is the locus of points that are equidistant from P and Q and equidistant from R and S? (3 possibilities)

Exercises 11–14 refer to figures in space. In each exercise, tell what the locus is. You need not draw the locus or describe it precisely.

Example: Given points G and H, what is the locus of points 4 cm from G and 4 cm from H? (3 possibilities)

Solution: The locus is a circle, one point, or no points.

11. Given point A in plane Z, what is the locus of points 5 cm from A and 3 cm from Z?

12. Given point A in plane Z, what is the locus of points 5 cm from A and d cm from Z? (3 possibilities)

13. Given $\overleftrightarrow{AB} \perp$ plane Q, what is the locus of points 2 cm from $\overleftrightarrow{AB}$ and 2 cm from Q?

14. Given point B and plane Y, what is the locus of points 2 cm from B and 2 cm from Y? (4 possibilities)

C **15.** Point P and intersecting lines j and k are all in a plane. What is the locus of points in the plane which are equidistant from j and k and a given distance from P? Draw 5 sketches. Then describe the five possibilities in one sentence.

16. Given three points, each 2 cm from the other two, draw a diagram to indicate the locus of points in the plane of the given points and not more than 2 cm away from any of them.

17. Describe the locus of points equidistant from the four points *V*, *A*, *B*, and *C*. The diagram indicates points in space.

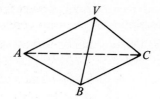

18. A mini-radio transmitter has been secured to a bear. Rangers at *D*, *E*, and *F* are studying the bear's movements. Rangers *D* and *E* can receive the bear's beep at distances up to 10 km, ranger *F* at distances up to 15 km. Draw a diagram showing where the bear might be at these times:

a. When all three rangers can receive the signal

b. When *F* suddenly detects the signal after a period of time during which only *D* and *E* could receive the signal

c. When station *D* is shut down, and *F* begins to detect the signal just as *E* loses it

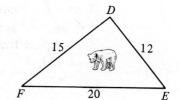

8-7 *Locus and Construction*

Sometimes the solution to a construction problem depends on finding a point that satisfies more than one condition. To locate the point, you may have to construct the locus of points satisfying one of the conditions.

Example: Given the angle and the segments shown, construct $\triangle ABC$ with $\angle A = n^\circ$, $AB = r$, and the altitude to $\overleftrightarrow{AB}$ equal to *s*.

Solution: It is easy to construct $\angle A$ and side $\overline{AB}$. Point *C* must satisfy two conditions: *C* must lie on $\overrightarrow{AZ}$, and *C* must be *s* units from $\overleftrightarrow{AB}$. The locus of points *s* units from $\overleftrightarrow{AB}$ is a pair of parallel lines. Only the upper parallel will intersect $\overrightarrow{AZ}$. We construct that parallel to $\overleftrightarrow{AB}$ as follows:

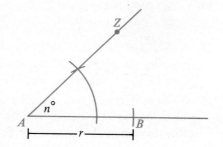

1. Construct the perpendicular to $\overleftrightarrow{AB}$ at any convenient point X.
2. Mark off s units on the perpendicular to locate point Y.
3. Construct the perpendicular to $\overleftrightarrow{XY}$ at Y. Call it $\overleftrightarrow{YW}$.

Note that all points on $\overleftrightarrow{YW}$ are s units from $\overleftrightarrow{AB}$. Thus the intersection of $\overleftrightarrow{YW}$ and $\overrightarrow{AZ}$ is the desired point C. To complete the solution, we simply draw $\overline{CB}$.

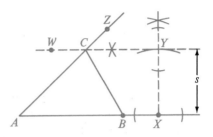

Classroom Exercises

1. The purpose of this exercise is to analyze the following construction problem:

   Given a circle and a segment with length k, inscribe in the circle an isosceles triangle RST with base $\overline{RS}$ k units long.

   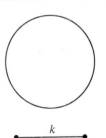

   a. Suppose R has been chosen. Where must S lie so that RS equals k? (In other words, what is the locus of points k units from R?)

   b. Suppose $\overline{RS}$ is fixed. Where must T lie so that $RT = ST$? (In other words, what is the locus of points equidistant from R and S?)

   c. Explain the steps of the construction shown.

   (1)   (2)   (3)

   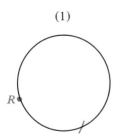 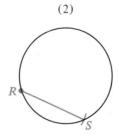 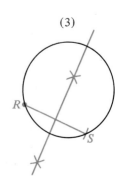

   d. Explain two different ways to finish the construction.

2. Two different solutions, both correct, are shown for the following construction problem. Analyze the diagrams and explain the solutions.

Given segments with lengths r and s, construct $\triangle ABC$ with $\angle C = 90°$, $AC = r$, and $AB = s$.

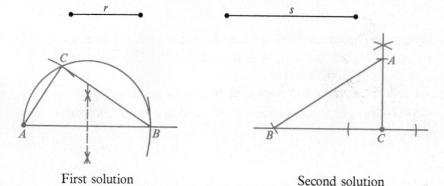

First solution   Second solution

Written Exercises

A 1. Draw a segment $\overline{AB}$ and a segment with length h. Construct the locus of points P such that for every $\triangle APB$ the altitude from P to $\overleftrightarrow{AB}$ would equal h.

2. Begin each part of this exercise by drawing a segment $\overline{DE}$. Then construct the locus of points P that meet the given condition.

   a. $\angle DEP$ is a right angle.

   b. $\angle DPE$ is a right angle. (*Hint:* Look at the first solution of Classroom Exercise 2.)

On your paper draw a segment roughly as long as the one shown:
Use it in Exercises 3–6.

3. Draw a rectangle $ABCD$. Find by construction the points that are equidistant from $\overrightarrow{BA}$ and $\overrightarrow{BC}$ and are v units from D. Use X to label each point.

4. Draw a rectangle $ABCD$. Find by construction the points that are equidistant from $\overleftrightarrow{AB}$ and $\overleftrightarrow{CD}$ and are v units from D.

5. Draw a figure roughly like the one shown. Then construct a circle, with radius v, that passes through N and is tangent to line k.

   $N\bullet$

   k _____

   (*Hint:* Construct the locus of points which would, as centers, be the correct distance from k. Also construct the locus of points which would, as centers, be the correct distance from N.)

6. Draw an angle XYZ. Construct a circle, with radius v, that is tangent to the sides of $\angle XYZ$.

On your paper draw an angle and three segments roughly like those shown. Use them in Exercises 7–15.

B **7.** Construct $\triangle ABC$ with $\angle A = \angle B = n°$ and the altitude to $\overline{AB}$ equal to s.

8. Construct $\triangle ABC$ with $\angle C = 90°$, $\angle A = n°$, and the altitude to $\overline{AB}$ equal to s.

9. Construct $\triangle ABC$ with $AB = s$, $AC = t$, and the altitude to $\overline{AB}$ equal to r.

10. Construct $\triangle ABC$ with $AB = t$, the median to $\overline{AB}$ equal to s, and the altitude to $\overline{AB}$ equal to r.

11. Construct an isosceles triangle with each leg equal to s and the altitude to each leg equal to r.

12. Construct a right triangle with the altitude to the hypotenuse equal to r and the median to the hypotenuse equal to s.

C **13.** Draw circle O roughly like the one shown. Construct a chord r units long that passes through X.

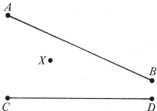

14. Construct an isosceles right triangle such that the radius of the inscribed circle is r.

15. Construct a right triangle such that the bisector of the right angle divides the hypotenuse into segments whose lengths are r and s.

16. Draw a figure roughly like the one shown. Without extending $\overline{AB}$ or $\overline{CD}$, construct a line through X that contains the intersection of $\overrightarrow{AB}$ and $\overrightarrow{CD}$.

Challenge

A mouse moves along $\overline{AJ}$. For any position M of the mouse, X and Y are such that $\overline{AX} \perp \overline{AJ}$ with $AX = AM$, and $\overline{JY} \perp \overline{AJ}$ with $JY = JM$. The cat is at C, the midpoint of $\overline{XY}$. Describe the locus of the cat as the mouse moves from A to J.

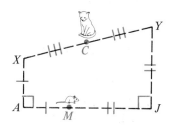

SELF-TEST

Describe briefly the locus of points that satisfy the conditions.

1. In the plane of an angle, and equidistant from the sides of the angle

2. In a plane, and 5 cm from a point in the plane

3. In the plane of a rectangle, and equidistant from the vertices of the rectangle

4. In the plane of equilateral triangle ABC, 4 cm from A, and 4 cm from $\overleftrightarrow{BC}$ (Each side of $\triangle ABC$ is 5 cm long.)

5. 5 cm from a given plane in space

6. 5 cm from each of two points in space which are 6 cm apart

7. Draw an acute angle and call it $\angle 1$. Draw a segment and label its length n. Construct $\square ABCD$ with $\angle A = \angle 1$, $AB = 2n$, and the altitude from D to $\overleftrightarrow{AB}$ equal to n.

Extra The Nine-Point Circle _____

Given any triangle ABC, let H be the intersection of the three altitudes. There is a circle that passes through these nine special points:

midpoints L, M, N of the three sides;

points R, S, T, where the three altitudes of the triangle meet the sides;

midpoints X, Y, Z, of $\overline{HA}$, $\overline{HB}$, $\overline{HC}$.

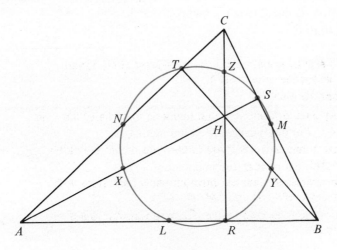

Key steps of proof:

1. *XYMN* is a rectangle.
2. The circle circumscribed about *XYMN* has diameters $\overline{MX}$ and $\overline{NY}$.
3. Because ∠*XSM* and ∠*YTN* are right angles, the circle contains points *S* and *T* as well as *X*, *Y*, *M*, and *N*.

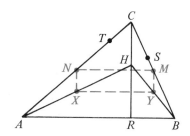

4. *XLMZ* is a rectangle.
5. The circle circumscribed about *XLMZ* has diameters $\overline{MX}$ and $\overline{LZ}$.
6. Because ∠*XSM* and ∠*ZRL* are right angles, the circle contains points *S* and *R* as well as *X*, *L*, *M*, and *Z*.

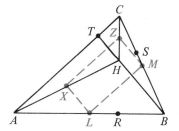

7. The circle of Steps 1–3 and the circle of Steps 4–6 must be the same circle, because $\overline{MX}$ is a diameter of both circles.
8. There is a circle that passes through the nine points *L*, *M*, *N*, *R*, *S*, *T*, *X*, *Y*, and *Z*. (See Steps 3 and 6.)

One way to locate the center of the circle is to locate points *X* and *M*, then the midpoint of $\overline{XM}$.

Exercises

1. Test your mechanical skill by constructing the nine-point circle for an acute triangle. (The larger the figure, the better.)
2. Repeat, but use an obtuse triangle.
3. Repeat, but use an equilateral triangle. What happens to some of the nine points?
4. Repeat, but use a right triangle. How many of the nine points are at the vertex of the right angle?
5. Prove that *XYMN* is a rectangle. Use the diagram shown for Steps 1–3 above. (*Hint:* Compare $\overline{NX}$ with $\overline{CR}$ and $\overline{NM}$ with $\overline{AB}$.)
6. How does the radius of the nine-point circle compare with the radius of the circumscribed circle?

Biographical Note Blaise Pascal

When Blaise Pascal (1623–1662) was twelve years old, he asked his father what geometry was about. He was told it consisted in constructing exact figures and in studying the relations between the parts. Intrigued by the subject, before long he had deduced several important properties of the triangle, including the fact that the sum of the angles is 180°, Theorem 2-10 in this book.

At the age of sixteen Pascal proved a theorem concerning certain collinear points, considered to be the greatest theorem of medieval geometry.

At twenty-five he abandoned mathematics to become a monk, devoting his attention to philosophy and religion. Later in life, however, Pascal was to have a final encounter with geometry. Bothered by a painful toothache, he found relief when he thought about the cycloid, a geometric curve traced out by the motion of a fixed point on the circumference of a wheel rolling along a straight line on a flat surface. Interpreting this relief as a divine remedy, he spent the next eight days discovering important properties of the cycloid.

Chapter Summary _____

1. Geometric constructions are made using only two instruments, a straightedge and a compass.

2. Basic constructions

    (1) A segment equal to a given segment

    (2) An angle equal to a given angle

    (3) The bisector of a given angle

    (4) A perpendicular to a line at a given point on the line

    (5) A perpendicular to a line from a given point outside the line

    (6) The perpendicular bisector of a given segment

    (7) A line parallel to a given line through a given point outside the line

    (8) A tangent to a given circle at a given point on the circle

    (9) A tangent to a given circle from a given point outside the circle

    (10) A circle circumscribed about a given triangle

    (11) A circle inscribed in a given triangle

    (12) Division of a given segment into a given number of equal parts

    (13) A segment of length x so that $\dfrac{a}{b} = \dfrac{c}{x}$ when segments of lengths a, b, and c are given

    (14) A segment whose length is the geometric mean between the lengths of two given segments

3. A locus is the set of all points, and only those points, that satisfy one or more conditions.

4. A locus which satisfies more than one condition is the intersection of the loci that satisfy the several conditions separately.

Chapter Review

In Exercises 1–9 and 12–14, draw a diagram which is similar to, but larger than, the one shown. Then do the constructions. Justifications are not required.

8-1 *What Construction Means*

1. Draw a line *l*. On *l* construct a segment equal to $\overline{AC}$.
2. Construct an angle equal to ∠ *B*.
3. Bisect ∠ *C*.

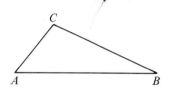

8-2 *Perpendiculars and Parallels*

4. Construct the perpendicular bisector of $\overline{DE}$.
5. Construct a perpendicular to $\overleftrightarrow{DE}$ at point *E*.
6. Construct the parallel to $\overleftrightarrow{DE}$ through *F*.
7. Construct the perpendicular to $\overleftrightarrow{DE}$ from *G*.

8-3 *Circles*

8. Construct a tangent to ⊙*O* from point *H*.
9. Construct the tangent to ⊙*O* at point *J*.
10. Draw an obtuse triangle and, by construction, inscribe a circle in it.
11. Draw an acute triangle and, by construction, circumscribe a circle about it.

8-4 *Special Segments*

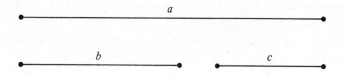

12. Construct a segment of length $\frac{1}{3}a$.

13. Construct a segment of length x so that $\dfrac{a}{c} = \dfrac{b}{x}$.

14. Construct a segment whose length is the geometric mean y between b and c.

8-5 *The Meaning of Locus*

15. What is the locus of points in a plane 4 cm from a given point P in that plane?

16. What is the locus of points in space which are equidistant from two parallel planes?

17. What is the locus of points in a plane which are equidistant from two points in that plane?

18. What is the locus of points in a plane at a distance of 3 cm from a line in that plane?

8-6 *Locus Problems*

19. Point P is 4 cm from line l. What is the locus of points, in the plane of P and l, 3 cm from P and 2 cm from l? Sketch the locus.

20. Two lines, j and k, intersect in point Q. What is the locus of points, in their plane, equidistant from j and k and 3 cm from Q? Sketch the locus.

21. What is the locus of points in space which are 3 cm from point A and 3 cm from a plane which does not contain A? Answer in a sentence.

8-7 *Locus and Construction*

22. Construct an isosceles right triangle in which the altitude to the hypotenuse is e units long.

23. Construct $\triangle RST$ so that $\overline{RS}$ has length f, $\overline{RT}$ has length g, and the altitude from T to $\overline{RS}$ has length e.

Chapter Test

In Exercises 1–5, draw a diagram which is similar to, but larger than, the one shown. Then do the required constructions. Justifications are not required.

1. Construct an isosceles triangle with each base angle equal to ∠*ABC*.

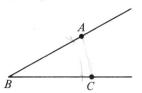

2. Construct the tangent to ⊙*O* at point *D*.

3. Construct the altitude from *H* to $\overline{JK}$.

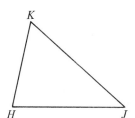

4. Circumscribe a circle about acute △*LMN*.

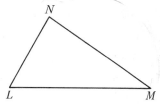

5. Construct a rectangle which has two sides of length *FG* and two sides of length ⅓*FG*.

6. Construct an isosceles right triangle.

7. Draw a segment, $\overline{AB}$, about 3 cm long. Construct a segment, $\overline{CD}$, which is twice the length of $\overline{AB}$. Construct the geometric mean *x* between *AB* and *CD*.

8. Construct an equilateral △*RST*. Through *T* construct the line *k* which is parallel to $\overline{RS}$.

9. In a plane, what is the locus of points that are 3 cm from a given point *T*? Answer in a sentence.

10. In space, what is the locus of points that are 3 cm from a given point *T*? Answer in a sentence.

For Exercises 11 and 12 draw two segments roughly like those shown.

11. Given points *P* and *Q* such that *PQ* = *s*, construct the locus of points, in a plane, equidistant from *P* and *Q*, and *r* units from $\overleftrightarrow{PQ}$.

12. Given points *X* and *Y* such that *XY* = *r*, construct the locus of points, in a plane, *s* units from $\overleftrightarrow{XY}$, and *s* units from point *Y*.

320

9

Areas of Plane Figures

Areas of Polygons

Objectives

1. Understand what is meant by the area of a polygon.
2. Understand the area postulates.
3. Know and use the formulas for the areas of rectangles, parallelograms, triangles, trapezoids, and regular polygons.

9-1 *Areas of Rectangles*

In everyday conversation people often refer to the *area* of a rectangle when what they really mean is the area of a rectangular region.

Rectangle   Rectangular Region

For the sake of simplicity, we shall continue this common practice. Thus, when we speak of the area of a triangle, we shall mean the area of the triangular region which includes the triangle *and* its interior.

In Chapter 1 we accepted postulates which enable us to express the lengths of segments and the measures of angles as positive numbers. Similarly, the areas of figures are positive numbers with properties given by the following area postulates.

Postulate 14 The area of a square is the square of the length of a side. $(A = s^2)$

Length: 1 unit

Area: 1 square unit

Area: 3^2, or 9, square units

Postulate 15 If two figures are congruent, then they have the same area. (Area Congruence Postulate)

Postulate 16 The area of a polygonal region is the sum of the areas of its non-overlapping parts. (Area Addition Postulate)

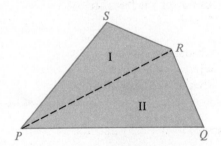

Area of *PQRS*:
Area I + Area II

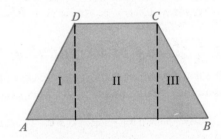

Area of *ABCD*:
Area I + Area II + Area III

The area formulas for a rectangle, a parallelogram, a triangle, and a trapezoid will be proved as theorems from the area postulates. These theorems will involve the ideas of *base* and *altitude*.

Any side of a rectangle or other parallelogram can be considered to be a base. The length of a base will be denoted by *b*. An altitude to a base is any segment perpendicular to the base line from any point on the opposite side. The length of an altitude is called the height (*h*). All the altitudes to a particular base have the same length.

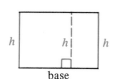 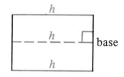 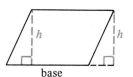 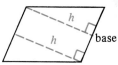

Theorem 9-1 The area of a rectangle equals the product of its base length and height. $(A = bh)$

Given: A rectangle with base length b and height h

Prove: $A = bh$

Proof:

Building onto the given rectangle, we can draw the large square shown. It consists of these non-overlapping parts:
  a smaller square with area b^2;
  a smaller square with area h^2;
  the given rectangle with area A;
  a congruent rectangle with area A.

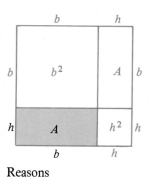

Statements	Reasons
1. Area of big square $= 2A + b^2 + h^2$	1. Area Addition Postulate
2. Area of big square $= (b + h)^2$, or $b^2 + 2bh + h^2$	2. $A = s^2$
3. $2A + b^2 + h^2 = b^2 + 2bh + h^2$	3. Substitution Property
4. $2A = 2bh$	4. Subtraction Postulate
5. $A = bh$	5. Division Postulate

Some common units of area are the square centimeter (cm^2) and the square meter (m^2).

Example: Find the area of each figure.

a.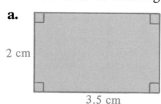

b.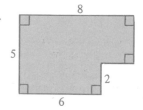

Solution: **a.** $A = 3.5 \cdot 2$
            $= 7$ (cm^2)

b. $A = (8 \cdot 5) - (2 \cdot 2) = 40 - 4$
      $= 36$ (square units)

Classroom Exercises

1. Tell what each letter represents in the formula $A = s^2$.
2. Tell what each letter represents in the formula $A = bh$.
3. Find the area and perimeter of a square with sides 5 cm long.
4. The perimeter of a square is 28 cm. What is the area?
5. The area of a square is 64 cm². What is the perimeter?

Exercises 6–13 refer to rectangles. Complete the table.

	6.	7.	8.	9.	10.	11.	12.	13.
b	8 cm	4 cm	12 m	?	$3\sqrt{2}$	$4\sqrt{2}$	$5\sqrt{3}$	$x + 3$
h	3 cm	1.2 cm	?	5 cm	2	$\sqrt{2}$	$2\sqrt{3}$	x
A	?	?	36 m²	55 cm²	?	?	?	?

14. **a.** What is the converse of the Area Congruence Postulate?
    b. Is this converse true or false? Explain.

15. **a.** Draw three rectangles each with perimeter 20 cm. Find the area of each rectangle.
    b. Of all rectangles having perimeter 20 cm, which one do you think has the greatest area? (Give its length and width.)
    c. What is the perimeter of the rectangle shown?
    d. What is the area?
    e. Sketch a graph showing how the area varies as x varies.

x ▭ $10 - x$

Written Exercises

Complete the tables. Exercises 1–16 refer to rectangles. p is perimeter.

A

	1.	2.	3.	4.	5.	6.	7.	8.
b	12 cm	8.2 cm	16 cm	?	$3\sqrt{2}$	$\sqrt{6}$	$2x$	$4k - 1$
h	5 cm	4 cm	?	8 m	$4\sqrt{2}$	$\sqrt{2}$	$x - 3$	$k + 2$
A	?	?	80 cm²	120 m²	?	?	?	?

	9.	10.	11.	12.	13.	14.	15.	16.
b	9 cm	10 cm	16 cm	$x + 5$	$a + 3$	$k + 7$	x	?
h	4 cm	?	?	x	$a - 3$	?	?	y
A	?	?	?	?	?	?	$x^2 - 3x$	$y^2 + 7y$
p	?	30 cm	42 cm	?	?	$4k + 20$	?	?

Consecutive sides of the figures below are perpendicular. Find the area of each figure.

B 17.

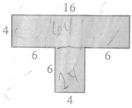

18.

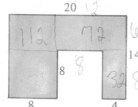

19.

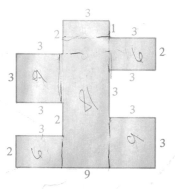

20.

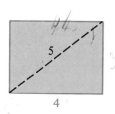

21.

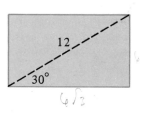

22.

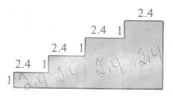

23. The diagonals of a square are $4\sqrt{2}$ cm long. What is the area?

24. The diagonals of a square are 10 cm long. What is the area?

25. The area of a rectangle is 360 cm². Find the length and width if their ratio is 5:2.

26. A walk 2 m wide surrounds a rectangular grass plot 40 m long and 20 m wide. Find the area of the walk.

27. For what value of x will a square with side length $3x - 9$ have area 81?

28. A rectangle with length $3x + 1$ and width $x - 2$ has area 48. Find the value of x.

C 29. Find the lengths of the sides of a rectangle of area 24 and perimeter 22.

30. If $2x^2 + 2x - 12$ represents the area of a polygon, what is the range of values for x?

31. If the length of a rectangle is increased by 25 percent, by what percent must the width be decreased if the area is to remain the same?

32. If rectangle $ABCD$ has area $36\sqrt{3}$, find the length and width.

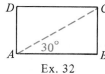

Ex. 32

9-2 Areas of Parallelograms and Triangles

Detailed formal proofs of most area theorems are lengthy and time consuming. For that reason, only key steps of proofs will accompany the theorems.

Theorem 9-2   The area of a parallelogram equals the product of its base length and height. (A = bh)

Given: $\square PQRS$

Prove: $A = bh$

Key steps of proof:

1. Draw altitudes $\overline{PV}$ and $\overline{QT}$, forming two rt. $\triangle$.
2. Area I = Area III
3. Area of $\square PQRS$ = Area II + Area I
   = Area II + Area III
   = Area of rect. $PQTV$
   = bh

Theorem 9-3   The area of a triangle equals half the product of its base length and height. ($A = \frac{1}{2}bh$)

Given: $\triangle XYZ$

Prove: $A = \frac{1}{2}bh$

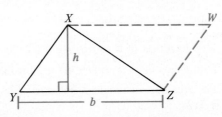

Key steps of proof:

1. Draw $\overline{XW} \parallel \overline{YZ}$ and $\overline{ZW} \parallel \overline{YX}$, forming $\square XYZW$.
2. $\triangle XYZ \cong \triangle ZWX$
3. Area of $\triangle XYZ = \frac{1}{2}$ Area of $\square XYZW$
   = $\frac{1}{2}bh$

Corollary   The area of a rhombus equals half the product of its diagonals. ($A = \frac{1}{2}d_1 d_2$)

The key steps of a proof of the corollary are left as Exercise 18.

Example 1:   Find the area of a triangle with sides 8, 8, and 6.

Solution: Draw the altitude to the 6 cm base. Since the triangle is isosceles, this altitude bisects the base.

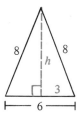

$$h^2 + 3^2 = 8^2 \quad \text{(Pythagorean Theorem)}$$
$$h^2 = 8^2 - 3^2 = 55$$
$$h = \sqrt{55}$$
$$A = \tfrac{1}{2}bh = \tfrac{1}{2} \cdot 6 \cdot \sqrt{55} = 3\sqrt{55}$$

Example 2: Find the area of an equilateral triangle with side 6.

Solution: Draw an altitude. Two 30-60-90 triangles are formed.

$$h = 3\sqrt{3}$$
$$A = \tfrac{1}{2}bh = \tfrac{1}{2} \cdot 6 \cdot 3\sqrt{3} = 9\sqrt{3}$$

Classroom Exercises

1. The area of the parallelogram can be found in two ways:

   a. $A = 8 \cdot \underline{\;?\;} = \underline{\;?\;}$

   b. $A = 4 \cdot \underline{\;?\;} = \underline{\;?\;}$

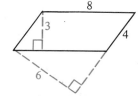

2. Find the area of $\triangle ABC$.
3. Find the area of $\triangle DBC$.
4. Find the area of $\triangle EBC$.

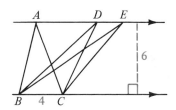

5. The area of the triangle can be found in 3 ways:

   a. $A = \tfrac{1}{2}(6) \cdot \underline{\;?\;} = \underline{\;?\;}$

   b. $A = \tfrac{1}{2}(10) \cdot \underline{\;?\;} = \underline{\;?\;}$

   c. $A = \tfrac{1}{2}(14) \cdot \underline{\;?\;} = \underline{\;?\;}$

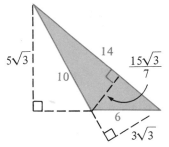

6. What is the area of a right triangle with legs having lengths a and b?
7. Find the area of a triangle with sides 5, 12, and 13 units long.
8. **a.** Find the area of a right triangle with legs 6 and 8 units long.

   b. How long is the altitude to the hypotenuse?

Written Exercises

Exercises 1–8 refer to triangles. Complete the table.

A

	1.	**2.**	**3.**	**4.**	**5.**	**6.**	**7.**	**8.**
b	8 cm	5 m	9	?	$3\sqrt{2}$	$6\sqrt{3}$	$5x$	?
h	7 cm	12 m	?	14	$2\sqrt{2}$	$3\sqrt{6}$	?	$3\sqrt{2}$
A	?	?	108	56	?	?	$15xy$	24

Find the area of each figure.

9.

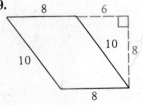

10.

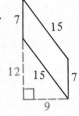

11.

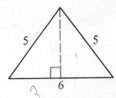

12. How much will it cost to blacktop the driveway shown if blacktopping costs $4.00 per square meter?

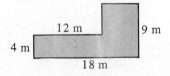

B 13. First find the value of x.
Then find the area of $PQRS$.

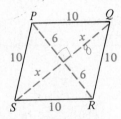

14. The area of $\triangle ABC$ is 240.
Find the area of $\triangle ABD$.

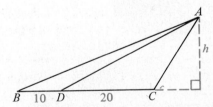

15. $\overline{AM}$ is a median of $\triangle ABC$. If the area of $\triangle ABC$ is 100, what is the area of $\triangle ABM$?

16. A parallelogram has sides 9 cm and 15 cm. If the shorter altitude is 3 cm, how long is the longer altitude?

17. a. What is the area of an equilateral triangle with side s?

b. If the area of an equilateral triangle is $\sqrt{3}$, what is the length of a side of the triangle?

18. Write the key steps of a proof for the corollary of Theorem 9-3 on page 326.

Find the area of each figure.

19. An isosceles triangle with legs 17 and base 16

20. A square with diagonal $6\sqrt{2}$

21. A 30-60-90 triangle with hypotenuse 14

22. An equilateral triangle with height 6

23. A regular hexagon with perimeter 60

24. A rectangle with length 12 inscribed in a circle with radius 6.5

Think of a parallelogram made with cardboard strips and hinged at each vertex so that the measure of $\angle C$ will vary. Find the area of the parallelogram for each measure of $\angle C$ given in Exercises 25-29.

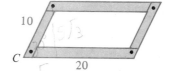

25. 30°   **26.** 45°   **27.** 60°   **28.** 90°   **29.** 120°

30. Approximate your answers to Exercises 26, 27, and 29 by using $\sqrt{2} \approx 1.4$ and $\sqrt{3} \approx 1.7$. Then record your answers to Exercises 25-29 on a graph like the one shown below.

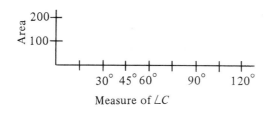

C 31. Draw any quadrilateral *ABCD*. Construct a triangle whose area is equal to the area of *ABCD*.

32. Draw any parallelogram and its diagonals. Write the key steps of a proof that the diagonals of a parallelogram form four triangles with equal areas.

33. a. Accurately draw or construct a large equilateral triangle. Choose any point inside the triangle and carefully measure the distances *x*, *y*, and *z*. Evaluate $x + y + z$.

b. Now choose another point on or inside the triangle and find $x + y + z$. What do you notice? Why does this happen?

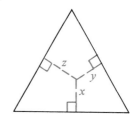

34. Two squares each with side 12 cm are placed so that a vertex of one lies at the center of the other. Find the area of the shaded region.

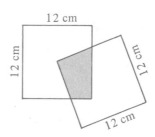

9-3 *Areas of Trapezoids*

An altitude of a trapezoid is any segment perpendicular to a base line from a point on the opposite base line. Since the bases are parallel, all altitudes have the same length, called the *height* (*h*) of the trapezoid.

Theorem 9-4 The area of a trapezoid equals half the product of the height and the sum of the bases. $A = \frac{1}{2}h(b_1 + b_2)$

Key steps of proof:

1. Draw diagonal $\overline{BD}$ of trap. $ABCD$, forming two triangular regions, I and II, each with height h.
2. Area of trapezoid = Area I + Area II
$$= \tfrac{1}{2}b_1h + \tfrac{1}{2}b_2h$$
$$= \tfrac{1}{2}h(b_1 + b_2)$$

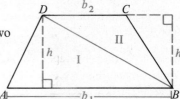

Example 1: Find the area of a trapezoid with height 7 and bases 12 and 8.

Solution: $A = \frac{1}{2}h(b_1 + b_2) = \frac{1}{2} \cdot 7 \cdot (12 + 8) = 70$

Example 2: Find the area of an isosceles trapezoid with sides 5, 6, 5, and 10.

Solution: When you draw the two altitudes shown, you get a rectangle and two congruent right triangles. The segments of the lower base must have lengths 6, 2, and 2. First find h:

$$h^2 + 2^2 = 5^2$$
$$h^2 = 21$$
$$h = \sqrt{21}$$

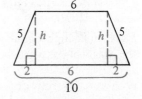

Then find the area: $A = \frac{1}{2}h(b_1 + b_2) = \frac{1}{2}\sqrt{21}(10 + 6) = 8\sqrt{21}$

Classroom Exercises

Find the area of each trapezoid.

1.

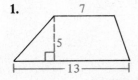

2.

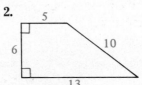

3.

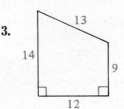

4. 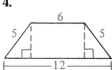   **5.**    **6.**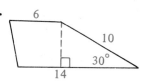

7. a. Find the lengths of the medians of the trapezoids in Exercises 1–3.

b. Explain why the area of a trapezoid can also be given by the formula
Area = height × median.

Written Exercises

Exercises 1–8 refer to trapezoids. Complete the table.

A

	1.	**2.**	**3.**	**4.**	**5.**	**6.**	**7.**	**8.**
b_1	12	6.8	$3\frac{1}{6}$	45	27	3	7	?
b_2	8	3.2	$4\frac{1}{3}$	15	9	?	?	$3k$
h	7	6.1	$1\frac{2}{5}$	?	?	3	$9\sqrt{2}$	$5k$
A	?	?	?	300	90	12	$36\sqrt{2}$	$45k^2$

9. Find the lengths of the medians of the trapezoids in Exercises 1–3.

10. A trapezoid has area 54 and height 6. How long is its median?

Find the area of each trapezoid.

11. 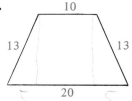   **12.** 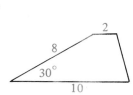   **13.**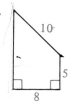

14. 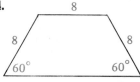   **15.** 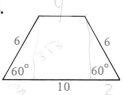   **16.**

B 17. a. The legs of an isosceles trapezoid are 10 cm and the bases are 9 cm and
21 cm. Find the area of the trapezoid.

b. Find the lengths of the diagonals.

18. An isosceles trapezoid has base lengths 12 and 28. The area is 300.

a. Find the height   **b.** Find the perimeter.

19. a. If the congruent trapezoids shown are slid together, what special quadrilateral is formed?

   b. Use your answer to derive the formula $A = \frac{1}{2}h(b_1 + b_2)$.

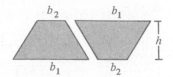

PQRS is a trapezoid.

20. Compare the areas of $\triangle PQS$ and $\triangle PQR$.

21. Compare the areas of $\triangle POS$ and $\triangle QOR$.

22. Compare the areas of $\triangle PSR$ and $\triangle QSR$.

23. If $\overline{PQ}$ is twice as long as $\overline{SR}$, compare the areas of $\triangle PSR$ and $\triangle PSQ$.

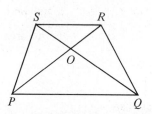

24. *ABCDEF* is a regular hexagon with side 12. Find the areas of the three regions formed when diagonals $\overline{AC}$ and $\overline{AD}$ are drawn.

C 25. An isosceles trapezoid with base lengths 12 and 16 is inscribed in a circle of radius 10. The center of the circle lies in the interior of the trapezoid. Find the area of the trapezoid.

26. Find the area of trapezoid *WXYZ* shown at the left below.

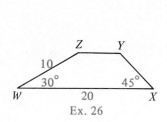

Ex. 26

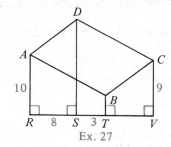

Ex. 27

27. If $DS = 16$, find the area of $\square ABCD$ shown at the right above.

COMPUTER KEY-IN

A formula for finding the area of a triangle, given the lengths of the sides, has been known for over two thousand years.

   Using *a*, *b*, and *c* to represent the lengths of the sides of a triangle and *s* to represent the semiperimeter,

$$s = \frac{a + b + c}{2}$$

The area *K* is then given by the following expression, known as Heron's Formula:

$$K = \sqrt{s(s - a)(s - b)(s - c)}$$

If $a = 6$, $b = 8$, and $c = 10$, we can calculate s and K as follows.

$$s = \frac{6 + 8 + 10}{2} = 12$$

$$K = \sqrt{12(12 - 6)(12 - 8)(12 - 10)} = \sqrt{576} = 24$$

Since 6, 8, and 10 are the sides of a right triangle with base 6 and height 8, we can check the value we found for K by using another method:

$$K = \tfrac{1}{2}bh = \tfrac{1}{2}(6)(8) = 24$$

Exercises

1. A triangle has sides 3, 13, and 14. Heron's Formula gives the area as $6\sqrt{10}$. If 3 is taken as the base, then $6\sqrt{10} = \tfrac{1}{2} \cdot 3h$, and the height, h, is $4\sqrt{10}$.

   a. If 13 is taken as the base, the height is ___?___ .

   b. If 14 is taken as the base, the height is ___?___ .

2. The computer print-out below gives the area of a triangle and the height when each side is taken as the base. For the print-out shown, 8, 9, and 11 were fed into the computer as the lengths of the sides.

   ```
   WHAT ARE THE LENGTHS OF THE SIDES
   OF THE TRIANGLE
   ?8
   ?9
   ?11
   THE AREA OF THE TRIANGLE IS   35.4965.

   IF  8 IS TAKEN AS THE BASE,
   THE HEIGHT IS  8.87412.

   IF  9 IS TAKEN AS THE BASE,
   THE HEIGHT IS  7.88811.

   IF  11 IS TAKEN AS THE BASE,
   THE HEIGHT IS  6.45391.
   ```

   Write a program which will compute and print the kind of information shown above for any triangle. Include steps which allow you to give the computer the lengths of the sides.

In Exercises 3–11, the lengths of the sides of a triangle are given. Use the program you wrote in Exercise 2 to find the area and heights of each triangle.

3. 5, 7, 9	**4.** 11, 13, 15	**5.** 6, 7, 8
6. 10.5, 11.5, 20	**7.** 12, 16, 27	**8.** 6.3, 7.2, 10.1
9. 8, 8, 10	**10.** 15, 16, 20	**11.** 5.5, 6.5, 10

9-4 *Areas of Regular Polygons*

Given any circle, you can inscribe in it a regular polygon of any number of sides. The diagrams below show how this can be done.

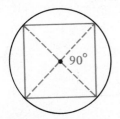
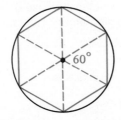
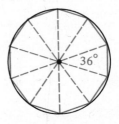

Square in circle: Draw four 90° central angles.

Regular hexagon in circle: Draw six 60° central angles.

Regular decagon in circle: Draw ten 36° central angles.

It is also true that if you are given any regular polygon, you can circumscribe a circle about it. This relationship between circles and regular polygons leads us to the following definitions:

The **center of a regular polygon** is the center of the circumscribed circle.

The **radius of a regular polygon** is the distance from the center to a vertex.

A **central angle of a regular polygon** is an angle formed by two radii drawn to consecutive vertices.

The **apothem of a regular polygon** is the (perpendicular) distance from the center of the polygon to a side.

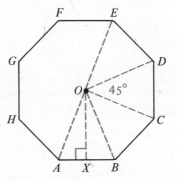

Center of regular octagon: O

Radius: OA, OB, OC, and so on

Central angle: $\angle AOB$, $\angle BOC$, and so on

Measure of central angle: $\dfrac{360°}{8} = 45°$

Apothem: OX

If you know the apothem of a regular polygon, you can use the next theorem to find the area of the polygon.

Theorem 9-5 The area of a regular polygon is equal to half the product of the apothem and the perimeter. $(A = \frac{1}{2}ap)$

Given: Regular n-gon $TUVW \cdots$; apothem a; side s; perimeter p; area A

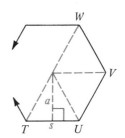

Prove: $A = \frac{1}{2}ap$

Key steps of proof:
1. If all radii are drawn, n congruent triangles are formed.
2. Area of each $\triangle = \frac{1}{2}sa$
3. $A = n(\frac{1}{2}sa) = \frac{1}{2}a(ns)$
4. Since $ns = p$, $A = \frac{1}{2}ap$.

Example 1: Find the radius and apothem of an equilateral triangle with side 6.

Solution: Use 30-60-90 $\triangle$ relationships.

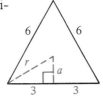

$$a = \frac{3}{\sqrt{3}} = \sqrt{3}$$
$$r = 2a = 2\sqrt{3}$$

Example 2: Find the area of a regular hexagon with a 9 cm apothem.

Solution: Use 30-60-90 $\triangle$ relationships.

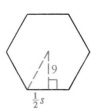

$$\frac{1}{2}s = \frac{9}{\sqrt{3}} = 3\sqrt{3}$$
$$s = 6\sqrt{3}; \; p = 36\sqrt{3}$$
$$A = \frac{1}{2}ap = \frac{1}{2} \cdot 9 \cdot 36\sqrt{3}$$
$$= 162\sqrt{3} \; (cm^2)$$

Classroom Exercises

Exercises 1–6 refer to square $PQRS$ with center O.

1. Name a radius.
2. Name a central angle.
3. The measure of a central angle is ___?___°.
4. OZ is the ___?___ of the polygon.
5. If the apothem is 3, the radius is ___?___.
6. If the radius is 8, the apothem is ___?___.

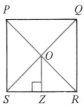

7. Find the area of a regular polygon with perimeter 40 and apothem 5.

8. Find the area of a regular pentagon with side length s and apothem a.

9. How large is a central angle of a regular polygon with 36 sides?

10. *ABCDE* is a regular pentagon.

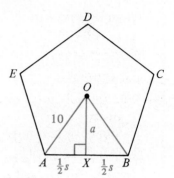

 a. $\angle AOB = \underline{\ ?\ }°$

 b. Explain why $\angle AOX = 36°$ and $\angle OAX = 54°$.

 Note: Parts (c)–(e) are optional exercises requiring use of the table of trigonometric ratios on page 225.

 c. $\sin 54° = \dfrac{a}{10}$. To the nearest tenth, $a \approx \underline{\ ?\ }$.

 d. $\cos 54° = \dfrac{\frac{1}{2}s}{?}$. To the nearest tenth, $s \approx \underline{\ ?\ }$.

 e. Find the perimeter and area of the pentagon.

Written Exercises

Copy and complete the tables for the regular polygons shown.

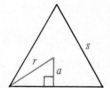

	r	a	s	p	A
A **1.**	?	?	12	?	?
2.	?	4	?	?	?
3.	8	?	?	?	?
4.	?	?	?	56	?
5.	?	?	?	?	25

	r	a	s	p	A
6.	6	?	?	?	?
7.	10	?	?	?	?
8.	?	4	?	?	?
9.	?	?	12	?	?
10.	?	?	?	10	?

11. A regular hexagon has perimeter 60. Find the radius and the apothem.

12. A regular hexagon has radius 12. Find the apothem and the perimeter.

Find the area of each polygon.

B 13. Equilateral triangle with radius $4\sqrt{3}$ 14. Square with radius $8k$

15. Regular hexagon with perimeter 72   16. Regular hexagon with apothem 4

Three regular polygons are inscribed in circles with radii 1. Find the apothem, the perimeter, and the area of each polygon. Use $\sqrt{3} \approx 1.73$ and $\sqrt{2} \approx 1.41$.

17.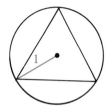

18.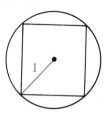

19.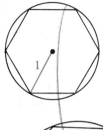

C 20. The regular octagon shown is inscribed in circle O with radius 1.

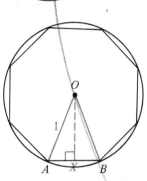

  a. Explain why $\angle AOX = 22.5°$.

  b. For trigonometric ratios of 22.5° use the average of the values for 22° and 23° in the table on page 225.

$$\sin 22.5° = \frac{AX}{1}, \text{ so } AX \approx \underline{\ ?\ }.$$

$$\cos 22.5° = \frac{OX}{1}, \text{ so } OX \approx \underline{\ ?\ }.$$

  c. Area of $\triangle AOB \approx \underline{\ ?\ }$ **d.** Area of octagon $\approx \underline{\ ?\ }$

SELF-TEST

Find the area of each figure.
1. A square with perimeter 20
2. A rectangle with base 12 and diagonal 15
3. A parallelogram with base 10 and height 7
4. A triangle with base 10 and height 7
5. An isosceles triangle with sides 4, 4, and 6
6. A rhombus with diagonals 6 and 8
7. A square with diagonal 4

Find the area of each trapezoid.

8.

9.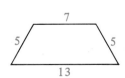

10.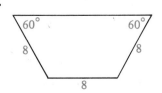

11. Each side of a regular hexagon is 4. Find the radius, the apothem, and the area.

Circles and Similar Figures

Objectives

1. **Understand how the area and perimeter formulas for regular polygons relate to the area and circumference formulas for circles.**
2. **Compute the circumferences and areas of circles.**
3. **Compute arc lengths and the areas of sectors of a circle.**
4. **Understand and apply the relationships between scale factors, perimeters, and areas of similar figures.**

9-5 *Circumference and Area of a Circle*

When you think of the perimeter of a figure, you probably think of the distance around the figure. Since the word "around" is not mathematically precise, perimeter is usually defined in other ways. For example, the perimeter of a polygon is defined as the sum of the lengths of its sides, which are segments. Since a circle has no sides, the perimeter of a circle must be defined differently. It is done as follows.

First consider a sequence of regular polygons inscribed in the circle. Four such polygons are shown below, but imagine more and more regular polygons having more and more sides.

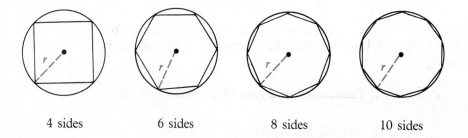

4 sides   6 sides   8 sides   10 sides

Now consider the perimeters of the polygons. As the drawings suggest, these perimeters give us a sequence of numbers that get closer and closer to a limiting number. This limiting number is defined to be the perimeter, or **circumference,** of the circle.

The area of a circle is defined in a similar way. The areas of the inscribed regular polygons get closer and closer to a limiting number, and this limit is defined to be the area of the circle.

The table below illustrates the preceding discussion. We have used trigonometry to find the perimeters and the areas of regular polygons inscribed in a circle with radius r.

Number of Sides of Polygon	Perimeter	Area
4	5.66r	2.00r^2
6	6.00r	2.60r^2
8	6.12r	2.83r^2
10	6.18r	2.93r^2
20	6.26r	3.09r^2
30	6.27r	3.12r^2
100	6.28r	3.14r^2

The results in the table suggest that the circumference and the area of a circle with radius r are approximately 6.28r and 3.14r^2. The exact values are given by the formulas below. (Proofs are suggested in Classroom Exercise 13 and Written Exercise 34.)

Circumference of circle: $2\pi \cdot$ radius $\qquad C = 2\pi r$

Area of circle: $\pi \cdot$ radius squared $\qquad A = \pi r^2$

These formulas involve a famous number denoted by the Greek letter π (*pi*). There isn't any decimal or fraction which expresses π exactly. Here are some common approximations for π:

$$3.14 \qquad \frac{22}{7} \qquad 3.1416$$

When you calculate the circumference and area of a circle, leave your answers in terms of π unless you are told to replace π by any of its approximations.

Example 1: Find the circumference and area of a circle with radius 6 cm.

Solution: $C = 2\pi r = 2\pi(6) = 12\pi$ (cm)
$A = \pi r^2 = \pi(6^2) = 36\pi$ (cm^2)

Example 2: Find the circumference of a circle with radius 10 cm. Use $\pi \approx 3.14$.

Solution: $C = 2\pi r \approx 2(3.14)(10) \approx 62.8$ (cm)

Example 3: Find the circumference of a circle if the area is 25π.

Solution: From $\pi r^2 = 25\pi$, we get $r^2 = 25$ and $r = 5$.
Thus, $C = 2\pi r = 2\pi(5) = 10\pi$.

Classroom Exercises

Complete the table.

	1.	2.	3.	4.	5.	6.	7.	8.
Radius	3	4	8	?	?	?	?	?
Circumference	?	?	?	10π	18π	?	?	?
Area	?	?	?	?	?	36π	49π	144π

Find the circumference and area to the nearest tenth. Use $\pi \approx 3.14$.

9. $r = 2$ **10.** $r = 6$ **11.** $r = \frac{1}{2}$ **12.** $r = 1.1$

13. The number π is defined to be the ratio of the circumference of a circle to the diameter. This ratio is constant from circle to circle. Supply the missing reasons in the key steps of a proof below.

Given: $\odot O$ and $\odot O'$ with circumferences
C and C' and diameters d and d'

Prove: $\dfrac{C}{d} = \dfrac{C'}{d'}$

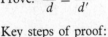

Key steps of proof:

Inscribe in each circle a regular polygon of n sides.
Let p and p' be the perimeters.

1. $p = ns$ and $p' = ns'$ (Why?)

2. $\dfrac{p}{p'} = \dfrac{ns}{ns'} = \dfrac{s}{s'}$ (Why?)

3. $\triangle AOB \sim \triangle A'O'B'$ (Why?)

4. $\dfrac{s}{s'} = \dfrac{r}{r'} = \dfrac{d}{d'}$ (Why?)

5. Thus, $\dfrac{p}{p'} = \dfrac{d}{d'}$ (Steps 2 and 4)

6. Steps 1–4 hold for any number of sides n. We can let n be so large that p is practically the same as C, and p' is practically the same as C'. In advanced courses, you learn that C and C' can be substituted for p and p' in Step 5. This gives $\dfrac{C}{C'} = \dfrac{d}{d'}$, or $\dfrac{C}{d} = \dfrac{C'}{d'}$.

Written Exercises

Find the circumference and area of each circle.

A   **1.** $r = 5$ **2.** $r = 11$ **3.** $d = 11$

    4. $r = 4\sqrt{2}$ **5.** $r = 6\sqrt{3}$ **6.** $r = 3k$

Find the circumference and area correct to the nearest tenth. Use $\pi \approx 3.14$.

7. $r = 3.2$ **8.** $d = 1$ **9.** $d = 5$

Find the circumference and area. Use $\pi \approx \frac{22}{7}$.

10. $r = 7$ **11.** $r = 35$ **12.** $r = 1\frac{3}{11}$

13. Find the radius of a circle with circumference 30π.

14. Find the radius of a circle with area 64π.

B **15.** The target below consists of four concentric circles with radii 1, 2, 3, and 4. Find the area of the bull's-eye and of each ring of the target. (Do you see a pattern to your answers?)

 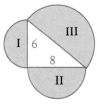

Ex. 15   Ex. 16

16. Semicircles are constructed on the sides of the right triangle shown above. Show that Area I + Area II = Area III.

In Exercises 17 and 18, the wheels of a racing bike are approximately 70 cm in diameter.

17. How far does the bike travel if the wheels make 10 revolutions? Use $\pi \approx \frac{22}{7}$.

18. About how many revolutions will the wheel make in a 22 km race? Recall that 1 km = 1000 m = 100,000 cm. Use $\pi \approx \frac{22}{7}$.

Find the area of each shaded region in terms of r.

19. 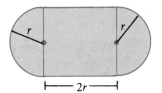   **20.**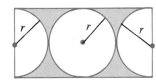

21.   **22.**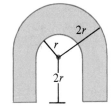

Find the area of a circle circumscribed about each polygon.

23. A square with side 8   **24.** An equilateral triangle with side 6

25. A regular hexagon with side 6   **26.** A right triangle with legs 5 and 12

27. An equilateral triangle has an inscribed circle with radius 3.

    a. What is the radius of the circumscribed circle?

    b. What is the ratio of the circumferences of these two circles?

    c. What is the ratio of the areas of these two circles?

28. The side of an equilateral triangle is 12. Find the areas of the inscribed and circumscribed circles.

Given a circle with radius 6. Find the areas of the inscribed polygons described.

29. An equilateral triangle

30. A square

31. A regular hexagon

For Exercises 32 and 33 use the table of trigonometric ratios on page 225 and find answers correct to the nearest tenth.

32. A regular decagon is inscribed in a circle with radius 6.

    a. Find $\angle AOB$ and $\angle AOX$.

    b. $\sin 18° = \dfrac{AX}{?}$. Thus $AX \approx \underline{?}$ and $AB \approx \underline{?}$.

    c. $\cos 18° = \dfrac{?}{?}$. Thus $OX \approx \underline{?}$.

    d. Area of $\triangle AOB \approx \underline{?}$

    e. Area of decagon $\approx \underline{?}$

    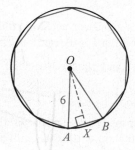

C 33. A regular polygon with 12 sides is inscribed in a circle with radius 6. Find the area enclosed between the circle and the polygon.

34. A regular polygon with apothem a is inscribed in a circle with radius r.

    a. As the number of sides increases, the value of a gets nearer to $\underline{?}$ and the perimeter of the polygon gets nearer to $2\pi r$.

    b. In the formula $A = \frac{1}{2}ap$, replace a by r, and p by $2\pi r$. What formula do you get?

35. Find the circumference of a circle inscribed in a rhombus with diagonals 12 cm and 16 cm.

36. Write a formula giving the area (A) of a circle in terms of the circumference (C) and the number π.

37. A circle is inscribed in a right triangle with sides 6, 8, and 10. Find the area of the circle.

38. Draw any circle O and any circle P. Construct a circle whose area equals the sum of the areas of circle O and circle P.

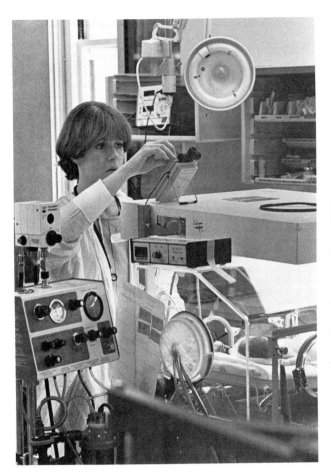

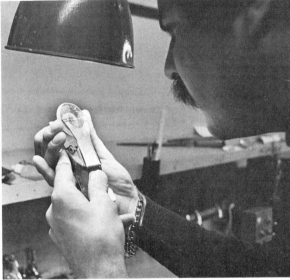

Career Notes
Biomedical Engineer

On an autumn morning in 1926, Philip Drinker stood in a laboratory observing a colleague work on a device to measure the breathing of a cat. The cat lay in a sealed metal box with its head outside and a rubber collar around its neck. This experience inspired Drinker's invention of the iron lung, a device which allows patients to breathe when their breathing muscles are paralyzed.

Biomedical engineers have used modern technology to design and develop other medical devices, too. Among the developments are artificial kidneys and hearts, pacemakers to regulate heartbeat, and eye surgery by means of lasers.

Biomedical engineers usually have a college degree in biomedical engineering, which requires a strong background in mathematics, science, and other branches of engineering.

9-6 *Areas of Sectors and Arc Lengths*

A **sector of a circle** is a region bounded by two radii and an arc of the circle. The shaded region of the diagram below is called sector AOB. The unshaded region is also a sector.

The length of $\overset{\frown}{AB}$ in circle O is part of the circumference of the circle. Since $\overset{\frown}{AB} = 60°$ and $\dfrac{60}{360} = \dfrac{1}{6}$,

$$\text{Length of } \overset{\frown}{AB} = \frac{1}{6}(2\pi \cdot 5) = \frac{5}{3}\pi.$$

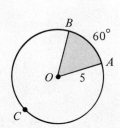

Similarly, the area of sector AOB is $\frac{1}{6}$ of the area of the circle. Thus,

$$\text{Area of sector } AOB = \frac{1}{6}(\pi \cdot 5^2) = \frac{25}{6}\pi.$$

In general, if $\overset{\frown}{AB} = n°$:

$$\text{Length of } \overset{\frown}{AB} = \frac{n}{360} \cdot 2\pi r$$

$$\text{Area of sector } AOB = \frac{n}{360} \cdot \pi r^2$$

Example 1: In $\odot O$ with radius 8, $\angle AOB = 120°$.
Find the length of $\overset{\frown}{AB}$ and the area of sector AOB.

Solution: Make a sketch as shown.
Since $\angle AOB = 120°$, $\overset{\frown}{AB} = 120°$.

$$\text{Length of } \overset{\frown}{AB} = \frac{120}{360} \cdot 2\pi \cdot 8 = \frac{16}{3}\pi$$

$$\text{Area of sector } AOB = \frac{120}{360} \cdot \pi \cdot 8^2 = \frac{64}{3}\pi$$

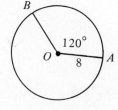

Example 2: Find the area of the shaded region.

Solution:

$$\text{Area of sector } XOY = \frac{90}{360} \cdot \pi \cdot 10^2 = 25\pi$$

$$\text{Area of } \triangle XOY = \frac{1}{2} \cdot 10 \cdot 10 = 50$$

$$\text{Area of shaded region} = 25\pi - 50$$

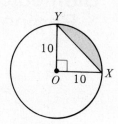

Classroom Exercises

1. If the length of an arc is $\frac{1}{4}$ of the circumference, what is the degree measure of the arc?

Find the arc length and area of each shaded sector.

2.

3.
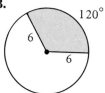
120°
6
6

4.
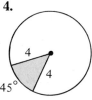
4
4
45°

5.

6 6
60°

6. In a circle with radius 6, $\widehat{AB} = 60°$. Find the area of the region bounded by $\overline{AB}$ and $\widehat{AB}$. First make a sketch.

Written Exercises

Sector AOB is described by giving $\angle AOB$ and the radius of circle O. Make a sketch and find the length of $\widehat{AB}$ and the area of sector AOB.

A

	1.	**2.**	**3.**	**4.**	**5.**	**6.**	**7.**	**8.**
$\angle AOB$	90°	60°	30°	45°	120°	240°	180°	270°
radius	10	12	12	4	3	3	5	8

Find the area of each shaded region. Point O marks the center of a circle.

B 9.

4
O

10.
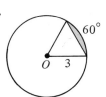
60°
O 3

11.

3   4

12.
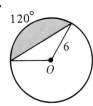
120°
6
O

13.
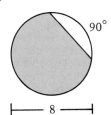
90°
|— 8 —|

14.

60°
O 6

Find the area of each shaded region. Point *O* marks the center of a circle.

15.

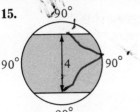

16.

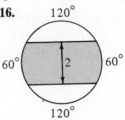

17.

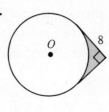

18. $\overline{PA}$ and $\overline{PB}$ are tangents to a circle. If $PA = 6$ and $\angle APB = 60°$, find the area of the region outside the circle but inside $\angle APB$.

C 19. $\overline{ST}$ represents a piece of string attached to the corner of square $SQAR$. If the string is held taut and rotated clockwise, the string will wrap once around the square. When this is done, how far does *T* move?

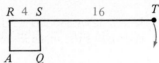

20. $\overline{AB}$ is a chord of a circle with radius 10. If $\overset{\frown}{AB} = 72°$, find the area of the region included between $\overline{AB}$ and $\overset{\frown}{AB}$, correct to the nearest tenth. (Use the table of trigonometric ratios on page 225.)

21. Three circles with radii 6 are tangent to each other. Find the area of the region enclosed between them.

22. $\overline{AB}$ is the common external tangent of the two circles below with radii 6 and 2. If the circles are tangent to each other, find the area of the shaded region. (*Hint:* What kind of figure is $AXYB$? How large is $\angle AXY$?)

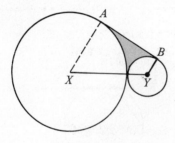

Ex. 22

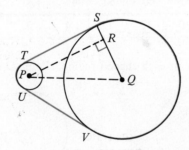

Ex. 23

23. The diagram at the right above shows a belt tightly stretched over two wheels with radii 5 cm and 25 cm. Find the length of this belt if the centers of the wheels are 40 cm apart.

24. Given: $\overset{\frown}{ACB}$, $\overset{\frown}{AXC}$, and $\overset{\frown}{CYB}$ are semicircles; $AC = CB$

Show that Area I + Area II = Area of $\triangle ABC$.

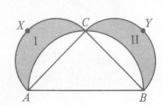

9-7 *Areas of Similar Figures*

The triangles shown below are similar. Notice that corresponding sides are in a 2:1 ratio and corresponding altitudes are in a 2:1 ratio also. Can you guess the ratio of the perimeters? the ratio of the areas?

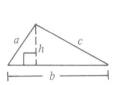
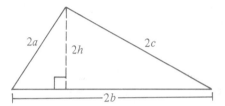

If you guessed that the ratio of the perimeters is 2:1, you are right.

$$\frac{\text{Perimeter of larger triangle}}{\text{Perimeter of smaller triangle}} = \frac{2a + 2b + 2c}{a + b + c} = \frac{2(a + b + c)}{a + b + c} = \frac{2}{1}$$

Many people guess that the ratio of the areas is 2:1 also, but it is 4:1.

$$\frac{\text{Area of larger triangle}}{\text{Area of smaller triangle}} = \frac{\frac{1}{2}(2b)(2h)}{\frac{1}{2}bh} = \frac{4bh}{bh} = \frac{4}{1}$$

The discussion above is an illustration of the following theorem.

Theorem 9-6   If the scale factor of two similar figures is $a:b$, then (1) the ratio of the perimeters is $a:b$ and (2) the ratio of the areas is $a^2:b^2$.

Example 1:   Find the ratio of the perimeters and the ratio of the areas of the two similar figures.

Solution: The scale factor is 8:12, or 2:3.
Therefore, the ratio of the perimeters is 2:3.
The ratio of the areas is $2^2:3^2$, or 4:9.

Example 2: *ABCD* is a trapezoid.

   a. Find the ratio of the areas of $\triangle COD$ and $\triangle AOB$.

   b. Find the ratio of the areas of $\triangle COD$ and $\triangle DOA$.

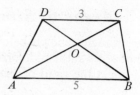

Solution: **a.** $\triangle COD \sim \triangle AOB$ (AA Similarity Postulate) with scale factor $3:5$.

$$\frac{\text{Area of } \triangle COD}{\text{Area of } \triangle AOB} = \frac{3^2}{5^2} = \frac{9}{25}$$

    b. $\triangle COD$ is *not* similar to $\triangle DOA$, so we cannot use Theorem 9-6. Instead, we consider $\overline{CO}$ and $\overline{AO}$ as bases of the two triangles, which then have the same height.

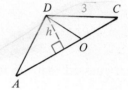

$$\frac{\text{Area of } \triangle COD}{\text{Area of } \triangle AOD} = \frac{\frac{1}{2} \cdot CO \cdot h}{\frac{1}{2} \cdot AO \cdot h} = \frac{CO}{AO}$$

To find $\dfrac{CO}{AO}$, we note that $\overline{CO}$ and $\overline{AO}$ are corresponding sides of similar triangles *COD* and *AOB* in the given trapezoid. Thus, $\dfrac{CO}{AO} = \dfrac{3}{5}$.

Classroom Exercises

The table refers to similar figures. Complete the table.

	1.	2.	3.	4.	5.	6.	7.	8.
Scale factor	1:3	1:5	3:4	6:9	?	?	?	?
Perimeter ratio	?	?	?	?	4:5	12:20	?	?
Area ratio	?	?	?	?	?	?	16:49	36:25

9. a. Are all circles similar?

  b. If two circles have radii 9 and 12, what is the ratio of the circumferences? of the areas?

10. The areas of two circles are 25π and 81π. What is the ratio of the circumferences?

11. *ABCD* is a parallelogram. Find each ratio.

  a. $\dfrac{\text{Area of } \triangle DEF}{\text{Area of } \triangle ABF}$

  b. $\dfrac{\text{Area of } \triangle DEF}{\text{Area of } \triangle CEB}$

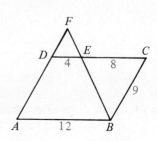

12. Consider triangles I, II, and III.

    a. Are any of these triangles similar?

    b. Can Theorem 9-6 be used?

    c. Find the ratio of the areas of triangles I and II.

    d. Find the ratio of the areas of triangles II and III.

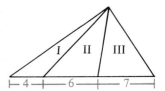

13. The figure is a trapezoid. Find the ratio of the areas of each pair of triangles.

    a. I and III

    b. I and II

    c. I and IV

    d. II and IV

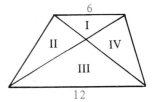

Written Exercises

The table refers to similar figures. Copy and complete the table.

A

	1.	**2.**	**3.**	**4.**	**5.**	**6.**	**7.**	**8.**
Scale factor	1:4	3:2	6:7	?	?	?	?	?
Ratio of perimeters	?	?	?	9:5	3:13	?	?	?
Ratio of areas	?	?	?	?	?	25:1	9:64	2:1

  9. Two circles have radii 7 and 11. What is the ratio of the areas?

10. The areas of two circles are 36π and 64π. What is the ratio of the circumferences?

11. L, M, and N are the midpoints of the sides of $\triangle ABC$. Find the ratio of the perimeters and the ratio of the areas of $\triangle LMN$ and $\triangle ABC$.

12. In the diagram below, $PQRS$ is a parallelogram. Find the ratio of the areas for each pair of triangles.

    a. $\triangle TOS$ and $\triangle QOP$ **b.** $\triangle TOS$ and $\triangle TQR$

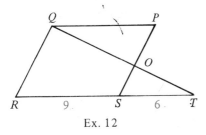

Ex. 12

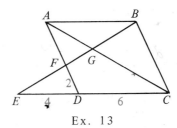

Ex. 13

B 13. In the diagram above, $ABCD$ is a parallelogram. Name *three pairs* of similar triangles and give the ratio of the areas for each pair.

14. *ABCD* is a trapezoid. Find the ratio of the areas for each pair of triangles.

    a. $\triangle AOB$ and $\triangle COD$

    b. $\triangle COD$ and $\triangle DOA$

    c. $\triangle COD$ and $\triangle COB$

    d. $\triangle DOA$ and $\triangle COB$

    e. $\triangle ABD$ and $\triangle DCB$

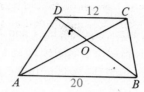

C 15. In $\triangle ABC$, the medians $\overline{BD}$ and $\overline{CE}$ meet at *G*. If the area of $\triangle DEG = 1$, find the area of each triangle.

    a. $\triangle BEG$ b. $\triangle DCG$ c. $\triangle BGC$ d. $\triangle ABC$

16. If you draw the three medians of a triangle, six small triangles are formed. Prove whatever you can about the areas of these six triangles.

17. *D*, *E*, *F*, and *G* are the midpoints of the sides of square *PQRS*. Show that the area of the shaded region is one fifth of the area of *PQRS*.

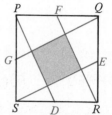

SELF-TEST

1. Find the circumference and area of a circle with radius 7.

2. The area of a circle is 36π. Find the circumference.

3. In the diagram, square *KLMN* has side 14. $\overset{\frown}{KPN}$ and $\overset{\frown}{LPM}$ are semicircles. Find the area of the shaded region. Use $\pi \approx \frac{22}{7}$.

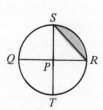

Ex. 3

4. In $\odot O$ with radius 9, $\overset{\frown}{AB} = 120°$.

    a. Find the length of $\overset{\frown}{AB}$.

    b. Find the area of sector *AOB*.

5. $\overline{QR}$ and $\overline{ST}$ are perpendicular diameters of a circle with radius 8.

    a. Find the length of $\overset{\frown}{SR}$.

    b. Find the area of the shaded region.

Ex. 5

6. Find the ratio of the areas of two circles with radii 3 and 5.

7. $\triangle ABC \sim \triangle DEF$. If $AB = 6$ and $DE = 9$, find the ratio of the areas of the triangles.

8. The areas of two similar triangles are 25 and 49. Find the ratio of the perimeters.

Challenge

Here $\overline{XY}$ has been divided into 5 equal parts and semicircles have been drawn. But suppose $\overline{XY}$ were divided into millions of equal parts and semicircles were drawn. What would the sum of the lengths of the arcs be?

A says, "XY, because all the points would be so close to $\overline{XY}$." B says, "A really large number, because there would be so many arc lengths to add up." What do you say?

Extra Networks

The Pregel River flows through the old city of Koenigsberg, now Kaliningrad. Once, seven bridges joined the shores and the two islands in the river as shown in the diagram at the left below. A popular problem of that time was to try to walk across all seven bridges without crossing any bridge more than once. Can you find a way to do it?

north shore

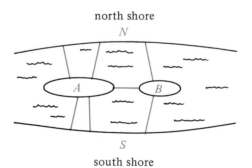

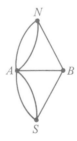

south shore

When the mathematician Leonard Euler analyzed this problem, he represented each land mass by a point, which he called a vertex, and each bridge by an arc, as in the diagram at the right above. A diagram like this is called a network, and Euler discovered which networks can be traced without backtracking, that is, without drawing over an arc twice. Euler classified all vertices as either odd or even. An odd vertex has an odd number of arcs coming from it; an even vertex has an even number. In the network above, *A* is an odd vertex. Are there any even vertices?

Exercises

Find the number of odd and even vertices in each network below. Imagine traveling each network to see whether it can be traced without backtracking. Record your findings in a table so that you can analyze your results.

1.

2.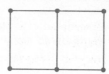

3.

4.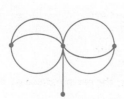

5.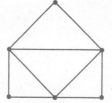

6.

7. Without actually trying to trace over a network, you can tell whether or not the network can be traced without backtracking. Do you see how? (The key lies in the number of odd vertices.) If not, read on.

8. In this exercise, you will follow Euler's steps to a solution of network problems.

   a. Suppose that a given network can be traced without backtracking. Consider a vertex that is neither the start nor end of a journey through this network. Is such a vertex odd or even? (Remember that every time you travel to such a vertex you must also depart from that vertex.)

   b. Now consider the two vertices at the start and finish of a journey through the network without backtracking. Can these vertices both be odd? (See the network in Exercise 2.) Must the vertices at the start and finish be odd vertices? (See the network in Exercise 5.)

   c. The first step of Euler's solution shows that all the vertices which are not at the start or finish of a journey without backtracking must be even. Also, the start and finish vertices can both be odd or both be even, but there is still the question of whether just one of these two can be odd. What do you think?

9. Tell why it is impossible to walk across the seven bridges without crossing any bridge more than once. Show that it is possible to add a bridge so that all can be traveled without backtracking. Also show that one bridge can be removed so that the resulting network can be traveled without backtracking.

Chapter Summary

1. If two figures are congruent, then they have the same area.
2. The area of a polygonal region is the sum of the areas of its non-overlapping parts.
3. Formulas for areas of polygons:

Square: $A = s^2$ Rhombus: $A = \frac{1}{2}d_1d_2$
Rectangle: $A = bh$ Trapezoid: $A = \frac{1}{2}h(b_1 + b_2)$
Parallelogram: $A = bh$ Regular Polygon: $A = \frac{1}{2}ap$
Triangle: $A = \frac{1}{2}bh$

4. Formulas for circles:

$$C = 2\pi r \qquad\qquad \text{Length of arc} = \frac{n}{360} \cdot 2\pi r$$

$$A = \pi r^2 \qquad\qquad \text{Area of sector} = \frac{n}{360} \cdot \pi r^2$$

5. If the scale factor of two similar figures is $a:b$, then (1) the ratio of the perimeters is $a:b$ and (2) the ratio of the areas is $a^2:b^2$.

Chapter Review

9-1 *Areas of Rectangles*

1. Find the area of a rectangle with base $4\sqrt{3}$ cm and height $2\sqrt{3}$ cm.
2. Find the perimeter of a rectangle 14 cm long and 6 cm wide.
3. Find the area of a square with sides 17 cm.
4. Find the perimeter of a square whose area is 9 cm².

9-2 *Areas of Parallelograms and Triangles*

Find the area of each figure.

5.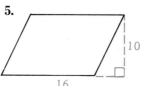

6.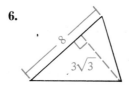

7.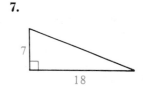

9-3 *Areas of Trapezoids*

Find the area of each trapezoid.

8.

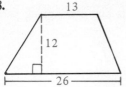

9.

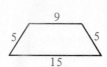

10.

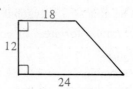

9-4 *Areas of Regular Polygons*

Find the area of each of the following regular polygons.

11. A square with radius $3\sqrt{2}$

12. An equilateral triangle with perimeter 24

13. A regular hexagon with apothem $5\sqrt{3}$

9-5 *Circumference and Area of a Circle*

14. Find the area of a circle with radius 5.

15. Find the circumference of a circle with radius 5.

16. The area of a circle is 81π. Find the circumference.

9-6 *Areas of Sectors and Arc Lengths*

17. In a circle of radius 6, a sector which has an arc of 120° has area   ? .

18. In a circle of radius 18, a 60° arc has length   ? .

19. In a circle with area 64π, a sector of area 8π has an arc of   ? °.

9-7 *Areas of Similar Figures*

20. If the ratio of the areas of two similar triangles is $9:49$, what is the ratio of two corresponding sides?

21. If the ratio of the circumferences of two circles is $2:5$, what is the ratio of the areas?

22. Two similar polygons have the scale factor $2:5$. The area of the smaller polygon is 36. Find the area of the larger polygon.

Chapter Test

Find the areas of the figures described.

1. A rectangle with length 17 and width 8
2. A circle with radius 11
3. A right triangle with legs 5 and 7
4. A square with perimeter 48
5. An equilateral triangle with perimeter 18
6. An isosceles triangle with base 12 and legs 10
7. A circle with circumference 16π
8. A rhombus with diagonals 12 and 7
9. A regular hexagon with sides 4
10. Sector AOB of circle O with radius 6 and $\overset{\frown}{AB} = 30°$
11. A square with diagonals $13\sqrt{2}$
12. An isosceles trapezoid with bases 20 and 26, and legs 5

Find the area of each shaded region.

13.

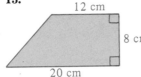

14.

15.

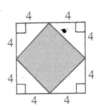

Find the ratio of the area of the smaller figure to the area of the larger similar figure.

16.

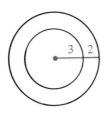

17.

18. What is the length of a 120° arc of a circle with radius 9?

Cumulative Review: Chapters 7–9 _____

True-False Exercises

Write T or F to indicate your answer.

A 1. Concentric circles are tangent to each other.

2. A minor arc has measure less than 180°.

3. If a right triangle is inscribed in a circle, the hypotenuse forms a diameter of the circle.

4. Two tangents drawn to a circle from the same point are equal in length.

5. The area of a circle is greater than the area of any polygon that is inscribed in the circle.

6. If two circles intersect in two points, the circles must be coplanar.

7. If we find the ratio of the circumference to the diameter for each of two circles, these ratios will be equal.

8. In $\odot O$, if sector AOB has area equal to sector COD, then $\overarc{AB} = \overarc{CD}$.

9. The locus of points in space 5 m from a given point is a circle.

10. If two rectangles have the same area, then they are congruent.

B 11. It is always possible to circumscribe a circle about a quadrilateral.

12. If a chord of a circle is equal to the radius of the circle, the arc of the chord measures 60°.

13. If two right triangles are inscribed in the same circle, they have equal areas.

14. The locus of points equidistant from two intersecting planes is a single plane.

15. The locus of points in a plane equidistant from three noncollinear points in the plane is a single point.

Construction Exercises

A 1. Construct a right triangle and inscribe a circle in it.

2. Draw a circle and a diameter of the circle. Construct a tangent to the circle which is parallel to the diameter you have drawn.

3. Construct an isosceles triangle with vertex angle of 120°.

4. Draw a triangle. Construct a similar triangle whose area is four times that of the triangle you have drawn.

B 5. Construct an equilateral triangle. By construction find the apothem of the triangle.

6. Draw a circle. Construct a regular inscribed octagon.

7. Draw a segment. Construct a square which has a diagonal equal to the segment you have drawn.

8. Draw a segment AB. By construction, find the point C on $\overline{AB}$ such that $\dfrac{AC}{BC} = \dfrac{1}{2}$.

9. Draw two unequal segments, with lengths labeled j and k. Construct the geometric mean between j and k.

10. Draw a segment. Construct an equilateral triangle whose altitude is equal to the segment which you have drawn.

Miscellaneous Exercises

State the areas of the figures described.

A 1. A right triangle with legs 4 cm and 6 cm.

2. A rectangle with a side 9 cm and diagonal 15 cm.

3. A trapezoid with bases 9 and 15 and altitude 5.

4. A circle with diameter 4.

5. A circle with circumference 10π.

6. An equilateral triangle with side 18.

Complete the statements.

7. A tangent to a circle is perpendicular to the ___?___ drawn to the point of tangency.

8. A minor arc of 40° has a central angle of ___?___°.

9. At three o'clock, the angle formed by the hands of a clock is ___?___°.

10. At five o'clock, the angle formed by the hands of a clock is ___?___°.

11. The measure of a semicircle is ___?___°.

12. An angle inscribed in a semicircle has measure ___?___°.

$\overline{DB}$ is tangent to the circle shown.

B 13. If $AB = AC$ and $\overset{\frown}{CB} = 68°$, $\angle ADB = $ ___?___°.

14. If $\angle ACB = 78°$ and $\angle CBD = 21°$, $\overset{\frown}{AC} = $ ___?___°.

15. If $\angle D = 25°$ and $\overset{\frown}{CB} = 76°$, $\angle ABC = $ ___?___°.

16. If $AD = 16$ and $AC = 7$, $DB = $ ___?___.

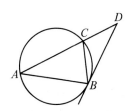

17. If $PQ = 13$, $ST = 4$, and $PT = 6$, the area of $\triangle QRS = $ ___?___.

18. If $PQ = 24$, $QR = 10$, and $RS = 14$, the area of trapezoid $PQST = $ ___?___.

19. If $PQ = 15$, $ST = 7$, and the area of $\triangle QRS = 24$, the area of rectangle $PQRT = $ ___?___.

C 20. If the radius of the circle inscribed in a 30-60-90 triangle is r, what is the area of the triangle?

10

Areas and Volumes of Solids

Important Solids

Objectives
1. Identify the parts of prisms, pyramids, cylinders, and cones.
2. Find the lateral area, total area, and volume of a right prism or pyramid.
3. Find the lateral area, total area, and volume of a right cylinder or cone.

10-1 *Prisms*

In this chapter you will be calculating surface areas and volumes of special solids. It is possible to begin with some postulates and then prove as theorems the formulas for areas and volumes of solids, as we did for plane figures. Instead, informal arguments will be given to show you that the formulas for solids are reasonable.

The first solid we shall study is the **prism.** The two shaded faces of the prism shown are its **bases.** Notice that the bases are congruent polygons lying in parallel planes. An **altitude** of a prism is a segment joining the two base planes and perpendicular to both. The length of an altitude is the *height (h)* of the prism.

The faces of a prism which are not its bases are called **lateral faces.** Adjacent lateral faces intersect in parallel segments called **lateral edges.**

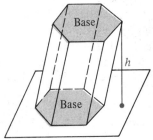

The lateral faces of a prism are parallelograms. If they are rectangles, the prism is a **right prism.** Otherwise it is an **oblique prism.** The diagrams below show that a prism is also classified by the shape of its base.

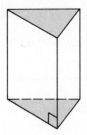 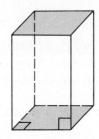 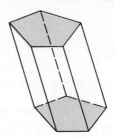

Right triangular prism   Right rectangular prism   Oblique pentagonal prism
                          (Rectangular solid)

The **lateral area** (L.A.) of a prism is the sum of the areas of its lateral faces. The **total area** (T.A.) is the sum of the areas of all its faces. Using B to denote the area of a base, we have the formula:

$$\textbf{T.A. = L.A. + 2B}$$

If a prism is a right prism, the next theorem gives us an easy way to get the lateral area.

Theorem 10-1 The lateral area of a right prism equals the perimeter of a base times the height of the prism. (L.A. = ph)

The formula above applies to any right prism. We'll use a right pentagonal prism to illustrate the development of the formula.

$$
\begin{aligned}
\text{L.A.} &= ah + bh + ch + dh + eh \\
&= (a + b + c + d + e)h \\
&= \text{perimeter} \cdot h \\
&= ph
\end{aligned}
$$

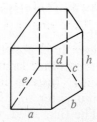

Prisms have *volume* as well as area. A rectangular solid with square faces is a cube. Since each edge of the shaded cube shown is 1 unit long, the cube is said to have a volume of 1 cubic unit. The larger rectangular solid has 3 layers of cubes, each layer containing $(4 \cdot 2)$ cubes. Hence its volume is $(4 \cdot 2) \cdot 3$, or 24 cubic units.

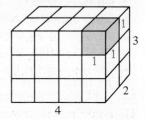

Volume: Base area × height
$(4 \cdot 2) \cdot 3$
24 cubic units

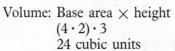

The same sort of reasoning is used to find the volume of any right prism. We shall list the result as a theorem without giving the details of the proof.

Theorem 10-2 The volume of a right prism equals the area of a base times the height of the prism. $(V = Bh)$

Some common units for measuring volume are the cubic centimeter (cm³) and the cubic meter (m³).

Example: A right trapezoidal prism is shown. Find (a) lateral area, (b) total area, and (c) volume.

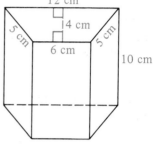

Solution: **a.** Lateral area
First find the perimeter of the base.
$p = 5 + 6 + 5 + 12 = 28$ (cm)
Now use the formula for lateral area.
L.A. $= ph = 28 \cdot 10 = 280$ (cm²)
b. Total area
First find the area of the base.
$B = \frac{1}{2} \cdot 4 \cdot (12 + 6) = 36$ (cm²)
Now use the formula for total area.
T.A. $=$ L.A. $+ 2B = 280 + 2 \cdot 36 = 352$ (cm²)
c. Volume
$V = Bh = 36 \cdot 10 = 360$ (cm³)

Classroom Exercises

Exercises 1–12 refer to the right prism shown.

1. What kind of polygons are the bases?

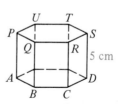

2. The prism is called a right ___?___ prism.

3. How many lateral faces are there?

4. What kind of figure is each lateral face?

5. Name two lateral edges.

6. Name an altitude.

7. The length of an altitude is called the ___?___ of the prism.

8. Suppose the bases are regular hexagons with 4 cm edges. Find the perimeter of a base.

9. If the height of the prism is 5 cm, find the lateral area.

10. Find the base area.

11. Find the total area.   12. Find the volume.

Written Exercises ⎯⎯⎯⎯⎯⎯⎯⎯⎯⎯⎯⎯⎯⎯⎯⎯⎯⎯⎯⎯⎯

Find the lateral area, the total area, and the volume of rectangular solids with the given dimensions.

A   **1.** $l = 6$ cm $w = 5$ cm $h = 3$ cm

       2. $l = 8$ mm $w = 4$ mm $h = 4$ mm

       3. $l = 1.5$ m $w = 1$ m $h = 1$ m

       4. $l = 2.5$ cm $w = 1.5$ cm $h = 2$ cm

       5. $l = 2$ m $w = 60$ cm $h = 30$ cm

       6. $l = 2$ cm $w = 4$ mm $h = 5$ mm

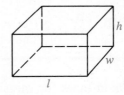

Find the total area and volume of cubes with the given edges.

     7. 3   **8.** 6   **9.** e **10.** $2e$

11. Use your answers to Exercises 9 and 10 to complete the following:
     If the edge of a cube is doubled, the total area is multiplied by _?_ and the volume is multiplied by _?_ .

Exercises 12–17 refer to cubes with edges e. Complete the table.

	12.	**13.**	**14.**	**15.**	**16.**	**17.**
T.A.	$6e^2$	600	486	216	?	?	?
V	e^3	?	?	?	8	64	125

18. Find the lateral area of a right pentagonal prism with height 13 and base edges 3.2, 5.8, 6.9, 4.7, and 9.4.

Facts about the base of a right prism and the height of the prism are given. Sketch each prism and find its lateral area, total area, and volume.

19. Equilateral triangle with side 8; $h = 10$

20. Triangle with sides 9, 12, 15; $h = 10$

B   **21.** Isosceles triangle with sides 13, 13, 10; $h = 7$

      22. Isosceles trapezoid with sides 10, 5, 4, 5; $h = 20$

      23. Regular hexagon with side 8; $h = 12$

      24. Rhombus with diagonals 6 and 8; $h = 9$

     25. A solid is shown submerged in a container having the shape of a rectangular solid. The solid has caused the water level to rise 2 cm. Find the volume of the solid.

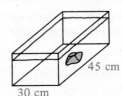

45 cm

30 cm

Find the total surface area and volume of each solid in terms of x and y.

26.

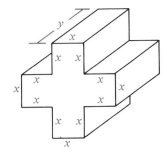

27.

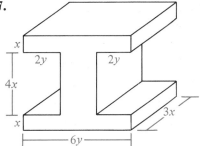

28. A rectangular solid has length 10, width 8, and height 6. Plane *WXYZ* intersects the solid as shown forming two trapezoidal prisms, one with base *AYXE* and the other with base *BYXF*.

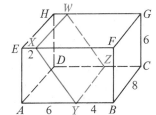

    a. Find the volumes of the two trapezoidal prisms.

    b. Find the total surface area of the prism with base *BYXF*.

C 29. The length of a rectangular solid is twice the width, and the height is three times the width. If the volume is 162 cm³, what are the dimensions of the solid?

30. A diagonal of a rectangular solid is defined to be a segment whose end points are two vertices not in the same face. If the diagonals of a cube are $4\sqrt{3}$ cm long, what is the volume?

31. All nine edges of a right triangular prism are equal. Find the length of these edges if the volume is $54\sqrt{3}$ cm³.

32. A right prism has height h and has bases which are regular hexagons with sides s. Show that the volume is $V = \frac{3}{2}\sqrt{3}\,s^2h$.

33. The diagonal of a box forms a 35° angle with the diagonal of the base shown. Use the Table of Trigonometric Ratios on page 225 to help you find the following:

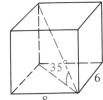

    a. The length of the diagonal of the box

    b. The volume of the box

Challenge

Given any two rectangles, draw a line that divides each rectangle into two parts of equal area.

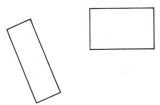

10-2 *Pyramids*

The diagram shows the pentagonal **pyramid** *V-ABCDE.*
Point *V* is the **vertex** of the pyramid and pentagon
ABCDE is the **base.** The perpendicular segment from the
vertex to the base is the **altitude** and its length is the *height*
(*h*) of the pyramid.

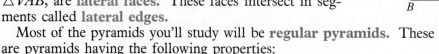

The five triangular faces with *V* in common, such as
△*VAB*, are **lateral faces.** These faces intersect in segments called **lateral edges.**

Most of the pyramids you'll study will be **regular pyramids.** These
are pyramids having the following properties:

(1) The base is a regular polygon.
(2) The altitude meets the base at its center.
(3) All lateral edges are equal.
(4) All lateral faces are congruent isosceles triangles. The
    height of a lateral face is called the **slant height** of the
    pyramid. It is denoted by *l.*

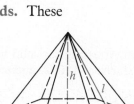

Regular hexagonal pyramid

To find the lateral area of a regular pyramid, you may use either of the
following methods:

Method 1 Find the area of one lateral face.
            Multiply this area by the number of lateral faces.

Method 2 Use the following formula, in which *p* denotes perimeter of
            the base. (See Classroom Exercise 1.)

$$\text{L.A.} = \tfrac{1}{2}pl$$

The prism and pyramid below have congruent bases and equal heights.
Since the volume of the prism is *Bh*, the volume of the pyramid must be
less than *Bh*. In fact, it is exactly $\tfrac{1}{3}Bh$.

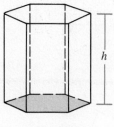

$V = Bh$

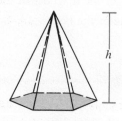

$V = \tfrac{1}{3}Bh$

Example: Given a regular square pyramid with base edge 10 and lateral
            edge 13, find (a) lateral area, (b) total area, and (c) volume.

Solution: **a.** $l = \sqrt{13^2 - 5^2} = \sqrt{144} = 12$
Perimeter of base $= p = 40$
L.A. $= \frac{1}{2}pl = \frac{1}{2} \cdot 40 \cdot 12 = 240$

b. Area of base $= B = 10^2 = 100$
T.A. $=$ L.A. $+ B = 240 + 100 = 340$

c. In rt. $\triangle VOM,$ $h = \sqrt{l^2 - 5^2}$
$= \sqrt{144 - 25} = \sqrt{119}$

$V = \frac{1}{3}Bh = \frac{1}{3} \cdot 100 \cdot \sqrt{119} = \dfrac{100\sqrt{119}}{3}$

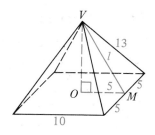

Classroom Exercises

1. The base of the regular pyramid shown is an n-sided polygon with edge a. Slant height is l.

   a. Perimeter of base $= p = $?

   b. Area of one lateral face $= $?

   c. Area of n lateral faces $= $?

   d. Express your answer to (c) in terms of p and l. What formula have you developed?

$V\text{-}ABCD$ is a regular square pyramid. Find numerical answers.

2. $OM = $?   3. $l = $?

4. Area of $\triangle VBC = $?   5. L.A. $= $?

6. Volume $= $?   7. $VC = $?

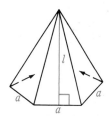
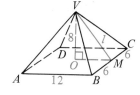

All edges of regular pyramid $V\text{-}XYZ$ are 6 cm long. Find numerical answers.

8. $XM = $?   9. $XO = $?

10. $h = $?   11. Base area $= $?

12. Volume $= $?   13. Slant height $= $?

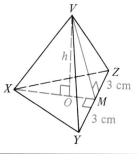

Written Exercises

Sketch each pyramid. Then find its lateral area.

A 1. A regular square pyramid with base edge 6 and slant height 7.

2. A regular triangular pyramid with base edge 4 and slant height 6.

3. A regular hexagonal pyramid with base edge 10 and lateral edge 13.

4. A regular square pyramid with base edge 16 and lateral edge 10.

Copy and complete the table below for the regular square pyramid shown.

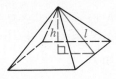

	5.	**6.**	**7.**	**8.**	**9.**	**10.**
height, h	4	12	24	?	?	15
slant height, l	5	13	?	12	5	?
base edge	?	?	14	?	8	?
lateral edge	?	?	?	15	?	17

Find the lateral area, total area, and volume of each regular square pyramid described.

11. base edge = 6, height = 4   **12.** base edge = 8, height = 3

B **13.** slant height = 20, height = 12   **14.** base edge = 10, slant height = 13

15. A pyramid and a prism both have height 8.2 cm and congruent hexagonal bases with area 22.3. Give the ratio of the volumes. (You do *not* need to calculate their volumes.)

16. The shaded pyramid is cut from a rectangular solid. How does the volume of the pyramid compare with the volume of the solid?

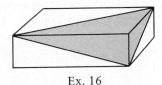

Ex. 16

17. Find the volume of a regular hexagonal pyramid with base edge 6 and lateral edge 10.

V-ABC is a regular triangular pyramid. Copy and complete the table.

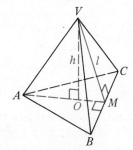

	AM	*AO*	*OM*	h	l	*VA*
18.	6	?	?	3	?	?
19.	9	?	?	8	?	?
20.	$6\sqrt{3}$	?	?	?	?	8
21.	$3\sqrt{3}$	?	?	?	$3\sqrt{3}$	?

Exs. 18-23

C **22.** Each edge of the pyramid shown above is 12 cm. Find the total area and the volume.

23. Suppose each edge of the pyramid shown above is *e* units long. Find the volume in terms of *e*.

24. The base of a rectangular pyramid has sides 12 and 30. The vertex is directly over the center of the base. The height is 8. Find the lateral area.

25. Show that the ratio of the volumes of the two regular square pyramids shown is $\dfrac{\tan 40°}{\tan 80°}$.

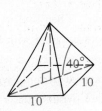
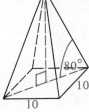

10-3 *Cylinders and Cones*

A cylinder is like a prism except that its bases are circles instead of polygons.

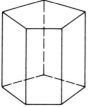 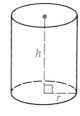 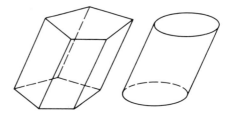

Right prism   Right cylinder   Oblique prism   Oblique cylinder

In a right cylinder, the segment joining the centers of the circular bases is an **altitude.** The length of an altitude is called the **height** h of the cylinder. A radius of a base is also called a **radius** r of the cylinder.

The four diagrams above show the relationship between prisms and cylinders. A similar relationship between pyramids and cones is shown in the four diagrams below.

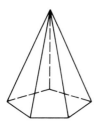 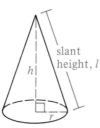  

Right pyramid   Right cone   Oblique pyramid   Oblique cone

In the discussion and exercises which follow, the words "cylinder" and "cone" will always refer to a right cylinder and a right cone.

The formulas for prisms also apply to cylinders, and the formulas for pyramids also apply to cones. Consequently you don't have to learn special formulas for cylinders and cones. Note that it is easy to write πr^2 for B and $2\pi r$ for p as shown below.

Cylinder $V = Bh = \pi r^2 h$ L.A. $= ph = 2\pi rh$
Cone $V = \frac{1}{3}Bh = \frac{1}{3}\pi r^2 h$ L.A. $= \frac{1}{2}pl = \frac{1}{2}(2\pi r)l = \pi rl$

Example 1:   Given a cylinder with radius 5 cm and height 4 cm, find
              (a) lateral area, (b) total area, and (c) volume.

Solution: **a.** L.A. $= 2\pi rh = 2\pi \cdot 5 \cdot 4 = 40\pi$ (cm²)
              b. T.A. $=$ L.A. $+ 2B = 40\pi + 2(\pi \cdot 5^2) = 90\pi$ (cm²)
              c. $V = \pi r^2 h = \pi \cdot 5^2 \cdot 4 = 100\pi$ (cm³)

Example 2: Find (a) lateral area, (b) total area, and (c) volume for the cone shown.

Solution: **a.** To find L.A., first find l.

$$l = \sqrt{6^2 + 3^2} = \sqrt{45} = 3\sqrt{5}$$
$$\text{L.A.} = \pi rl = \pi \cdot 3 \cdot 3\sqrt{5} = 9\pi\sqrt{5}$$

b. T.A. = L.A. + B = $9\pi\sqrt{5} + \pi \cdot 3^2 = 9\pi\sqrt{5} + 9\pi$

c. $V = \frac{1}{3}\pi r^2 h = \frac{1}{3}\pi \cdot 3^2 \cdot 6 = 18\pi$

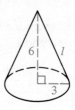

Classroom Exercises

1. a. When the label of a soup can is cut off and laid flat, it is a rectangular piece of paper. (See diagram below.) How are the length and width of this rectangle related to r and h?

b. What is the area of this rectangle?

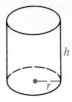

Find the lateral area, total area, and volume of each cylinder.

2. $r = 2$	**3.** $r = 3$	**4.** $r = 2$	**5.** $r = 4$
$h = 3$	$h = 2$	$h = 5$	$h = 2$

6. One container of bath powder is twice as wide as another but only half as tall. Which container holds more powder, or do they hold the same amount? Guess first and then calculate the ratio of their volumes.

Complete the table for the cone shown.

	r	h	l	L.A.	T.A.	V
7.	3	4	?	?	?	?
8.	5	12	?	?	?	?
9.	6	?	10	?	?	?

Written Exercises

Find the lateral area, total area, and volume of each cylinder.

A **1.** $r = 4$
$\quad\;\; h = 5$

2. $r = 8$
$\quad\;\; h = 10$

3. $r = 4$
$\quad\;\; h = 3$

4. $r = 8$
$\quad\;\; h = 6$

5. The volume of a cylinder is 64π. If $r = h$, find r.

6. The lateral area of a cylinder is 18π. If $h = 6$, find r.

7. The volume of a cylinder is 72π. If $h = 8$, find L.A.

8. The total area of a cylinder is 100π. If $r = h$, find r.

The table below refers to a cone. Copy and complete the table.

	r	h	l	L.A.	T.A.	V
9.	4	3	?	?	?	?
10.	8	6	?	?	?	?
11.	5	?	13	?	?	?
12.	?	$3\sqrt{3}$	6	?	?	?
13.	?	?	10	60π	?	?
14.	3	?	?	?	?	27π

15. In the first two rows of the preceding table, the ratio of the radii is $\dfrac{4}{8}$, or $\dfrac{1}{2}$,

and the ratio of the heights is $\dfrac{3}{6}$, or $\dfrac{1}{2}$. Use your answers from these two rows
of the table to determine the ratios of the following:

  a. lateral areas   **b.** total areas   **c.** volumes

16. A cone and a cylinder both have height 48 and radius 15. Give the ratio of
their volumes without actually calculating the two volumes.

B **17. a.** Guess which contains more, the can or the bottle? (Assume the top part of the bottle is a complete cone.)

  b. See if your guess is right by finding the volumes of both.

18. A pipe is 2 meters long and has inside radius 5 cm and
outside radius 6 cm. How many cubic centimeters of metal
are in the pipe?

19. In rectangle $ABCD$, $AB = 10$ and $AD = 6$.

  a. If the rectangle is revolved in space about $\overline{AB}$, what is the volume of the
space through which it moves?

  b. Answer part (a) if the rectangle is revolved about $\overline{AD}$.

20. a. The segment joining $(0, 0)$ and $(4, 3)$ is rotated about the x-axis, forming the lateral surface of a cone. Find the lateral area and the volume of this cone.

b. Make a sketch showing the cone that would be formed if the segment had been rotated about the y-axis. Find the lateral area and the volume of this cone.

c. Are your answers to parts (a) and (b) the same?

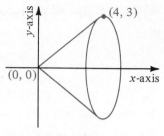

21. If the radius and height of a cylinder are both multiplied by 3, the lateral area is multiplied by __?__ and the volume is multiplied by __?__.

22. The total area of a cylinder is 40π. If $h = 8$, find r.

23. The total area of a cylinder is 90π. If $h = 12$, find r.

24. A square pyramid with base edge 4 is inscribed in a cone with height 6. What is the volume of the cone?

25. A square pyramid is inscribed in a cone with radius 4 and height 4.

a. What is the volume of the pyramid?

b. Find the slant heights of the cone and the pyramid.

Exs. 24, 25

C 26. A regular hexagonal pyramid with base edge 6 and height 8 is inscribed in a cone. Show that the lateral area of the cone is 60π and the lateral area of the pyramid is $18\sqrt{91}$.

27. In $\triangle ABC$, $AB = 15$, $AC = 20$, and $BC = 25$. If the triangle is rotated in space about $\overline{BC}$, what is the volume of the space through which it moves?

28. A $120°$ sector is cut out of a circular piece of tin with radius 6 and bent to form the lateral surface of a cone. What will be the volume of the cone?

SELF-TEST

1. The volume of a cube is 125 cm^3. Find the total area.

2. The lateral area of a cylinder is 120π. If the height of the cylinder is 10, find the radius.

Find the volume and total area of each figure.

3. A rectangular solid with dimensions 3 cm, 4 cm, and 5 cm.

4. A cylinder with radius 3 and height 6.

5. A regular square pyramid with base edge 10 and height 12.

6. A cone with radius 8 and slant height 10.

7. A right prism whose height is 6 and whose base is an equilateral triangle with side 3.

Similar Solids

Objectives

1. **Find the area and the volume of a sphere.**
2. **State and apply the properties of similar solids.**

10-4 *Spheres*

The area and the volume of a sphere are given by the formulas below. After some examples showing how these formulas are used, we shall see how they may be derived.

$$A = 4\pi r^2 \qquad V = \tfrac{4}{3}\pi r^3$$

Example 1:   Find the area and the volume of a sphere with radius 2.

Solution: $A = 4\pi r^2 = 4\pi \cdot 2^2 = 16\pi$

$$V = \frac{4}{3}\pi r^3 = \frac{4}{\cdot 3}\pi \cdot 2^3 = \frac{32\pi}{3}$$

Example 2:   The area of a sphere is 400π. Find the volume.

Solution: To find the volume, we must first find the radius.

(1) $A = 400\pi = 4\pi r^2$

$$100 = r^2$$
$$10 = r$$

(2) $V = \dfrac{4}{3}\pi r^3 = \dfrac{4}{3}\pi \cdot 10^3$

$$= \frac{4000\pi}{3}$$

    The next example shows how to find the area of the circle formed when a sphere is cut by a plane.

Example 3:   A plane passes 4 cm from the center of a sphere with radius 7 cm. Find the area of the circle of intersection.

Solution: Let x = radius of the circle.

$$x = \sqrt{7^2 - 4^2} = \sqrt{33}$$
$$A = \pi x^2 = \pi(\sqrt{33})^2 = 33\pi \ (\text{cm}^2)$$

Deriving the Volume Formula (Optional)

The sphere, the cylinder, and the double cone below all have radius r and height $2r$. Imagine a thin circular disc inscribed in each solid at a distance x units above its center. Suppose these three discs all have thickness h. Study the calculations below the figures.

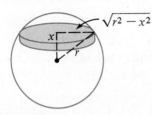

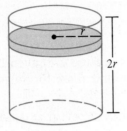

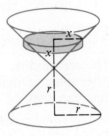

Disc volume:

$$\pi(\sqrt{r^2 - x^2})^2 h = \pi(r^2 - x^2)\,h$$
$$= \pi r^2 h - \pi x^2 h$$

Disc volume: $\pi r^2 h$

Disc volume: $\pi x^2 h$

The calculations show that no matter what the distance x is, the volume of the first disc equals the difference between the volumes of the other two discs. This means that if you inscribe many discs of the same thickness in the three solids (as in the diagrams below) the total volume of all the discs in the first solid will equal the difference between the total volumes of all the discs in each of the other two solids.

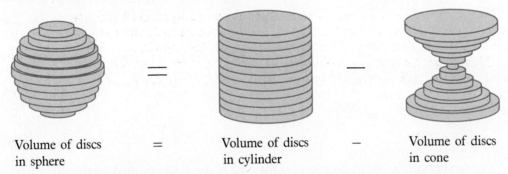

| Volume of discs in sphere | = | Volume of discs in cylinder | − | Volume of discs in cone |

The equation above holds if there are just a few discs inscribed in each solid or very many discs. If there are very many discs inscribed in a solid, their total volume will be practically the same as the volume of the solid. It follows that we can find the volume of the sphere by subtracting the volume of the double cone from the volume of the cylinder.

$$\text{Volume of sphere} = \pi r^2 \cdot 2r - 2(\tfrac{1}{3}\pi r^2 \cdot r)$$
$$= 2\pi r^3 \qquad - \tfrac{2}{3}\pi r^3$$
$$= \tfrac{4}{3}\pi r^3$$

Deriving the Area Formula (Optional)

Imagine a rubber ball with inner radius r and rubber thickness t. To find the volume of the rubber, we can use the formula for the volume of a sphere. We just subtract the volume of the inner sphere from the volume of the outer sphere.

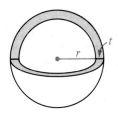

$$\text{Exact vol. of rubber} = \tfrac{4}{3}\pi(r + t)^3 - \tfrac{4}{3}\pi r^3$$
$$= \tfrac{4}{3}\pi[(r + t)^3 - r^3]$$
$$= \tfrac{4}{3}\pi[r^3 + 3r^2t + 3rt^2 + t^3 - r^3]$$
$$= 4\pi r^2 t + 4\pi rt^2 + \tfrac{4}{3}\pi t^3$$

The volume of the rubber can be found in another way as well. If we think of a small piece of the rubber ball, its approximate volume would be its outer area A times its thickness t. The same thing is true for the whole ball.

$$\text{Volume of rubber} \approx \text{Surface area} \times \text{thickness}$$
$$V \approx At$$

Now let us use both results for the volume of the rubber.

$$At \approx 4\pi r^2 t + 4\pi rt^2 + \tfrac{4}{3}\pi t^3$$

If we divide both sides of the equation by t, we get the result:

$$A \approx 4\pi r^2 + 4\pi rt + \tfrac{4}{3}\pi t^2$$

This approximation for A gets better and better as the layer of rubber gets thinner and thinner. As t gets near zero, the last two terms in the formula above also get near zero. The limiting result is the formula $A = 4\pi r^2$.

This is exactly what we would expect, since the surface area of a ball clearly does not depend at all on the thickness of the rubber, but only on the size of the radius.

Classroom Exercises

Copy and complete the table for spheres.

	1.	2.	3.	4.	5.	6.
Radius	1	3	6	?	?	?
Area	?	?	?	64π	100π	?
Volume	?	?	?	?	?	$\dfrac{4000\pi}{3}$

A plane passes h cm from the center of a sphere with radius r cm. Find the area of the circle of intersection, shaded in the diagram, for the given values.

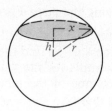

7. $r = 5$
$\quad h = 3$

8. $r = 17$
$\quad h = 8$

9. $r = 7$
$\quad h = 6$

Written Exercises

Copy and complete the table for spheres.

A

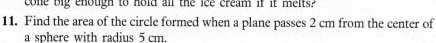

	1.	2.	3.	4.	5.	6.	7.	8.
Radius	4	5	$\frac{1}{2}$	$\frac{3}{4}$	?	?	$\sqrt{2}$	?
Area	?	?	?	?	36π	324π	?	?
Volume	?	?	?	?	?	?	?	288π

9. A sphere has radius 2 and a hemisphere has radius 4. Compare their volumes.

10. A scoop of ice cream with diameter 6 cm is placed in an ice-cream cone with diameter 5 cm and height 10 cm. Is the cone big enough to hold all the ice cream if it melts?

Ex. 10

11. Find the area of the circle formed when a plane passes 2 cm from the center of a sphere with radius 5 cm.

12. Find the area of the circle formed when a plane passes 6 cm from the center of a sphere with radius 7 cm.

B 13. A silo of a barn consists of a cylinder capped by a hemisphere as shown. Find the volume of the silo.

14. Two cans of paint cover the hemispherical dome of the silo shown. Estimate how many cans are needed to paint the rest of the silo's exterior.

Exs. 13, 14

15. A sphere with radius r is inscribed in a cylinder. Find the volume of the cylinder in terms of r.

16. A sphere is inscribed in a cylinder. Show that the area of the sphere equals the lateral area of the cylinder.

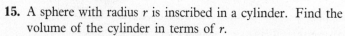

Exs. 15, 16

17. If you double the radius of a sphere, the area of the sphere is multiplied by __?__ and the volume is multiplied by __?__ .

18. If you triple the radius of a sphere, the area of the sphere is multiplied by __?__ and the volume is multiplied by __?__ .

19. A double cone is inscribed inside the cylinder shown. Find the volume of the space inside the cylinder but outside the double cone.

Ex. 19

20. A hollow rubber ball has outer radius 11 cm and inner radius 10 cm.

    a. Find the exact volume of the rubber.

    b. The volume of the rubber can be approximated by the formula:

    $V \approx$ inner surface area $\times$ thickness of rubber

    Use this formula to approximate V and compare your answer with part (a).

    c. Is the approximation method used in part (b) better for a ball with a thick layer of rubber or a ball with a thin layer?

Ex. 20

21. A cylinder with height 12 is inscribed in a sphere with radius 10. Find the volume of the cylinder.

C 22. A cylinder with height $2x$ is inscribed in a sphere with radius 10.

    a. Show that the volume of the cylinder, V, is $2\pi x(100 - x^2)$.

    b. By using calculus, one can show that V is maximum when $x = \dfrac{10\sqrt{3}}{3}$. If you substitute this value for x, the maximum volume $V = \underline{\ ?\ }$.

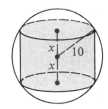

Exs. 21, 22

    c. (Optional computer exercise) Use a computer to evaluate $V = 2\pi x(100 - x^2)$ for various values of x between 0 and 10. What approximate value of x gives you the largest volume V?

23. A cone is inscribed in a sphere with radius 10 as shown.

    a. Show that the volume of the cone, V, is $\frac{1}{3}\pi(100 - x^2)(10 + x)$.

    b. By using calculus, one can show that V is maximum when $x = \frac{10}{3}$. If you substitute this value for x, the maximum volume $V = \underline{\ ?\ }$.

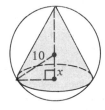

    c. (Optional computer exercise) Use a computer to evaluate $V = \frac{1}{3}\pi(100 - x^2)(10 + x)$ for various values of x between 0 and 10. Show that the maximum volume V occurs when x is $\frac{10}{3}$.

24. a. Find the volume of a sphere inscribed in a cube with edges 6 cm long.

    b. Show that the volume of the sphere is a little more than half the volume of the cube.

25. A sphere is inscribed in a cone whose radius is 6 and whose height is 8. Find the radius of the sphere.

10-5 *Areas and Volumes of Similar Solids*

Similar solids are solids which have the same shape but not necessarily the same size. It's easy to see that all spheres are similar. To decide whether two solids are similar, just see whether bases are similar and corresponding lengths are proportional.

Right cylinders   Regular square pyramids

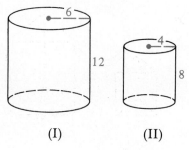

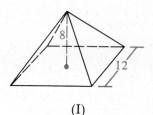

  (I)   (II)   (I)   (II)

Similar because $\dfrac{6}{4} = \dfrac{12}{8}$ Similar because $\dfrac{12}{6} = \dfrac{8}{4}$

Scale factor: $\dfrac{3}{2}$ Scale factor: $\dfrac{2}{1}$

The table below shows the ratios of the perimeters, areas, and volumes for both pairs of similar solids shown above. What do you notice about the ratios for the cylinders? for the pyramids?

	Cylinders I and II	Pyramids I and II
Scale Factor	$\dfrac{3}{2}$	$\dfrac{2}{1}$
$\dfrac{\text{Base Perimeter (I)}}{\text{Base Perimeter (II)}}$	$\dfrac{2\pi \cdot 6}{2\pi \cdot 4} = \dfrac{6}{4}$, or $\dfrac{3}{2}$	$\dfrac{4 \cdot 12}{4 \cdot 6} = \dfrac{48}{24}$, or $\dfrac{2}{1}$
$\dfrac{\text{L.A. (I)}}{\text{L.A. (II)}}$	$\dfrac{2\pi \cdot 6 \cdot 12}{2\pi \cdot 4 \cdot 8} = \dfrac{9}{4}$, or $\dfrac{3^2}{2^2}$	$\dfrac{\frac{1}{2} \cdot 48 \cdot 10}{\frac{1}{2} \cdot 24 \cdot 5} = \dfrac{4}{1}$, or $\dfrac{2^2}{1^2}$
$\dfrac{\text{Volume (I)}}{\text{Volume (II)}}$	$\dfrac{\pi \cdot 6^2 \cdot 12}{\pi \cdot 4^2 \cdot 8} = \dfrac{27}{8}$, or $\dfrac{3^3}{2^3}$	$\dfrac{\frac{1}{3} \cdot 12^2 \cdot 8}{\frac{1}{3} \cdot 6^2 \cdot 4} = \dfrac{8}{1}$, or $\dfrac{2^3}{1^3}$

The results shown in the table above are generalized in the following theorem. (See Exercises 15–20 for proofs.)

Theorem 10-3 If the scale factor of two similar solids is $a:b$, then
(1) the ratio of corresponding perimeters is $a:b$;
(2) the ratios of the base areas, of the lateral areas, and of the total areas are $a^2:b^2$; and
(3) the ratio of the volumes is $a^3:b^3$.

Example: For the similar solids shown, find the ratios of (a) base perimeters, (b) lateral areas, and (c) volumes.

Solution: The scale factor is $6:10$, or $3:5$.
a. Ratio of base perimeters $= 3:5$
b. Ratio of lateral areas $= 3^2:5^2 = 9:25$
c. Ratio of volumes $= 3^3:5^3 = 27:125$

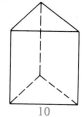

Theorem 10-3 above is the three-dimensional counterpart of Theorem 9-6 on page 347. (Take a minute to compare these theorems.) There is a similar relationship between the two results shown below.

In two dimensions:
If $\overline{XY} \| \overline{AB}$, then
$\triangle VXY \sim \triangle VAB$.

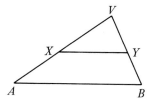

In three dimensions:
If plane $XYZ \|$ plane ABC, then $V\text{-}XYZ \sim V\text{-}ABC$.

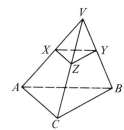

Classroom Exercises

Tell whether the solids in each pair are similar.

1.

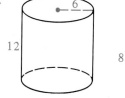

Right Cylinders

2.

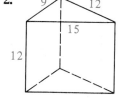

Right Prisms

3. a. What is the ratio of the lateral areas of the prisms in Exercise 2?

   b. What is the ratio of the total areas of the prisms?

   c. What is the ratio of the volumes of the prisms?

4. Two spheres have diameters 24 and 36.

   a. What is the ratio of the areas?

   b. What is the ratio of the volumes?

5. Two spheres have volumes 2π and 16π. Find the ratios of the following:

   a. volumes   **b.** radii   **c.** areas

Complete the table below, which refers to two similar pyramids.

	6.	**7.**	**8.**	**9.**	**10.**	**11.**
Scale factor	3:4	5:7	?	?	?	?
Ratio of base perimeters	?	?	2:1	?	?	?
Ratio of slant heights	?	?	?	6:9	?	?
Ratio of lateral areas	?	?	?	?	4:9	?
Ratio of total areas	?	?	?	?	?	?
Ratio of volumes	?	?	?	?	?	8:125

12. Plane *PQR* is parallel to the base of the pyramid and bisects the altitude. Find the ratios.

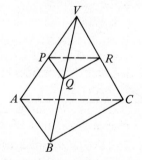

   a. The perimeter of $\triangle PQR$ to the perimeter of $\triangle ABC$

   b. The lateral area of the top part of the pyramid to the lateral area of the whole pyramid

   c. The lateral area of the top part of the pyramid to the lateral area of the bottom part

   d. The volume of the top part of the pyramid to the volume of the bottom part

13. Find the ratios in Exercise 12 (a)–(c) if the height of the top pyramid is 3 and the height of the whole pyramid is 5.

Challenge

Each face of a pyramid is an equilateral triangle. Each edge of the pyramid has length 6. Find the length of (a) the altitude of the pyramid and (b) the radius of the sphere that contains the four vertices of the pyramid.

Written Exercises

A 1. Two cones have radii 6 and 9. The heights are 10 and 15. Are the cones similar?

2. The heights of two right prisms are 18 and 30. The bases are squares with sides 8 and 15. Are the prisms similar?

3. Two similar cylinders have radii 3 and 4. Find the ratios of the following:
   a. heights   **b.** base circumferences   **c.** lateral areas   **d.** volumes

4. Two similar pyramids have heights 12 and 18. Find the ratios of the following:
   a. base areas   **b.** lateral areas   **c.** total areas   **d.** volumes

5. Assume that the earth and the moon are smooth spheres with diameters 12,800 km and 3,200 km. Find the ratios of the following:
   a. lengths of their equators   **b.** areas   **c.** volumes

6. Two similar cylinders have lateral areas 81π and 144π. Find the ratios of the following:
   a. heights   **b.** total areas   **c.** volumes

7. Two similar cones have volumes 8π and 27π. Find the ratios of the following:
   a. radii   **b.** lateral heights   **c.** lateral areas

8. Two similar pyramids have volumes 3 and 375. Find the ratios of the following:
   a. heights   **b.** base areas   **c.** total areas

B 9. A certain kind of string is sold in a ball 6 cm in diameter and in a ball 12 cm in diameter. The smaller ball costs $.50 and the larger one costs $3.00. Which is the better buy?

10. Two similar prisms have heights 10 and 25. If the volume of the larger is 500, what is the volume of the smaller?

11. Two similar pyramids have lateral areas 8 and 18. If the volume of the smaller is 32, what is the volume of the larger?

12. A plane parallel to the base of a cone divides the cone into two pieces. Find the ratios of the following:
   a. The areas of the shaded circles
   b. The lateral area of the top piece to that of the whole cone
   c. The lateral area of the top piece to that of the bottom piece
   d. The volume of the top piece to that of the whole cone
   e. The volume of the top piece to that of the bottom piece

9 cm

3 cm

13. Redraw the figure for Exercise 12 changing the 9 cm and 3 cm dimensions to 10 cm and 4 cm. Then find the five ratios described in Exercise 12.

14. A pyramid with height 15 cm is separated into two pieces by a plane parallel to the base and 6 cm above it. What are the volumes of these two pieces if the volume of the original pyramid is 250 cm³?

The purpose of Exercises 15–20 is to prove Theorem 10-3 for some similar solids.

15. Two spheres have radii a and b. Prove that the ratio of the areas is $a^2:b^2$.

16. Two spheres have radii a and b. Prove that the ratio of the volumes is $a^3:b^3$.

17. Two similar cones have radii r_1 and r_2 and heights h_1 and h_2. Prove that the ratio of the volumes is $h_1{}^3:h_2{}^3$.

18. Two similar cones have radii r_1 and r_2 and lateral heights l_1 and l_2. Prove that the ratio of the lateral areas is $r_1{}^2:r_2{}^2$.

19. The bases of two similar prisms are regular pentagons with base edges e_1 and e_2 and base areas B_1 and B_2. The heights are h_1 and h_2. Prove that the ratio of the lateral areas is $e_1{}^2:e_2{}^2$.

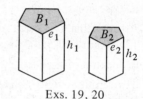

20. Refer to Exercise 19 and prove that the ratio of the volumes of the prisms is $e_1{}^3:e_2{}^3$.

Exs. 19, 20

C 21. A plane parallel to the base of a pyramid separates the pyramid into two pieces with equal volumes. If the height of the pyramid is 12, find the height of the top piece.

SELF-TEST

1. A sphere has diameter 6. What is the volume?

2. A sphere has area 100π. What is the radius?

3. A plane passes 4 cm from the center of a sphere with radius 5 cm. What is the area of the circle of intersection?

4. Two similar cones have radii 6 and 12. What is the ratio of their total areas? What is the ratio of their volumes?

5. The lateral areas of two similar cylinders are 10π and 90π. What is the ratio of their radii?

Challenge _____

Think of the earth as a perfect sphere. A band fits snugly around the earth at the equator. The band is then stretched one meter and is placed so that its points are equidistant from the earth. Which of the following animals could crawl under the band—an ant, a mouse, a kitten, a large dog?

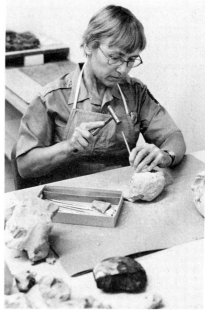

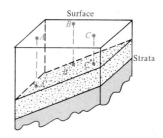

Surface

Strata

Career Notes

Geologist

Rock formations often occur in *strata,* or layers, beneath the surface of the earth. Earthquakes occur at *faults,* breaks in the strata. In search of a fault, how would you determine the position of a stratum of rock buried deep beneath the surface of the earth?

A geologist might start by picking three noncollinear points, *A, B,* and *C,* on the surface and drilling holes to find the depths of points *A', B',* and *C'* on the stratum. These three points determine the plane of the surface of the stratum.

Geologists may work for industry, searching for oil or minerals. They may work in research centers, developing ways to predict earthquakes. Today, geologists are trying to locate sources of geothermal energy, generated by the earth's internal heat.

A career in geology usually requires a graduate degree in geology, preferably a Ph.D.

CALCULATOR KEY·IN

The diagrams show two rectangles inscribed in an isosceles triangle with legs 5 and base 6. There are many more such rectangles. The question is, Which one has the greatest area?

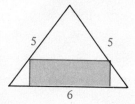

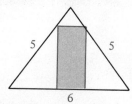

To solve the problem, let *CDEF* represent any rectangle inscribed in isosceles △*ABV* with legs 5 and base 6. If we let *OD* = *x* and *ED* = *y*, the area of the rectangle is 2*xy*. Our goal is to express this area in terms of *x* alone. Then we can find out how the area changes as *x* changes.

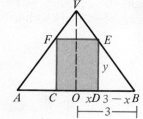

1. In right △*VOB*, *OB* = 3 and *VB* = 5. Thus *VO* = 4.

2. △*EDB* ~ △*VOB*   (Why?)

3. $\dfrac{ED}{VO} = \dfrac{DB}{OB}$ (Why?)

4. $\dfrac{y}{4} = \dfrac{3 - x}{3}$ (By substitution in Step 3.)

5. $y = \dfrac{4}{3}(3 - x)$ (Multiplication Postulate)

6. Area of rectangle: $A = 2xy = 2x \cdot \dfrac{4}{3}(3 - x) = \dfrac{8x(3 - x)}{3}$

We can use the formula found in Step 6 and a calculator to find the area for many values of *x*. It is easiest to calculate 3 − *x* first, then multiply by *x*, then multiply by 8, and finally divide by 3.

x	Area
0	0
0.25	1.83333
0.5	3.33333
0.75	4.5
1	5.33333
1.25	5.83333
1.5	6
1.75	5.83333
2	5.33333
2.25	4.5
2.50	3.33333
2.75	1.83333
3	0

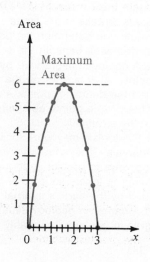

The table was used to make a graph showing how the area varies with *x*. Both the table and the graph suggest that the greatest area, 6, occurs when *x* = 1.5.

Exercises

Suppose the original triangle had sides 5, 5, and 8 instead of 5, 5, and 6.

1. Draw a diagram like the third diagram on page 382 and show that $A = \dfrac{3x(4 - x)}{2}$.

2. Find the value of x for which the greatest area occurs.

COMPUTER HEY-IN

The diagrams at the left below show two cylinders inscribed in a cone with diameter 6 and lateral height 5. There are many more such cylinders. The question is, Which one has the greatest volume?

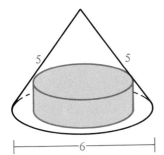 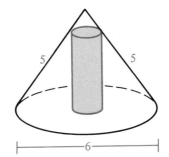 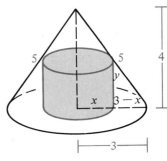

The diagram at the right above shows a typical inscribed cylinder. By using similar triangles, we can write the proportion $\dfrac{y}{4} = \dfrac{3 - x}{3}$. Thus $y = \dfrac{4}{3}(3 - x)$. The volume of the cylinder is as follows:

$$V = \pi x^2 y = \pi x^2 \cdot \frac{4}{3}(3 - x) \approx \frac{4}{3}(3.14159)x^2(3 - x)$$

The program in BASIC below will evaluate V for various values of x.

```
10   PRINT "X", "VOLUME"
20   FOR X=0 TO 3 STEP .25
30   LET V=4/3*3.14159*X↑2*(3-X)
40   PRINT X, V
50   NEXT X
60   END
```

Exercises

1. If your computer uses a language other than BASIC, write a similar program in that language to evaluate V for various values of x.

2. a. Run the program. Make a graph that shows how the volume varies with x.

   b. For what value of x did you find the greatest volume?

Chapter Summary

1. Formulas for areas and volumes of solids:

   Right Prism
   L.A. $= ph$
   $V = Bh$

   Regular Pyramid
   L.A. $= \frac{1}{2}pl$
   $V = \frac{1}{3}Bh$

   Right Cylinder
   L.A. $= 2\pi rh$
   $V = \pi r^2h$

   Right Cone
   L.A. $= \pi rl$
   $V = \frac{1}{3}\pi r^2h$

   Prism or cylinder: Total area $=$ L.A. $+ 2B$
   Pyramid or cone: Total area $=$ L.A. $+ B$
   Sphere: Area $= 4\pi r^2$ Volume $= \frac{4}{3}\pi r^3$

2. If the scale factor of two similar solids is $a:b$, then
   (a) the ratio of corresponding lengths is $a:b$;
   (b) the ratio of corresponding areas is $a^2:b^2$; and
   (c) the ratio of the volumes is $a^3:b^3$.

Chapter Review

10-1 *Prisms*

1. How many lateral faces does a pentagonal prism have?
2. What is the lateral area of a right pentagonal prism with height 10 and base edges 5, 3, 6, 4, and 8?
3. What is the total area of a cube whose edges are 3 cm?
4. What is the height of a right prism with base area 6 cm² and volume 24 cm³?

10-2 *Pyramids*

Use the regular square pyramid shown.

5. L.A. $= \underline{\ ?\ }$
6. T.A. $= \underline{\ ?\ }$
7. $V = \underline{\ ?\ }$
8. $l = \underline{\ ?\ }$

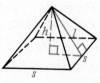

10-3 *Cylinders and Cones*

9. A cylinder with radius 4 cm and height 5 cm has lateral area $\underline{\ ?\ }$ cm².
10. A cone with radius 4 cm and height 3 cm has total area $\underline{\ ?\ }$ cm².
11. A cone with radius 9 cm and height 10 cm has volume $\underline{\ ?\ }$ cm³.
12. A cylinder with radius 5 cm and height 8 cm has volume $\underline{\ ?\ }$ cm³.

10-4 *Spheres*

13. A sphere with radius 8 cm has area _?_ cm².

14. A sphere with radius 9 cm has volume _?_ cm³.

15. A sphere with volume 36π cm³ has radius _?_ cm.

16. A sphere with area 9π cm² has volume _?_ cm³.

10-5 *Areas and Volumes of Similar Solids*

17. A cone has radius 21 cm and height 28 cm. A larger cone has radius 27 cm and height 36 cm. Are the cones similar?

18. Two spheres have radii 12 cm and 42 cm. What is the ratio of the volumes?

19. Two similar cylinders have radii 10 cm and 15 cm. What is the ratio of the total areas?

20. Two similar cones have lateral areas of 3 cm² and 48 cm². What is the ratio of the volumes?

Chapter Test

1. Find the volume of a rectangular solid with length 7 cm, width 6 cm, and height 3 cm.

2. Find the total area of the rectangular solid in Exercise 1.

3. If a cube has volume 27 cm³, what is the total area?

4. A regular square pyramid has base edge 16 cm and slant height 12 cm. What is the lateral area?

5. Find the volume of the pyramid in Exercise 4.

6. A cylinder has height 12 cm and radius 5 cm. What is the lateral area?

7. Find the volume of the cylinder in Exercise 6.

8. The volume of a cylinder is 288π cm³. If the height is 8 cm, what is the radius?

9. The height of a cone is equal to the radius. If the volume is 9π cm³, what is the radius?

10. A cone has radius 7 cm and lateral area 84π cm². What is the slant height?

11. Find the total area of the cone in Exercise 10.

12. A sphere has radius 6 cm. What is the area?

13. What is the volume of the sphere in Exercise 12?

14. Two cylinders have radii 9 cm and 16 cm. If the heights are 12 cm and 24 cm, are the cylinders similar?

15. Two pyramids are similar. If the ratio of their volumes is 64:1000, what is the ratio of their lateral areas?

16. A prism has a base which is an equilateral triangle. If each base edge is 12 cm and the height of the prism is $5\sqrt{3}$, what is the volume?

11

Coordinate Geometry

Using the Distance Formula

Objectives
1. Specify points in the coordinate plane by means of their coordinates.
2. State and apply the distance formula.
3. State and apply the general equation of a circle.
4. State and apply the midpoint formula.

11-1 *The Distance Formula*

Some of the terms you have used in your study of graphs are reviewed below.

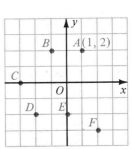

Origin: Point O
Axes: x-axis and y-axis
Quadrants: Regions I, II, III, IV
Coordinate plane: The plane of the x-axis and y-axis

The arrowhead on each axis shows the positive direction.

The **x-coordinate** of A is 1.
The **y-coordinate** of A is 2.

The **coordinates** of A are 1 and 2. Point A, or point (1, 2), is the **graph** of the *ordered pair* (1, 2). Note that the order of the coordinates has significance in locating the point in the coordinate plane. The points shown are:

$B(-1, 2)$ $C(-3, 0)$ $D(-2, -2)$
$E(0, -2)$ $F(2, -3)$ $O(0, 0)$

It is easy to find the distance between two points that lie on a horizontal line or on a vertical line. Recall that distance is always a positive number.

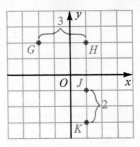

The distance between G and H is 3.
Using the x-coordinates of G and H:
$|1 - (-2)| = 3$, or $|-2 - 1| = 3$.

The distance between J and K is 2.
Using the y-coordinates of J and K:
$|-3 - (-1)| = 2$, or $|-1 - (-3)| = 2$.

If two points do *not* lie on a horizontal or vertical line, the distance between the points can be found by using the Pythagorean Theorem.

Example 1: Find the distance between points $A(-1, -1)$ and $B(3, 2)$.

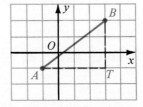

Solution: Draw the horizontal and vertical segments shown. The coordinates of T are $(3, -1)$. Then $AT = 4$, and $BT = 3$. Thus, $(AB)^2 = 4^2 + 3^2 = 25$, and $AB = 5$.

Using a method suggested by Example 1, you can find a formula for the distance between points $P_1(x_1, y_1)$ and $P_2(x_2, y_2)$. First draw a right triangle as shown. The coordinates of T are (x_2, y_1).

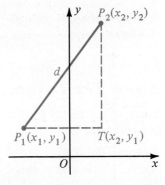

$$P_1T = |x_2 - x_1|; \quad P_2T = |y_2 - y_1|$$
$$d^2 = (P_1T)^2 + (P_2T)^2$$
$$= |x_2 - x_1|^2 + |y_2 - y_1|^2$$
$$= (x_2 - x_1)^2 + (y_2 - y_1)^2$$
$$d = \sqrt{(x_2 - x_1)^2 + (y_2 - y_1)^2}$$

Theorem 11-1 The distance d between points (x_1, y_1) and (x_2, y_2) is given by:
$$d = \sqrt{(x_2 - x_1)^2 + (y_2 - y_1)^2}$$

Example 2: Find the distance between points $(-2, 1)$ and $(4, -3)$.

Solution 1: Let (x_1, y_1) be $(-2, 1)$, and (x_2, y_2) be $(4, -3)$.
$$d = \sqrt{(x_2 - x_1)^2 + (y_2 - y_1)^2}$$
$$d = \sqrt{(4 - (-2))^2 + (-3 - 1)^2} = \sqrt{36 + 16} = \sqrt{52}, \text{ or } 2\sqrt{13}$$

Solution 2: Draw a right triangle. The legs have lengths 6 and 4.

$$d^2 = 6^2 + 4^2 = 36 + 16 = 52$$
$$d = \sqrt{52} = 2\sqrt{13}$$

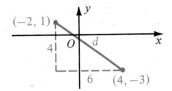

Classroom Exercises

1. State the coordinates of A, B, C, D, E, F, G, and O.

2. What is the x-coordinate of every point that lies on a vertical line through G?

3. Which of the following points lie on a horizontal line through G?

$(0, -3)$ $(3, -3)$ $(79, -3)$
$(2, 3)$ $(-3, 0)$ $(-13, -3)$

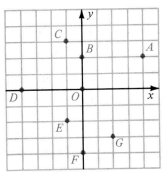

4. Each dashed segment shown is either horizontal or vertical. State the coordinates of point T.

a.

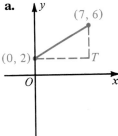

b.

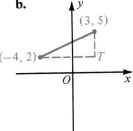

c.

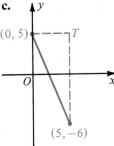

d.

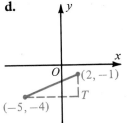

e.

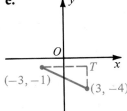

f.

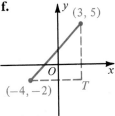

5. State the lengths of the legs of the right triangles shown in Exercise 4.

6. a. To use the distance formula to find the distance between points $(1, 2)$ and $(3, 7)$, you can let $P_1 = (1, 2)$ and $P_2 = (3, 7)$. Then:

$$(x_2 - x_1)^2 = \underline{?} \qquad\qquad (y_2 - y_1)^2 = \underline{?} \qquad\qquad d = \underline{?}$$

b. On the other hand, to find the distance between points $(1, 2)$ and $(3, 7)$, you can let $P_1 = (3, 7)$ and $P_2 = (1, 2)$. Then:

$$(x_2 - x_1)^2 = \underline{?} \qquad\qquad (y_2 - y_1)^2 = \underline{?} \qquad\qquad d = \underline{?}$$

7. Find the distance between the points named. Write answers in simplified radical form.

    a. $(0, 0)$ and $(7, -1)$ **b.** $(-2, -2)$ and $(4, -4)$ **c.** $(-3, 2)$ and $(8, 2)$

Written Exercises

Find the distance between the two points. If necessary, you can draw graphs, but you shouldn't need to use the distance formula.

A   **1.** $(-3, 0)$ and $(4, 0)$ **2.** $(2, 2)$ and $(2, 6)$

    3. $(4, -1)$ and $(4, 4)$ **4.** $(-5, -2)$ and $(-1, -2)$

    5. $(-4, 5)$ and $(4, 5)$ **6.** $(3, 8)$ and $(3, 1)$

Use the distance formula to find the distance between the two points.

    7. $(0, 0)$ and $(3, 4)$ **8.** $(-6, 8)$ and $(0, 0)$

    9. $(1, 1)$ and $(4, 4)$ **10.** $(3, 6)$ and $(4, -1)$

  11. $(-1, -2)$ and $(-4, -4)$ **12.** $(5, -7)$ and $(0, 3)$

Find the distance between the points named. Use any method you prefer.

  13. $(4, 3)$ and $(0, 0)$ **14.** $(0, 0)$ and $(8, -6)$

  15. $(2, 2)$ and $(5, 5)$ **16.** $(-3, 0)$ and $(0, 4)$

  17. $(5, -2)$ and $(2, -8)$ **18.** $(-3, 1)$ and $(-5, 0)$

Write an expression for the distance between the points named.

  19. (a, b) and (c, d) **20.** $(j, -k)$ and $(-m, -n)$

B **21.** There are twelve points, each with integer coordinates, that are 10 units from the origin. List the points. (*Hint:* Recall the 6-8-10 right triangle.)

  22. List twelve points, each having integer coordinates, that are 5 units from point $(2, -1)$.

In Exercises 23–27 you will want to find and then compare lengths of segments.

  23. Show that the triangle with vertices $A(-3, 1)$, $B(2, -2)$, and $C(1, 4)$ is a scalene triangle.

  24. Show that quadrilateral $RSTU$ has equal diagonals when the vertices are $R(1, 2)$, $S(0, -4)$, $T(-1, -3)$ and $U(-2, 1)$.

25. Discover and prove something about a triangle with vertices $D(3, 0)$, $E(11, -2)$, and $F(5, -8)$.

26. Discover and prove something about quadrilateral $GHIJ$ when the vertices are $G(-4, -2)$, $H(-5, -5)$, $I(1, 1)$ and $J(2, 4)$.

C 27. Discover and prove something about the triangle with vertices $D(-4, -2)$, $E(2, 3)$, and $F(12, -9)$.

28. It is known that $\triangle RUN$ is isosceles; R is point $(-1, -5)$; U is point $(1, 1)$; the x-coordinate of N is 3. Find all possible values for the y-coordinate of N.

29. The point (j, k) is equidistant from the three points $(8, 7)$, $(10, 1)$ and $(4, -5)$. Find the values of j and k.

30. Use an indirect proof to show that there isn't any point that is equidistant from the three points $(-5, 2)$, $(-2, 3)$, and $(4, 5)$.

11-2 *Circles*

You can use the distance formula to develop the equation of a circle with center at the origin and with radius 5.

Let $P(x, y)$ represent any point on the circle. The distance between $O(0, 0)$ and $P(x, y)$ is 5.

$$\sqrt{(x - 0)^2 + (y - 0)^2} = 5$$
$$\sqrt{x^2 + y^2} = 5$$
$$x^2 + y^2 = 25$$

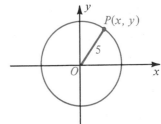

You can proceed in much the same way to develop the equation of a circle with center $Q(a, b)$ and radius r.

Let $P(x, y)$ represent any point on the circle. The distance between $Q(a, b)$ and $P(x, y)$ is r.

$$\sqrt{(x - a)^2 + (y - b)^2} = r$$
$$(x - a)^2 + (y - b)^2 = r^2$$

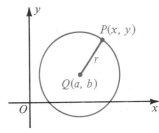

The development proves the theorem stated below.

Theorem 11-2 The circle with center (a, b) and radius r has the equation $(x - a)^2 + (y - b)^2 = r^2$.

Example: Find the center and radius of the circle with equation
$(x - 5)^2 + (y + 3)^2 = 81$.

Solution: Rewrite the equation in the form $(x - a)^2 + (y - b)^2 = r^2$.
$(x - 5)^2 + (y - (-3))^2 = 9^2$
The center is point $(5, -3)$ and the radius is 9.

Classroom Exercises

1. State the coordinates of the center of each circle.

   a. $(x - 2)^2 + (y - 5)^2 = 10^2$ **b.** $(x - 2)^2 + (y + 5)^2 = 7^2$

   c. $(x - 0)^2 + (y - 12)^2 = 3^2$ **d.** $x^2 + (y - 4)^2 = 1$

   e. $(x + 4)^2 + y^2 = 25$ **f.** $(x + 1)^2 + (y + 2)^2 = 26$

2. State the radius of each circle in Exercise 1.

3. State the equation of the circle that has the given center and radius.

   a. Center $(3, 6)$; radius 4   **b.** Center $(0, -5)$; radius 2

   c. Center $(-4, 1)$; radius r **d.** Center (j, k); radius n

4. Write an inequality that describes the points (x, y) that are less than 4 units from the origin. See the diagram below.

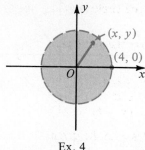

Ex. 4

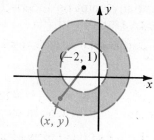

Ex. 5

5. Refer to the diagram above. Complete the statement to describe points (x, y) that are more than 3 units but less than 6 units from point $(-2, 1)$.

$$3^2 < (\underline{?})^2 + (\underline{?})^2 < 6^2$$

6. Describe the locus of points whose coordinates satisfy the pair of statements: $x^2 + y^2 = 100$ and $y \geq 0$.

Written Exercises

Find the center and radius of each circle.

A **1.** $x^2 + (y - 3)^2 = 8^2$ **2.** $(x + 2)^2 + y^2 = 3^2$

    3. $(x - 7)^2 + (y - 4)^2 = 25$ **4.** $(x + 1)^2 + (y + 9)^2 = \frac{49}{4}$

    5. $(x - j)^2 + (y - k)^2 = c^2$ **6.** $(x - d)^2 + (y + e)^2 = 13$

Write the equation of the circle that has the center and the radius named.

	7.	8.	9.	10.	11.	12.
Center	$(0, 5)$	$(-4, 0)$	$(3, 3)$	$(-2, -2)$	$(4, -7)$	(p, q)
Radius	3	11	1	$\frac{1}{2}$	q	v

In Exercises 13–16, find the equation of the circle described.

B **13.** The circle has center $(6, 6)$ and is tangent to both axes.

14. The circle has center (c, d) and is tangent to the y-axis.

15. The circle has center $(4, 0)$ and passes through point $(12, 6)$.

16. The circle with diameter $\overline{AB}$ if A is $(0, 0)$ and B is $(6, 0)$.

17. Two points on the circle $(x + 2)^2 + (y - 4)^2 = 25$ both have x-coordinate 1. What are the y-coordinates of those two points?

18. Find the equation of the locus of the centers of all circles that have radius 3 and pass through point $(5, -1)$.

C **19. a.** Draw the four circles obtained by using all combinations of signs in the equation $(x \pm 5)^2 + (y \pm 5)^2 = 25$.

   b. Write the equation of a circle that surrounds the four given circles and is tangent to each of them.

   c. Write the equation of another circle that is tangent to each of the four given circles.

20. Find the center and radius of the circle $x^2 - 6x + y^2 - 10y = 2$. (*Hint:* Fill in the blanks in $(x^2 - 6x + \underline{\,?\,}) + (y^2 - 10y + \underline{\,?\,}) = 2 + \underline{\,?\,}$ in such a way that you can proceed to an equation in the form $(x - a)^2 + (y - b)^2 = r^2$.)

Use algebra to find the intersection points of each pair of circles with the given equations. (*Suggestions:* Write each equation in expanded form; eliminate x^2 and y^2 by subtraction; solve the resulting equation for y in terms of x; substitute for y in one of the equations; solve for x.)

21. $(x + 5)^2 + (y + 4)^2 = 25$ and $(x - 3)^2 + y^2 = 65$

22. $(x + 4)^2 + (y - 2)^2 = 50$ and $(x - 4)^2 + (y - 10)^2 = 18$

Challenge

Use only a compass to construct the diagram shown. If the radius of the circle is 6, find the area and the perimeter of the shaded region.

COMPUTER KEY-IN

The graph shows a quarter-circle inscribed in a square with area 1. If points are picked at random inside the square, some of them will also be inside the quarter-circle. Let n be the number of points picked inside the square and let q be the number of these points that fall inside the quarter-circle. If many, many points are picked at random inside the square, the following ratios are approximately equal:

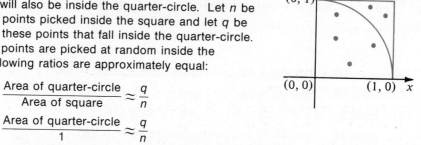

$$\frac{\text{Area of quarter-circle}}{\text{Area of square}} \approx \frac{q}{n}$$

$$\frac{\text{Area of quarter-circle}}{1} \approx \frac{q}{n}$$

$$\text{Area of whole circle} \approx 4 \times \frac{q}{n}$$

Any point (x, y) in the square region has coordinates such that $0 < x < 1$ and $0 < y < 1$. (Note that this restriction excludes points on the boundaries of the square.) A computer can pick a random point inside the unit square by choosing two random numbers x and y between 0 and 1. We let d be the distance from O to the point (x, y). By the Pythagorean Theorem, $d = \sqrt{x^2 + y^2}$. Do you see that if $d < 1$, the point lies inside the quarter-circle?

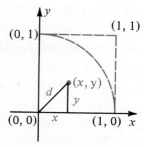

Exercises

1. Write a computer program to do all of the following:

   a. Choose n random points (x, y) inside the unit square.

   b. Test each point chosen to see whether it is also inside the quarter-circle.

   c. Keep count of the number of points (q) which *do* lie inside the quarter-circle.

   d. Print out the value of $4 \times \dfrac{q}{n}$.

2. Run the program you have written for $n = 100$, $n = 500$, and $n = 1000$. What famous number have you approximated?

Challenge

In the square with area 1 shown above, suppose that a curve drawn from $(0,0)$ to $(1,1)$ is part of the graph of the equation $y = x^2$. Write and run a program to approximate the area under the curve and inside the square. (*Hint:* Refer to Exercise 1 above. Test each point chosen in part (a) to see whether $y < x^2$. What ratio should you print out?)

11-3 *The Midpoint Formula*

You can see that the midpoint M of $\overline{AB}$ has coordinate 3, the average of 1 and 5. On page 34 you learned that if A and B have coordinates a and b, then the midpoint of $\overline{AB}$ has coordinate $\dfrac{a + b}{2}$.

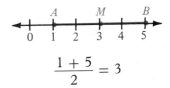

$$\frac{1 + 5}{2} = 3$$

You can use this idea twice to find the coordinates of the midpoint of a slanting segment with endpoints $P_1(x_1, y_1)$ and $P_2(x_2, y_2)$. In the diagram below, M is the midpoint of $\overline{P_1P_2}$. Horizontal and vertical segments are drawn as shown. Then T is point (x_2, y_1).

Since $\overline{MJ} \parallel \overline{P_2T}$, it follows that $\overline{MJ}$ divides $\overline{P_1P_2}$ and $\overline{P_1T}$ proportionally. Since $P_1M = \frac{1}{2}P_1P_2$, then $P_1J = \frac{1}{2}P_1T$.

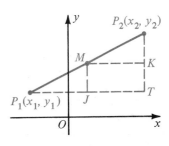

J, the midpoint of $\overline{P_1T}$, has x-coordinate $\dfrac{x_1 + x_2}{2}$.

Therefore M has x-coordinate $\dfrac{x_1 + x_2}{2}$.

By similar reasoning, M has y-coordinate $\dfrac{y_1 + y_2}{2}$.

The discussion above leads to the following theorem.

Theorem 11-3   The midpoint of the segment that joins points (x_1, y_1) and (x_2, y_2) is the point $\left(\dfrac{x_1 + x_2}{2}, \dfrac{y_1 + y_2}{2} \right)$.

Classroom Exercises

Find the coordinates of the midpoint of the segment that joins the given points.

1. $(0, 5)$ and $(6, 1)$

2. $(1, 8)$ and $(3, 8)$

3. $(-2, -4)$ and $(-8, 0)$

4. $(-1, 6)$ and $(-5, -6)$

5. $(4, 4)$ and $(6, -6)$

6. (j, k) and (r, s)

7. P_1 is point $(0, 3)$; point $(4, 7)$ is the midpoint of $\overline{P_1P_2}$. Find the coordinates of P_2.

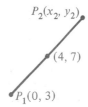

Written Exercises

Find the coordinates of the midpoint of the segment that joins the given points.

A **1.** $(-1, 4)$ and $(5, 6)$ **2.** $(0, 8)$ and $(-6, -8)$

    3. $(2, 2)$ and $(-6, 5)$ **4.** $(1.2, 2.9)$ and $(1.6, 4.5)$

    5. (a, b) and (c, d) **6.** $(a + 2, \frac{1}{5})$ and $(a - 2, \frac{7}{5})$

In Exercises 7–10, M is the midpoint of $\overline{P_1P_2}$. The coordinates of P_1 and M are given. Find the coordinates of P_2.

    7. $P_1(0, 5)$; $M(3, 7)$ **8.** $P_1(-3, 1)$; $M(-1, 0)$

    9. $P_1(8, -5)$; $M(8, 1)$ **10.** $P_1(c, d)$; $M(e, f)$

B **11.** The vertices of quadrilateral $OABC$ have the coordinates shown.

     a. Show that $OA = CB$ and $OC = AB$.

     b. What special kind of quadrilateral is $OABC$?

     c. The midpoint of $\overline{OB}$ has coordinates $(\underline{?}, \underline{?})$.

     d. The midpoint of $\overline{AC}$ has coordinates $(\underline{?}, \underline{?})$.

     e. Note that $\overline{OB}$ and $\overline{AC}$ have the same midpoint. State a theorem, from Chapter 4, that is suggested by this exercise.

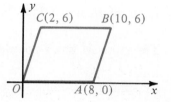

12. The vertices of right triangle TOP have the coordinates shown. M is the midpoint of $\overline{TP}$.

     a. M has coordinates $(\underline{?}, \underline{?})$.

     b. Find, and compare, the lengths MT, MP, and MO.

     c. State a theorem, from Chapter 4, suggested by this exercise.

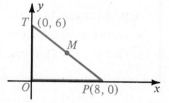

13. The vertices of quadrilateral $OABC$ have the coordinates shown. R, S, T, and U are the midpoints of the sides of the quadrilateral $OABC$. Find the midpoint of $\overline{TR}$ and the midpoint of $\overline{SU}$. What does your answer prove about quadrilateral $RSTU$?

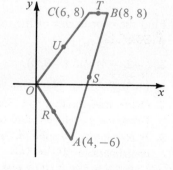

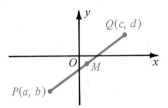

C 14. According to Theorem 11-3, the midpoint M of $\overline{PQ}$ has coordinates $\left(\dfrac{a+c}{2}, \dfrac{b+d}{2}\right)$. Verify the coordinates by using the distance formula and showing that $PM = MQ = \frac{1}{2}PQ$.

15. Given points $A(0, 3)$ and $B(5, 13)$, find the coordinates of a point T on $\overline{AB}$ such that $\dfrac{AT}{TB} = \dfrac{2}{3}$.

16. Given points $A(j, k)$ and $B(r, s)$, with B lying above and to the right of A, find the coordinates of a point T on $\overline{AB}$ such that $\dfrac{AT}{TB} = \dfrac{m}{n}$.

SELF-TEST

For each pair of points find (a) the distance between the two points and (b) the midpoint of the segment that joins the two points.

1. $(2, 3)$ and $(8, 3)$ **2.** $(1, -5)$ and $(1, 7)$

3. $(0, 0)$ and $(-6, 8)$ **4.** $(-5, 6)$ and $(3, -9)$

Write the equation of the circle described.

5. Center at the origin; radius 7

6. Center $(3, -2)$; radius 2

7. Find the center and the radius of the circle $x^2 + (y + 1)^2 = 25$.

Lines

Objectives

1. **Find the slope of the line containing two given points.**
2. **Determine whether two lines are parallel, perpendicular, or neither.**
3. **Draw the graph of a line specified by a given equation.**
4. **Write an equation of a line when given either one point and the slope of the line or two points on the line.**
5. **Determine the intersection of two lines.**

11-4 *Slope of a Line*

The slope of a line is a measure of the "steepness" of the line. In each of the diagrams below, you can think of getting from A to B by moving horizontally and then vertically. Informally, the slope is the quotient:

$$\frac{\text{change in } y}{\text{change in } x}$$

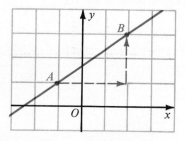 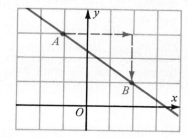

$$\text{Slope of } \overleftrightarrow{AB} = \frac{\text{change in } y}{\text{change in } x} = \frac{2}{3} \qquad \text{Slope of } \overleftrightarrow{AB} = \frac{\text{change in } y}{\text{change in } x} = \frac{-2}{3} = -\frac{2}{3}$$

You see that the slope of $\overleftrightarrow{AB}$ is positive in the first diagram and negative in the second.

The slope m of a line through $P_1(x_1, y_1)$ and $P_2(x_2, y_2)$, where $x_1 \neq x_2$, is defined as follows:

$$m = \frac{y_2 - y_1}{x_2 - x_1}$$

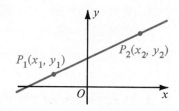

This is also the slope of $\overline{P_1 P_2}$, a segment of $\overleftrightarrow{P_1 P_2}$.

When you are given several points on a line, you can use any two of them to compute the slope. See Classroom Exercise 6.

Example: Find the slope of the segment that joins points $(-2, 4)$ and $(3, 1)$.

Solution 1: Let $(-2, 4)$ be (x_1, y_1) and $(3, 1)$ be (x_2, y_2).

$$m = \frac{y_2 - y_1}{x_2 - x_1} = \frac{1 - 4}{3 - (-2)} = \frac{-3}{5} = -\frac{3}{5}$$

Solution 2: Let $(3, 1)$ be (x_1, y_1) and $(-2, 4)$ be (x_2, y_2).

$$m = \frac{y_2 - y_1}{x_2 - x_1} = \frac{4 - 1}{-2 - 3} = \frac{3}{-5} = -\frac{3}{5}$$

The two solutions show that you can compute the slope of a segment that joins two points by choosing either point to be (x_1, y_1).

Consider any horizontal line. Note that y_1 and y_2 must be equal. Thus $m = \dfrac{y_2 - y_1}{x_2 - x_1} = \dfrac{0}{x_2 - x_1} = 0$.

The slope of any horizontal line equals zero.

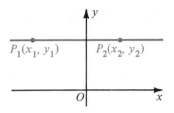

Consider any vertical line. Note that x_1 and x_2 must be equal. Since the denominator of $\dfrac{y_2 - y_1}{x_2 - x_1}$ is then 0, the expression for m has no meaning.

Slope is not defined for vertical lines.

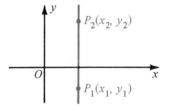

Classroom Exercises

1. Find the slope of the line.

a.

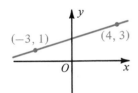

b.

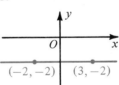

c.

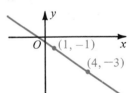

2. Is the expression positive or negative for the line below that rises to the right?

a. $y_2 - y_1$ **b.** $x_2 - x_1$ **c.** $\dfrac{y_2 - y_1}{x_2 - x_1}$

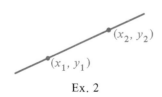

Ex. 2

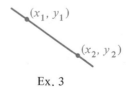

Ex. 3

3. Repeat Exercise 2 for the line above that falls to the right.

4. Does the slope of the line appear to be positive, negative, zero, or not defined?

5. a. What value does the slope of $\overleftrightarrow{RS}$ have?

b. What value does tan $n°$ have?

c. Consider the statement: If a line that rises to the right makes an acute angle of $n°$ with the x-axis, then the slope of the line is tan $n°$. Do you think this statement is true or false? Explain.

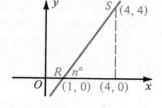

6. The purpose of this exercise is to prove that you can use any two points on a line to determine the slope of the line. Horizontal and vertical segments have been drawn as shown.

1. $\angle 1 = \angle 2$ (Why?)

2. $\angle S = \angle T$ (Why?)

3. $\triangle ASB \sim \triangle CTD$ (Why?)

4. $\dfrac{BS}{DT} = \dfrac{AS}{CT}$, or $\dfrac{BS}{AS} = \dfrac{DT}{CT}$ (Why?)

5. But slope of $\overline{AB} = \dfrac{BS}{AS}$ and slope of $\overline{CD} = \dfrac{DT}{CT}$.

6. Slope of $\overline{AB}$ = slope of $\overline{CD}$ (Why?)

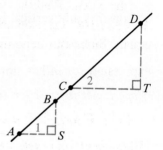

Written Exercises

Find the slope of the line through the points named. If a line is vertical, write *slope not defined.*

A

1. $(1, 3)$; $(4, 5)$ **2.** $(1, 3)$; $(-4, -5)$ **3.** $(2, -5)$; $(0, 1)$

4. $(-2, 5)$; $(0, 1)$ **5.** $(-3, 4)$; $(8, 4)$ **6.** $(2, 1)$; $(2, 8)$

7. $(-8, 8)$; $(8, -8)$ **8.** $(4, -3)$; $(-1, -1)$ **9.** $(0, 0)$; (a, b)

10. $(-c, d)$; (e, d) **11.** (f, g); (h, i) **12.** $(0, k)$; $(j, 0)$

Find the missing coordinate.

B **13.** A line with slope $\dfrac{2}{3}$ passes through points $(1, 5)$ and $(7, \underline{\ ?\ })$.

14. A line with slope $-\dfrac{4}{3}$ passes through points $(0, 5)$ and $(\underline{\ ?\ }, 13)$.

15. A line with slope k passes through points (a, b) and $(c, \underline{\ ?\ })$. Answer in terms of a, b, c, and k.

In each of Exercises 16–19, A, B, and C are vertices of a right triangle with right angle at B. Find the slopes of the legs of each triangle. What do you notice?

16. $A(1, 2)$ **17.** $A(0, -2)$ **18.** $A(0, 4)$ **19.** $A(-1, -1)$

     $B(3, 4)$ $B(1, 1)$ $B(2, 0)$ $B(1, 2)$

     $C(4, 3)$ $C(-2, 2)$ $C(0, -1)$ $C(7, -2)$

20. Given $A(-3, -4)$, $B(2, -2)$, $C(11, 2)$, and $D(21, 6)$, a student computed the slopes of $\overline{AB}$ and $\overline{CD}$ and then decided that the four given points were collinear. Was the conclusion correct? Explain.

C 21. A line passes through points $(-3, -2)$ and $(10, 5)$. Where does the line intersect the x-axis? the y-axis?

22. A line through $A(2, 3)$ and $B(4, k)$ rises to the right and makes a $60°$ angle with the x-axis. Find the value of k.

23. In $\triangle ABC$, the midpoint of $\overline{AB}$ is $D(5, 4)$, the midpoint of $\overline{BC}$ is $E(8, 10)$, and the midpoint of $\overline{AC}$ is $F(3, 6)$. Find the coordinates of A, B, and C.

11-5 *Parallel and Perpendicular Lines*

When you look at two parallel lines, you probably believe that the lines have equal slopes. You can use properties from Chapters 2, 6, and 11 to show that for two lines that rise to the right:

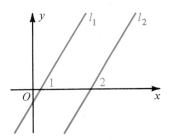

$l_1 \| l_2$ if and only if $\angle 1 = \angle 2$;

$\angle 1 = \angle 2$ if and only if $\tan \angle 1 = \tan \angle 2$;

$\tan \angle 1 = \tan \angle 2$ if and only if slope of $l_1 = $ slope of l_2.

It follows that $l_1 \| l_2$ if and only if slope of $l_1 = $ slope of l_2.

This result can also be proved for two lines that fall to the right. When the lines are parallel to the x-axis, both have slope zero.

Theorem 11-4 Two nonvertical lines are parallel if and only if their slopes are equal.

In Exercises 16–19 of the previous section, you may have noticed that perpendicular lines, too, have slopes that are related in a special way. See Classroom Exercise 9 and Exercise 16 for proofs of the following theorem.

Theorem 11-5 Two nonvertical lines are perpendicular if and only if the product of their slopes is -1.

$$m_1 \cdot m_2 = -1, \text{ or } m_1 = -\frac{1}{m_2}$$

Classroom Exercises

1. Line j has slope $\frac{3}{7}$. State the slope of every line that is:

   a. parallel to j **b.** perpendicular to j

2. Line k has slope -2, or $-\frac{2}{1}$. State the slope of every line that is:

   a. parallel to k **b.** perpendicular to k

Tell whether two lines with the given slopes are parallel, perpendicular, or neither.

3. $m_1 = \frac{2}{3}$; $m_2 = \frac{10}{15}$ 4. $m_1 = 1$; $m_2 = -1$

5. $m_1 = 2$; $m_2 = -2$ 6. $m_1 = 2$; $m_2 = -\frac{1}{2}$

7. $m_1 = 0$; $m_2 = -1$ 8. $m_1 = 0$; m_2 is not defined.

9. In this exercise you will prove the statement: If two nonvertical lines are perpendicular, the product of their slopes is -1.

   Given: l_1 has slope m_1; l_2 has slope m_2;
   $$l_1 \perp l_2$$

   Key steps of proof:

   1. Draw the vertical segment shown.

   2. $m_1 = \frac{v}{u}$ (Why?)

   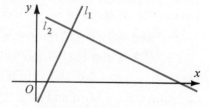

   3. $m_2 = -\frac{v}{w}$ (Explain)

   4. $m_1 \cdot m_2 = \frac{v}{u} \cdot \left(-\frac{v}{w}\right)$ (Why?)

   5. $\frac{u}{v} = \frac{v}{w}$ (State the corollary from Chapter 6.)

   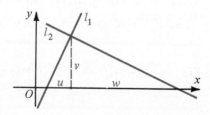

   6. $m_1 \cdot m_2 = \frac{v}{u} \cdot \left(-\frac{u}{v}\right)$ (Why?)

   $$= -1$$

Written Exercises

A   1. Given $P(-5, 0)$ and $R(2, 4)$, find the slope of each line described.

      a. $\overleftrightarrow{PR}$ **b.** any line parallel to $\overleftrightarrow{PR}$ **c.** any line perpendicular to $\overleftrightarrow{PR}$

  2. Given $S(-4, 4)$ and $T(5, -2)$, find the slope of each line described.

      a. $\overleftrightarrow{ST}$ **b.** any line parallel to $\overleftrightarrow{ST}$ **c.** any line perpendicular to $\overleftrightarrow{ST}$

3. In the left diagram below, $OABC$ is a parallelogram. What is the slope of $\overline{OC}$? of $\overline{OA}$? of $\overline{AB}$? of $\overline{CB}$?

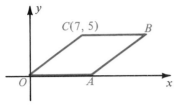

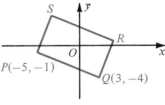

Ex. 3   Ex. 4

4. In the diagram at the right above, $PQRS$ is a rectangle. What is the slope of $\overline{PQ}$? of $\overline{QR}$? of $\overline{RS}$? of $\overline{PS}$?

5. a. What is the slope of $\overline{EF}$? of $\overline{HG}$?

   b. $\overline{HG} \| \overline{EF}$. Why?

   c. What is the slope of $\overline{EH}$? of $\overline{FG}$?

   d. $\overline{EH}$ is not parallel to $\overline{FG}$. Why?

   e. What special kind of quadrilateral is $EFGH$?

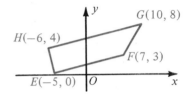

6. Quadrilateral $OGHJ$ is known to be a parallelogram.

   a. What is the slope of $\overline{HJ}$? of $\overline{OJ}$?

   b. $\overline{HJ} \perp \overline{OJ}$. Why?

   c. $\square OGHJ$ is a rectangle. Why?

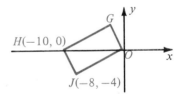

Find the slope of each side and each altitude of $\triangle ABC$.

B   **7.** $A(2, 0)$ $B(5, -1)$ $C(3, 2)$ **8.** $A(0, 0)$ $B(-5, 0)$ $C(-7, 4)$

Decide what you can prove about quadrilateral $DEFG$. Then prove it.

  9. $D(0, 0)$ $E(8, 0)$ $F(9, 5)$ $G(2, 5)$

10. $D(-3, 0)$ $E(-2, -2)$ $F(4, 1)$ $G(3, 3)$

11. $D(-2, 5)$ $E(-7, 2)$ $F(-2, -1)$ $G(3, 2)$

12. $D(0, 0)$ $E(3, 1)$ $F(1, 4)$ $G(-2, 3)$

13. Point $A(4, -3)$ lies on the circle $x^2 + y^2 = 25$. What is the slope of the line that is tangent to the circle at A? (*Hint:* Draw a diagram; also draw $\overline{OA}$.)

14. Point $B(9, 8)$ lies on the circle $(x - 3)^2 + y^2 = 100$. What is the slope of the line that is tangent to the circle at B?

C 15. Prove that the quadrilateral formed by joining, in order, the midpoints of the sides of quadrilateral $OABC$ below is a parallelogram. Find a value of l, expressed in terms of g, h, j, and k, for which the parallelogram will be a rectangle.

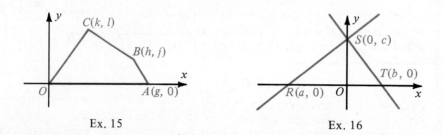

Ex. 15   Ex. 16

16. Here is another way to prove Theorem 11-5. Refer to the diagram above.

a. Use the Pythagorean Theorem to prove:

When $\overleftrightarrow{RS} \perp \overleftrightarrow{ST}$, the product of the slopes of $\overleftrightarrow{RS}$ and $\overleftrightarrow{ST}$ equals -1.

That is, prove $\dfrac{c}{-a} \cdot \dfrac{c}{-b} = -1$.

b. Use the converse of the Pythagorean Theorem to prove:

When $\dfrac{c}{-a} \cdot \dfrac{c}{-b} = -1$, then $\overleftrightarrow{RS} \perp \overleftrightarrow{ST}$.

Application _____

Vectors

The diagram illustrates three jets cruising on the same course and speed. In one hour, the jets travel from points A, B, and C to points A', B', and C'. The flight path of the jet traveling from A to A' is represented by the arrow from A to A' and is denoted $\overrightarrow{AA'}$ (not to be confused with ray $\overrightarrow{AA'}$).

The flight path $\overrightarrow{AA'}$ can also be represented by the pair of numbers (3, 4) because the jet has traveled 3 units east and 4 units north. Thus, we can write $\overrightarrow{AA'} = (3, 4)$. Since the other two jets are flying on the same course and speed, we can also write $\overrightarrow{BB'} = (3, 4)$ and $\overrightarrow{CC'} = (3, 4)$. When we recognize the familiar 3-4-5 right triangle, we see that each jet has traveled 5 units.

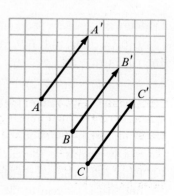

If positive numbers are used for travel east and north, negative numbers are used for movement west and south. Thus, referring to the diagram, $\overrightarrow{EE'} = (-3, -1)$. This jet has traveled $\sqrt{(-3)^2 + (-1)^2}$ units, or $\sqrt{10}$ units. What ordered pairs describe $\overrightarrow{FF'}$ and $\overrightarrow{GG'}$? How far has each of these jets traveled?

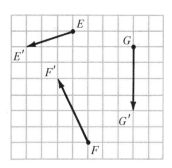

When an ordered pair is used to represent a flight path, or anything else having both magnitude and direction, the ordered pair is called a **vector.** Vectors can be added by the following simple rule:

$$(a, b) + (c, d) = (a + c, b + d)$$

To see an application of adding vectors, suppose that a jet travels from P to Q and then changes its course from Q to R. A succession of these two flight paths will move the jet to the same position as the single flight path from P to R. We abbreviate this fact by writing

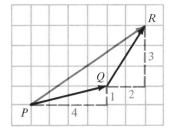

$$\overrightarrow{PQ} + \overrightarrow{QR} = \overrightarrow{PR}$$
$$(4, 1) + (2, 3) = (6, 4)$$

Because forces, like flight paths, have direction and magnitude, they too can be represented by vectors. For example, a force of 5 units acting in a northerly direction is represented by the vector $(0, 5)$ or by the arrow $\overrightarrow{KX}$ as shown at the left below. Likewise, $\overrightarrow{KY} = (3, -3)$ represents a force acting in a southeasterly direction. The magnitude of this force is $\sqrt{3^2 + (-3)^2}$ units, or $3\sqrt{2}$ units.

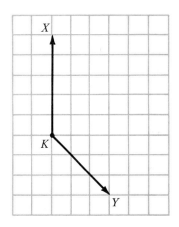
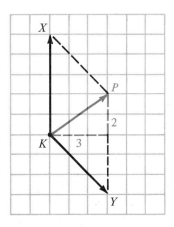

Suppose the given forces $\overrightarrow{KX}$ and $\overrightarrow{KY}$ are simultaneously acting on an object at point K. Then the single force which has the same effect as these two forces can be found by adding vectors: $\overrightarrow{KX} + \overrightarrow{KY} = (0, 5) + (3, -3) = (3, 2)$. Notice in the diagram at the right above that this vector sum is also represented by the arrow $\overrightarrow{KP}$ where P is the fourth vertex of the parallelogram with sides $\overrightarrow{KX}$ and $\overrightarrow{KY}$.

Exercises

1. Use a grid and any starting points you choose. Draw arrows to represent the following vectors:

    a. $(3, 1)$ **b.** $(4, -4)$ **c.** $(-5, 0)$ **d.** $(-3, -4)$

2. Find the magnitude of each vector in Exercise 1.

3. Find the following vector sums:

    a. $(2, 1) + (4, 3)$ **b.** $(-4, 7) + (3, -2)$ **c.** $(4, -9) + (-4, 6)$

4. A girl rides her bicycle from A to B and then from B to C. Show her trip on a coordinate grid if $\overrightarrow{AB} = (2, 9)$ and $\overrightarrow{BC} = (6, -3)$. How far is C from A?

5. The vector $(-5, 5)$ represents a force. In what compass direction is the force acting? What is the magnitude of the force?

6. Make a drawing showing an object being pulled by the two forces $\overrightarrow{KX} = (-1, 5)$ and $\overrightarrow{KY} = (7, 3)$. What single force has the same effect as the two forces acting together?

7. Repeat Exercise 6 for the forces $\overrightarrow{KX} = (2, -3)$ and $\overrightarrow{KY} = (-2, 3)$.

8. If $ABCD$ is a parallelogram, does $\overrightarrow{AB} = \overrightarrow{DC}$? Does $\overrightarrow{BC} = \overrightarrow{AD}$?

Challenge _____

Two forces act on an object. They are represented by arrows which are in the first quadrant of the coordinate plane and which start at the origin. One force, of magnitude j, is directed at an $e°$ angle with respect to the x-axis. The other force, of magnitude k, is directed at an $f°$ angle with respect to the x-axis. Show that the single force which has the same effect as these two forces combined is given by the vector:

$$(j \cos e° + k \cos f°, j \sin e° + k \sin f°).$$

11-6 *Equation of a Line*

You may remember from algebra that the graph of an equation such as $3x + 2y = 12$ is a line. We state a theorem, but omit the proof.

Theorem 11-6   The graph of any equation that can be written in the form $ax + by = c$, with a and b not both zero, is a line.

When you are given the equation of a line, it is easy to draw the line.

Example 1:   Draw the graph of $3x + 2y = 12$.

Solution: Two points determine a line. We assign two values to x or y. (We can use *any* values, but zero is often convenient.) Substitute in the given equation.

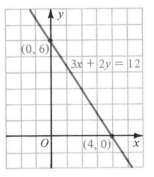

$$\text{Let } x = 0 \qquad\qquad \text{Let } y = 0$$
$$3(0) + 2y = 12 \qquad\qquad 3x + 2(0) = 12$$
$$y = 6 \qquad\qquad\qquad x = 4$$

Plot the points $(0, 6)$ and $(4, 0)$. Draw the line.

As a check, find a third point. Let $x = 2$;
$3(2) + 2y = 12$; $y = 3$.
Does $(2, 3)$ lie on the line?

The converse of Theorem 11-6 is also true. For any particular line, you can write an equation in the form $ax + by = c$. Sometimes, however, it is convenient to write an equation in another form.

Example 2:   Write an equation of the line that passes through the point $(1, 4)$ and has slope $-\dfrac{2}{5}$.

Solution: Let (x, y) represent any other point on the line. Then the slope of the line is $\dfrac{y - 4}{x - 1}$. Because the slope is given as $-\dfrac{2}{5}$,

$$\frac{y - 4}{x - 1} = -\frac{2}{5}, \text{ or } y - 4 = -\frac{2}{5}(x - 1).$$

The final equation in Example 2 is said to be in *point-slope form*. You might guess that the equation $y + 3 = 2(x - 6)$, which is in this form, describes the line that passes through point $(6, -3)$ and has slope 2. That guess would be correct. By following the method of Example 2, the next theorem can be proved.

Theorem 11-7   The equation of the line that passes through point (x_1, y_1) and has slope m is $y - y_1 = m(x - x_1)$.

Example 3: Find an equation of the line containing points (1, 3) and (4, −4).

Solution: Let (1, 3) be (x_1, y_1) and (4, −4) be (x_2, y_2).

First find the slope: $m = \dfrac{-4 - 3}{4 - 1} = -\dfrac{7}{3}$

An equation of the line through (1, 3) with slope $-\dfrac{7}{3}$ is

$$y - 3 = -\frac{7}{3}(x - 1).$$

In Example 3, if we let (4, −4) be (x_1, y_1), we would get the equation $y + 4 = -\dfrac{7}{3}(x - 4)$. This equation and $y - 3 = -\dfrac{7}{3}(x - 1)$ are equivalent. See Exercise 20.

When you are given the equations of two lines, you can graph each line and find the coordinates of the intersection point. On the other hand, you can use algebra to find the coordinates.

Example 4: Find the intersection point of the lines $2x + 3y = 11$ and $3x + y = 13$.

Solution 1: Addition or Subtraction Method: First we choose a letter to eliminate. Here we choose y:

$$
\begin{array}{rl}
2x + 3y = 11 & \text{(First equation)} \\
9x + 3y = 39 & \text{(Second equation} \times 3) \\
\hline
-7x = -28 & \text{(Subtract to eliminate } y.) \\
x = 4 & \\
2(4) + 3y = 11 & \text{(Substitute in first equation.)} \\
y = 1 &
\end{array}
$$

The point of intersection is (4, 1).

Solution 2: Substitution Method: First we choose one equation to solve for one letter.

$$
\begin{array}{ll}
3x + y = 13 & \text{(Second equation)} \\
y = 13 - 3x & \text{(Solve for } y.) \\
2x + 3(13 - 3x) = 11 & \text{(Substitute in the other equation.)} \\
-7x = -28 & \\
x = 4 & \\
y = 13 - 3(4) = 1 &
\end{array}
$$

The point of intersection is (4, 1).

Classroom Exercises

Complete the table of values for each equation.

x	y
0	?
?	0
?	2

1. $x + y = 6$
2. $x + 2y = 6$
3. $3x - y = 6$
4. $3x - 2y = 6$
5. $y = 4x$
6. $y - 3 = 2(x + 1)$

State an equation of the line through point P and having slope m.

7. $P(2, 5)$

$m = \dfrac{1}{3}$

8. $P(3, -1)$

$m = \dfrac{5}{2}$

9. $P(4, 0)$

$m = -2$

10. $P(0, 3)$

$m = \dfrac{1}{4}$

Suppose the pair of equations is to be solved by the addition or subtraction method.

a. Tell which letter you would eliminate.

b. State the two equations you would write before you actually added or subtracted.

11. $2x + 3y = 4$
    $5x + y = 10$

12. $7x + 2y = 9$
    $5x - 4y = 1$

13. $10x + 2y = 13$
    $5x + 3y = 14$

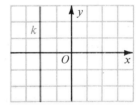

14. It is easy to see that the coordinates of all points on line k satisfy the equation:

$$x + 0y = -2, \text{ or } x = -2.$$

What is the equation of each line shown below?

a.

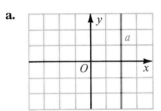

b.

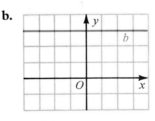

c.

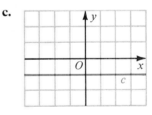

15. **a.** On one set of axes sketch the graphs of $y = \dfrac{1}{2}x$; $y = x$; $y = 2x$;

$y = -\dfrac{1}{2}x$; $y = -x$; and $y = -2x$.

b. Find the slope of each line.

c. State the slopes of these lines: $y = 3x$; $y = -3x$; $y = \dfrac{2}{5}x$.

16. **a.** On one set of axes sketch the graphs of $y = 2x$; $y = 2x + 2$; $y = 2x + 4$; and $y = 2x - 3$.

b. What is the point at which the line $y = mx + k$ intersects the y-axis?

Written Exercises

Draw the graph of each equation.

A **1.** $x + 2y = 4$ **2.** $x - y = 5$ **3.** $2x - 3y = 6$

    4. $4x + y = 8$ **5.** $y = \dfrac{1}{2}x + 3$ **6.** $3x = 4y$

Write, in point-slope form, the equation of the line that passes through *P* and has slope *m*.

7. $P(1, 4); \; m = \dfrac{2}{3}$ **8.** $P(5, 2); \; m = -\dfrac{1}{2}$

9. $P(0, -2); \; m = 3$ **10.** $P(-3, 0); \; m = \dfrac{5}{2}$

Write an equation of the line that contains the given points. (See Example 3.)

11. $(0, 2)$ and $(3, 3)$ **12.** $(3, 0)$ and $(4, 7)$

13. $(-1, 1)$ and $(3, -5)$ **14.** $(0, -4)$ and $(-3, 5)$

Find the intersection point of the two lines. (See Example 4.)

15. $x + y = 10$ **16.** $3x + 4y = 10$
      $x - y = 2$ $4x + y = 9$

17. $3x + 5y = 7$ **18.** $2x + 3y = 21$
      $x + 4y = 7$ $y = 2x - 1$

Show that the equations are equivalent by writing each in the form $ax + by = c$.

B **19.** $y - 3 = \dfrac{1}{2}(x - 4)$ **20.** $y - 3 = -\dfrac{7}{3}(x - 1)$

      $y - 1 = \dfrac{1}{2}x$ $y + 4 = -\dfrac{7}{3}(x - 4)$

In Exercises 21–28, write an equation of each line described.

21. The line through $(4, 5)$ and parallel to $\overleftrightarrow{OC}$

22. The line through $(4, 5)$ and perpendicular to $\overleftrightarrow{OC}$

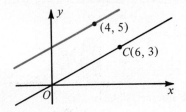

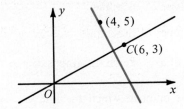

In Exercises 23–28, write an equation of each line described.

23. The line through A and the midpoint of $\overline{BC}$

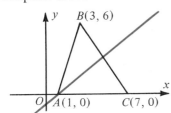

24. The perpendicular bisector of $\overline{JK}$

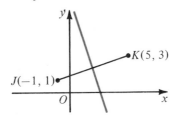

25. The line through $(2, 7)$ parallel to the x-axis

26. The line through $(1, -5)$ parallel to the y-axis

27. The x-axis   **28.** The y-axis

C **29.** Write an equation of the line that is tangent to the circle $(x - 2)^2 + (y + 7)^2 = 25$ at $(-1, -3)$.

30. The vertices of $\triangle ORS$ are $O(0, 0)$, $R(8, 0)$ and $S(2, 6)$.

    a. Find the equations of the three lines that contain the altitudes of $\triangle ORS$.

    b. Find, algebraically, the intersection point, H, of two of the altitude lines.

    c. Show that the third altitude also passes through H. You can do this by showing that the coordinates of H satisfy the equation of the third altitude line.

31. The vertices of $\triangle OAB$ are $O(0, 0)$, $A(18, 0)$ and $B(6, 6)$. The medians of $\triangle OAB$ are $\overline{OM}$, $\overline{AN}$, and $\overline{BP}$.

    a. Find the equations of $\overleftrightarrow{OM}$, $\overleftrightarrow{AN}$, and $\overleftrightarrow{BP}$.

    b. Find, algebraically, the intersection point, G, of two of the medians.

    c. Show that the third median passes through G.

    d. Find BG and BP, and show that $BG = \frac{2}{3}BP$.

32. Find the equation of the circle circumscribed about the triangle whose vertices are $P(3, 6)$, $Q(5, 2)$, and $R(-3, -6)$.

Challenge

The area of the shaded region is 12. Find the equation of line l.

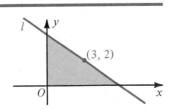

SELF-TEST

Find the slope of the line through the points named.

1. $(0, 0)$ and $(2, 5)$ **2.** $(-3, 1)$ and $(2, -3)$

3. For which kind of line is slope *not* defined, a horizontal line or a vertical line?

4. The slope of line k is $-\dfrac{7}{2}$. Find the slope of any line that is:

    a. parallel to k **b.** perpendicular to k

5. A line has the equation $y + 3 = \dfrac{1}{2}(x - 4)$.

    a. Find the slope of the line.

    b. Find the coordinates of a point on the line.

6. Draw the graph of the line $x - 2y = 4$.

7. Find the intersection point of the two lines $x + y = 3$ and $2x - y = 9$.

Coordinate Geometry Proofs

Objectives

1. Place axes on a given polygon in a convenient way.

2. Prove statements by using coordinate geometry methods.

11-7 *Organizing Coordinate Proofs*

We shall illustrate coordinate geometry proofs by proving a property studied earlier: *The midpoint of the hypotenuse of a right triangle is equidistant from the three vertices.*

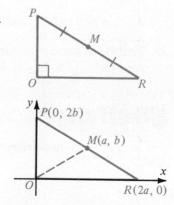

Proof:

Let $\overleftrightarrow{OR}$ and $\overleftrightarrow{OP}$ be the *x*-axis and *y*-axis.
Let R and P have the coordinates shown.
Then the coordinates of M are (a, b).

$MO = \sqrt{(a - 0)^2 + (b - 0)^2} = \sqrt{a^2 + b^2}$

$MP = \sqrt{(a - 0)^2 + (b - 2b)^2} = \sqrt{a^2 + b^2}$

$MR = \sqrt{(a - 2a)^2 + (b - 0)^2} = \sqrt{a^2 + b^2}$

$MO = MP = MR$

If you have a right triangle, such as △*POR* above, the most convenient place to put the *x*-axis and *y*-axis is along the legs of the triangle. If a triangle is not a right triangle, the two most convenient ways to place your axes are shown below.

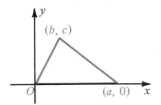

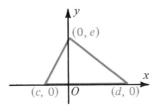

The most common ways of placing other special figures on the coordinate axes are shown below.

△*TON* is isosceles: *TO = TN*
Then *T* can be labeled (*a*, *b*).

△*GTK* is isosceles: *GT = GK*
Then *T* can be labeled (−*a*, 0).

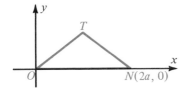

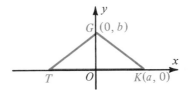

ROST is a rectangle.
Then *T* can be labeled (*a*, *b*).

ROST is a parallelogram.
Then *T* can be labeled (*a* + *b*, *c*).

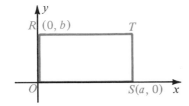

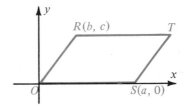

ROST is a trapezoid.
Then *T* can be labeled (*d*, *c*).

ROST is an isosceles trapezoid.
Then *T* can be labeled (*a* − *b*, *c*).

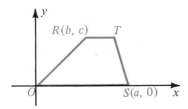

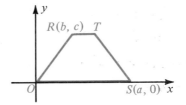

Classroom Exercises

Supply the missing coordinates without introducing any new letters.

1. Rt. △*TON* is isosceles.

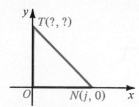

2. *POST* is a square.

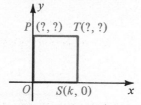

3. *NOPQ* is a rectangle.

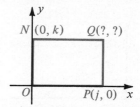

4. △*VOW* is isosceles.

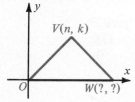

5. *MORU* is a trapezoid.

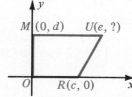

6. *BOLT* is a parallelogram.

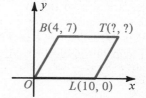

Written Exercises

Copy the figure. Supply the missing coordinates without introducing any new letters.

A **1.** Rectangle

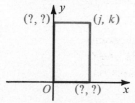

2. Square

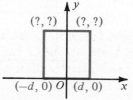

3. Parallelogram

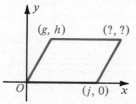

4. Parallelogram

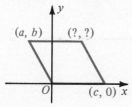

5. Isosceles trapezoid

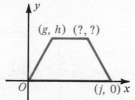

6. Isosceles triangle

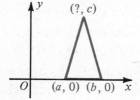

B **7.** An equilateral triangle is shown below. Express the missing coordinates in terms of *a*.

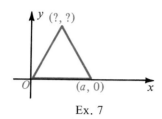

Ex. 7

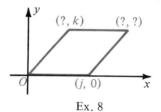

Ex. 8

8. A rhombus is shown above. Express the missing coordinates in terms of *j* and *k*.

Use slope properties to prove each statement in Exercises 9 and 10.

Example: In a plane, two lines perpendicular to the same line are parallel.

Proof:

Let the lines be $\overleftrightarrow{AB}$, $\overleftrightarrow{CD}$, and $\overleftrightarrow{EF}$;
$\overleftrightarrow{AB} \perp \overleftrightarrow{EF}$; $\overleftrightarrow{CD} \perp \overleftrightarrow{EF}$.

Let the slope of $\overleftrightarrow{EF}$ be *j*.

The slope of $\overleftrightarrow{AB} = -\dfrac{1}{j}$. (Why?)

The slope of $\overleftrightarrow{CD} = -\dfrac{1}{j}$.

$\overleftrightarrow{AB} \| \overleftrightarrow{CD}$. (Why?)

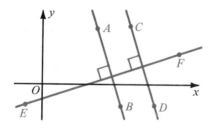

9. In a plane, two lines parallel to the same line are parallel to each other.

10. In a plane, a line perpendicular to one of two parallel lines is perpendicular to the other one also.

11. Supply the missing coordinates to prove: The segments that join the midpoints of opposite sides of any quadrilateral bisect each other. Let *N*, *H*, *M*, and *G* be the midpoints of the sides of quadrilateral *PORT*. Choose axes and coordinates as shown.

a. *M* has coordinates (_?_, _?_).

b. *N* has coordinates (_?_, _?_).

c. The midpoint of $\overline{MN}$ has coordinates (_?_, _?_).

d. *G* has coordinates (_?_, _?_).

e. *H* has coordinates (_?_, _?_).

f. The midpoint of $\overline{GH}$ has coordinates (_?_, _?_).

g. Because (_?_, _?_) is the midpoint of both $\overline{MN}$ and $\overline{GH}$, $\overline{MN}$ and $\overline{GH}$ bisect each other.

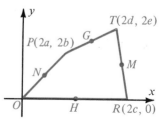

416 / *Geometry*

C 12. Draw a regular octagon. Select axes and label the coordinates of the vertices in terms of a single letter.

13. Repeat Exercise 12, but use a regular hexagon.

14. Given isosceles trapezoid *GOLD* and the axes and coordinates shown. Using the definitions of trapezoid and isosceles trapezoid, prove that $v = c$ and $u = a - b$.

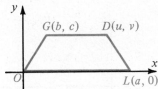

11-8 *Properties of Triangles and Quadrilaterals*

You can now use coordinate geometry to prove some theorems stated in Chapter 4.

Example: Prove Theorem 4-15: The segment that joins the midpoints of two sides of a triangle has two properties:

(1) It is parallel to the third side.
(2) Its length equals half the length of the third side.

Proof: Given $\triangle TOP$, we choose axes and coordinates as shown. Midpoint M has coordinates (b, c). Midpoint N has coordinates $(a + b, c)$.

Slope of $\overline{OP} = 0$

Slope of $\overline{MN} = \dfrac{c - c}{(a + b) - b} = 0$

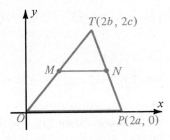

Since $\overline{MN}$ and $\overline{OP}$ have equal slopes, $\overline{MN} \parallel \overline{OP}$.

$MN = (a + b) - b = a$
$OP = 2a - 0 = 2a$
$MN = a = \dfrac{1}{2}(2a) = \dfrac{1}{2}OP$

Classroom Exercises

The purpose of Exercises 1–8 is to prove Theorem 4-18: The lines that contain the altitudes of a triangle intersect in a point.

Given $\triangle RON$ with altitude lines j, k, and l, we choose axes and coordinates as shown.

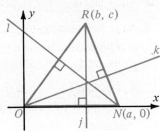

1. The equation of line j is $x = \underline{\ ?\ }$.

2. Since the slope of $\overline{RN} = \dfrac{c}{b - a}$, the slope of $k = \underline{\ ?\ }$.

3. The equation of line k is $y = \left(\dfrac{a - b}{c}\right)x$. (Explain)

4. Lines j and k intersect where $x = b$ and $y = \left(\dfrac{a - b}{c}\right)b$, or

$y = \dfrac{ab - b^2}{c}$. (Explain)

5. Since the slope of $\overline{OR} = \dfrac{c}{b}$, the slope of l is ? .

6. The equation of line l is $y = -\dfrac{b}{c}(x - a)$. (Explain)

7. Lines j and l intersect where $x = b$ and $y = -\dfrac{b}{c}(b - a)$, or

$y = \dfrac{ab - b^2}{c}$. (Explain)

8. The three altitude lines intersect in a point. Name the coordinates of that point.

Written Exercises

In Exercises 1–5, use coordinate geometry to prove the statement.

A **1.** The diagonals of a rectangle are equal.

    2. The diagonals of a parallelogram bisect each other.

B **3.** The diagonals of an isosceles trapezoid are equal.

    4. The medians drawn to the legs of an isosceles triangle are equal.

    5. A triangle is formed by joining the midpoints of the sides of an isosceles triangle. What can you prove about the smaller triangle?

C **6.** The median of a trapezoid has two properties:
(1) It is parallel to the bases.
(2) Its length equals half the sum of the lengths of the bases. (Theorem 4-14)
Write a coordinate geometry proof.

    7. The midpoints of the sides of an isosceles trapezoid are joined, in order. What can you prove about the quadrilateral formed? Write a coordinate geometry proof.

    8. What two properties can you prove for the segment that joins the midpoints of the diagonals of a trapezoid? Write coordinate geometry proofs.

    9. Use axes and coordinates as shown to prove: The medians of a triangle intersect in a point that is two thirds of the distance from each vertex to the midpoint of the opposite side. (Theorem 4-19)
(*Hints:* Find the coordinates of the midpoints; then the slopes of the medians; then the equations of the lines containing the medians.)

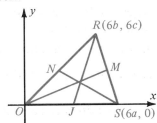

10. The diagram shows axes and coordinates that can be used for any triangle *ROS*.

C, the intersection point of the perpendicular bisectors of the sides, has coordinates $\left(3a, \dfrac{3b^2 + 3c^2 - 3ab}{c}\right)$.

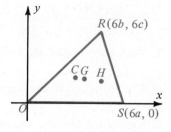

G, the intersection point of the medians, has coordinates $(2a + 2b, 2c)$. (See Exercise 9.)

H, the intersection point of the altitudes, has coordinates $\left(6b, \dfrac{6ab - 6b^2}{c}\right)$. (See the Classroom Exercises.)

Prove each statement.

a. Points C, G, and H are collinear. The line containing these points is called *Euler's Line*. (*Hint:* One way to prove this is to show that slope of $\overline{CG}$ = slope of $\overline{GH}$.)

b. $CG = \frac{1}{3}CH$

SELF-TEST

State the coordinates of point T without introducing any new letters.

1. Isosceles triangle   **2.** Parallelogram   **3.** Isosceles trapezoid

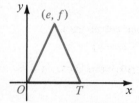   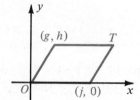   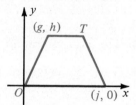

4. Show that the triangle whose vertices are $(2, 0)$, $(19, 0)$, and $(10, 15)$ is an isosceles triangle.

5. The vertices of a quadrilateral are $R(3, 7)$, $O(0, 0)$, $N(9, 0)$ and $D(12, 7)$. Show that quad. *ROND* is a parallelogram.

Chapter Summary

1. We can represent ordered pairs of numbers by points located with respect to the *x*- and *y*-axes in a coordinate plane. For any ordered pair (a, b), a is the *x*-coordinate and b is the *y*-coordinate.

2. The distance between points (x_1, y_1) and (x_2, y_2) is $\sqrt{(x_2 - x_1)^2 + (y_2 - y_1)^2}$. The midpoint of the segment joining these points is the point $\left(\dfrac{x_1 + x_2}{2}, \dfrac{y_1 + y_2}{2}\right)$.

3. The circle with center (a, b) and radius r has the equation $(x - a)^2 + (y - b)^2 = r^2$.

4. The slope m of a line through two points (x_1, y_1) and (x_2, y_2) is defined as follows: $m = \dfrac{y_2 - y_1}{x_2 - x_1}$.

   The slope of a horizontal line is zero. Slope is not defined for vertical lines.

5. Two nonvertical lines with slopes m_1 and m_2 are:

   a. parallel if and only if $m_1 = m_2$.

   b. perpendicular if and only if $m_1 \cdot m_2 = -1$.

6. The graph of any equation that can be written in the form $ax + by = c$, with a and b not both zero, is a line. The equation of the line through point (x_1, y_1) with slope m is $y - y_1 = m(x - x_1)$.

7. To prove theorems using coordinate geometry, proceed as follows:

   a. Place x- and y-axes in a convenient position with respect to a figure.

   b. Use known properties of the figure in assigning coordinates to points of the figure.

   c. Use the distance formula, the midpoint formula, and the slope properties of parallel and of perpendicular lines to prove theorems.

Chapter Review ─────────────────────

11-1 *The Distance Formula*

1. Draw the graph of each of the following points using one set of axes.

   $A(1, 3)$ $B(-1, 0)$ $C(0, 0)$ $D(0, 2)$ $E(-2, -2)$

2. The distance between points (x_1, y_1) and (x_2, y_2) is given by the formula $d = \underline{\ ?\ }$.

3. The distance between $(-4, 2)$ and $(5, 2)$ is $\underline{\ ?\ }$.

4. The distance between $(3, 3)$ and $(-2, -1)$ is $\underline{\ ?\ }$.

11-2 *Circles*

Find the center and radius of each circle.

5. $(x - 2)^2 + (y - 3)^2 = 64$ 6. $x^2 + (y + 1)^2 = 25$

Write the equations of the circles described.

7. Center $(5, 1)$ and radius 2

8. Center $(2, 5)$ and tangent to the x-axis

11-3 *The Midpoint Formula*

Find the coordinates of the midpoint of the segment that joins the given points.

9. $(5, 3)$ and $(9, 7)$ 10. $(0, 0)$ and $(-8, 6)$ 11. $(-6, 2)$ and $(2, -2)$

12. $M(-1, 0)$ is the midpoint of $\overline{AB}$. If A is $(-4, 1)$, then B is $(\underline{\ ?\ }, \underline{\ ?\ })$.

11-4 *Slope of a Line*

13. What is the slope of the line through $(-6, 3)$ and $(3, 10)$?

14. What is the slope of the line through $(4, 8)$ and $(13, 2)$?

15. A line with slope 2 passes through $(-3, 2)$ and $(1, \underline{\ ?\ })$.

16. What is the slope of a line which is parallel to the x-axis?

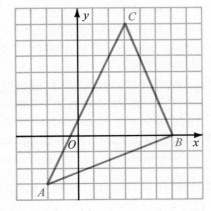

11-5 *Parallel and Perpendicular Lines*

17. What is the slope of a line parallel to $\overleftrightarrow{BC}$?

18. What is the slope of the altitude to $\overline{AC}$?

19. What is the slope of $\overleftrightarrow{AB}$?

20. Are $\overleftrightarrow{AB}$ and $\overleftrightarrow{BC}$ perpendicular?

11-6 *Equation of a Line*

21. Draw the graph of $3x + 2y = 12$.

22. Write an equation of the line that passes through $(-2, 4)$ and has slope $\dfrac{1}{2}$.

23. Write an equation of the line passing through $(-4, 2)$ and $(2, -1)$.

24. Find the intersection point of the lines $x - 3y = 3$ and $3x + y = -11$.

11-7 *Organizing Coordinate Proofs*

25. A rectangle is shown. Find the coordinates of points A and C without introducing any new letters.

26. The base of an equilateral triangle lies on the x-axis with its midpoint at the origin. If one vertex is point $(a, 0)$, then the other vertices are $(\underline{\ ?\ }, \underline{\ ?\ })$ and $(\underline{\ ?\ }, \underline{\ ?\ })$.

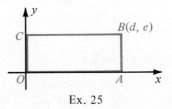

Ex. 25

11-8 *Properties of Triangles and Quadrilaterals*

Use coordinate geometry to prove each statement.

27. If rectangle $DOBC$ has $OB = 2 \cdot BC$ and E is the midpoint of $\overline{DC}$, then $\overline{OE}$ is perpendicular to $\overline{BE}$.

28. In rectangle $DOBC$, if E is the midpoint of $\overline{DC}$, then $\triangle OBE$ is isosceles.

29. The lengths of the diagonals of a square are equal.

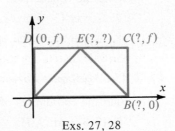

Exs. 27, 28

Chapter Test

Given point $A(4, -1)$ and point $B(1, 3)$, complete the statements.

1. The y-coordinate of point A is __?__.
2. The distance between A and B is __?__.
3. The midpoint of $\overline{AB}$ is __?__.
4. A line parallel to $\overleftrightarrow{AB}$ has slope __?__.
5. A line perpendicular to $\overleftrightarrow{AB}$ has slope __?__.
6. An equation of $\overleftrightarrow{AB}$ is __?__.
7. An equation of a line through $C(1, -1)$ and perpendicular to $\overleftrightarrow{AB}$ is __?__.
8. The circle with center B and radius 4 has the equation __?__.

9. If $L(2, -1)$ is the midpoint of $\overline{HK}$, and H is $(-5, -4)$, find the coordinates of K.
10. If three vertices of a rectangle are $(1, 3)$, $(1, -2)$, and $(6, -2)$, then the fourth vertex is $(\underline{\,?\,}, \underline{\,?\,})$.

What kind of special quadrilateral is quadrilateral *DEFG*?
11. $D(-2, -1)$; $E(4, 1)$; $F(1, 9)$; $G(-5, 7)$
12. $D(-4, 5)$; $E(2, -3)$; $F(6, 0)$; $G(0, 8)$

Find the point where the two lines intersect.
13. $3x + 2y = 4$
    $x - 2y = 4$
14. $4x - 3y = 15$
    $2x + 5y = 27$

15. Isosceles trapezoid *RSTU* and the coordinate axes are as shown below. The midpoint of the base of the trapezoid is the origin. What are the coordinates of points T and U?

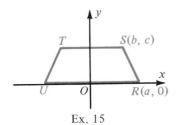

Ex. 15

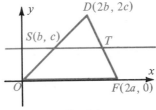

Ex. 16

16. Use coordinate geometry to prove: A line bisecting one side of a triangle and parallel to a second side bisects the third side. Use the diagram at the right above.

12

Transformations

Some Basic Mappings

Objectives

1. Recognize and use the terms *mapping, image, preimage, transformation,* and *isometry.*
2. Locate images of figures by translation, reflection, and rotation.
3. Recognize the properties of the basic mappings.

12-1 *Isometries*

Have you ever wondered how maps of the round earth can be made on flat paper? The diagram illustrates the idea behind a *polar map* of the northern hemisphere. A plane is placed tangent to a globe of the earth at its north pole N. Every point P of the globe is projected straight upward to a point P' in the plane. P' is called the **image** of P, and P is called the **preimage** of P'. The diagram shows the images of two points P and Q on the globe's equator. It also shows D', the image of a point above the equator.

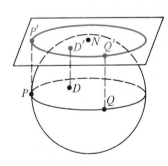

This correspondence between points of the globe's northern hemisphere and points in the plane is an example of a **mapping.** Most of the mappings we shall discuss will be correspondences between the points in a plane. These special mappings are called *transformations* and are defined as follows:

A **transformation of the plane** is a mapping with two properties:
(1) every point of the plane has exactly one image, and
(2) every point of the plane has exactly one preimage.

Suppose a transformation S maps each point with coordinates (x, y) to an image point with coordinates $(4x, y)$.

We write $S:(x, y) \rightarrow (4x, y)$.
We say S *maps* (x, y) *to* $(4x, y)$.

Transformation S maps $\triangle ABC$ to its image $\triangle A'B'C'$ as shown in the diagram.

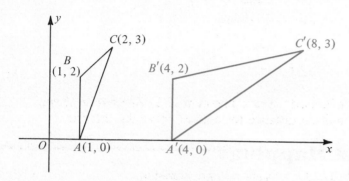

Notice that the transformation S distorts figures. $\triangle ABC$ is distorted into $\triangle A'B'C'$. Notice that distance BC is distorted into the longer distance $B'C'$.

Some transformations do *not* distort distances. These transformations are called **isometries.** If an isometry maps P to P' and Q to Q', then distance PQ equals distance $P'Q'$.

The first kind of isometry we shall study is a **translation.** The idea behind a translation is that all points of the plane slide the same distance in the same direction. For example, consider a translation T in which every point slides 7 units right and 2 units up. To indicate this, we write $T:(x, y) \rightarrow (x + 7, y + 2)$. The diagram below shows how $\triangle PQR$ is mapped by T to $\triangle P'Q'R'$. Use the distance formula to check that $PQ = P'Q'$.

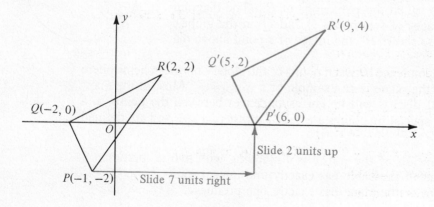

Definition of a translation, using coordinates:

Let a and b be any two real numbers. Then a transformation which maps (x, y) to $(x + a, y + b)$ is a translation.

Definition of a translation, without coordinates:

A translation is a transformation that maps any two points P and Q to P' and Q' in such a way that $PP' = QQ'$ and $PQ = P'Q'$.

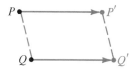

Either definition can be used to prove Theorem 12-1. In Exercise 17 you will use coordinates and the distance formula to prove the theorem.

Theorem 12-1 A translation is an isometry.

Classroom Exercises

Transformation S maps $\triangle ABC$ to $\triangle A'B'C'$.

1. What is the image of A? of B?

2. Is the transformation S an isometry? Explain your answer.

3. What is the preimage of C'?

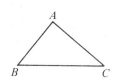
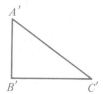

4. Use the transformation $T:(x, y) \rightarrow (x + 1, y - 2)$ in this exercise.

   a. Plot each of the following points and their images on the chalkboard.

      $A(0, 0)$ $B(3, 4)$ $C(5, 1)$ $D(-1, -3)$

   b. Find the distances AB and $A'B'$.

   c. Find the distances CD and $C'D'$.

   d. Is this transformation an isometry?

   e. What is the preimage of $(0, 0)$? of $(4, 5)$?

5. A translation maps $(2, 3)$ to $(5, 4)$.

   a. The change in x is $5 - 2$, or 3. The change in y is _?_.

   b. Find the image of $(0, 0)$.

   c. Find the image of (x, y).

Exercises 6 and 7 refer to the globe on page 423.

6. What is the image of point N?

7. Is the distance between N and P on the globe the same as the corresponding distance on the polar map? Is this mapping an isometry?

8. An isometry maps $\triangle PQR$ to $\triangle P'Q'R'$.

    a. What distances must be equal?

    b. Tell why $\triangle PQR \cong \triangle P'Q'R'$. (*Note:* An isometry is also called a *congruence mapping.*)

9. Use the mapping $(x, y) \rightarrow (x, 0)$ in this exercise.

    a. Find the images of $(2, 4)$ and $(2, 5)$.

    b. Name several preimages of $(2, 0)$.

    c. Is this mapping a transformation? Explain your answer.

Written Exercises

In Exercises 1–4, use the translation $T{:}(x, y) \rightarrow (x + 2, y - 1)$.

A **1.** Plot the following points and their images.

    $A(3, 1)$ $B(1, 5)$ $C(-5, 3)$ $D(-2, -1)$

2. Find the distances AB and $A'B'$.

3. Find the distances CD and $C'D'$.

4. Find the preimage of $(0, 0)$.

5–8. Repeat Exercises 1–4 using $T{:}(x, y) \rightarrow (x - 2, y - 4)$.

9. Use $S{:}(x, y) \rightarrow (2x, y + 1)$ in this exercise.

    a. Plot the points $A(1, 3)$ and $B(5, -1)$ and their images A' and B'.

    b. Compare the distances AB and $A'B'$. Is S an isometry?

    c. Find the coordinates of M, the midpoint of $\overline{AB}$.

    d. Find the image of M.

    e. Is the image of M the midpoint of $\overline{A'B'}$?

10. Repeat Exercise 9 using $S{:}(x, y) \rightarrow (-x, -y)$.

In Exercises 11 and 12, use $S{:}(x, y) \rightarrow (x + y, y)$.

B **11. a.** Plot the points $A(1, 1)$, $B(5, 1)$, and $C(2, 6)$ and their images A', B', and C'.

    b. Draw $\triangle ABC$ and $\triangle A'B'C'$ and find the area of each.

12. a. Plot the points $P(0, 0)$, $Q(-4, 0)$, and $R(0, -6)$ and their images P', Q', and R'.

    b. Draw $\triangle PQR$ and $\triangle P'Q'R'$ and find the area of each.

A cylindrical piece of paper is wrapped around a globe of the earth as shown. O is the center of the earth and a point P of the globe is projected along $\overrightarrow{OP}$ to a point P' of the cylinder.

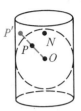

13. Describe the image of the globe's equator.

14. Is the image of the Arctic Circle the same size as the image of the equator?

15. Are distances near the equator distorted more or less than distances near the Arctic Circle?

16. Does the North Pole (point N) have an image?

C 17. Consider the translation $T:(x, y) \rightarrow (x + a, y + b)$. Let $P(x_1, y_1)$ and $Q(x_2, y_2)$ be any two points. Use the distance formula to prove that $PQ = P'Q'$.

12-2 *Reflections*

When you stand before a mirror, your image appears to be as far behind the mirror as you are in front of it. The diagram shows a transformation in which a line acts like a mirror. Points P and Q are reflected in line j to their images P' and Q'. This transformation is called a *reflection*. Line j is called the *mirror* of the reflection.

A **reflection** maps every point P to a point P' such that:
(1) If P is not on the mirror, then the mirror is the perpendicular bisector of $\overline{PP'}$.
(2) If P is on the mirror, then $P' = P$.

To abbreviate *reflection in line j*, we write M_j. To abbreviate the statement M_j *maps P to P'*, we write $M_j:P \rightarrow P'$. This may also be read as P *is mapped by* M_j *to* P', or P *is reflected in line j to* P'.

Theorem 12-2   A reflection in a line is an isometry.

Theorem 12-2 can be proved with or without the use of coordinates. If coordinates are not used, we must show that $PQ = P'Q'$ for the four cases shown below. In Exercises 19–21 you will prove Cases 2–4. Use the fact that the mirror line is the perpendicular bisector of $\overline{PP'}$ and $\overline{QQ'}$.

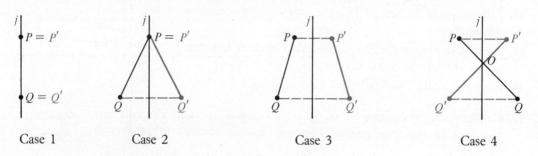

Case 1   Case 2   Case 3   Case 4

To prove the theorem by using coordinates, we assign coordinates in the plane so that the mirror line becomes the *y*-axis. Although the diagram shows P and Q on the same side of the *y*-axis, you should realize that the coordinates $x_1, y_1, x_2,$ and y_2 can be positive, negative, or zero, thereby covering the four cases shown above. In Exercise 37 you will use the distance formula to prove that $PQ = P'Q'$.

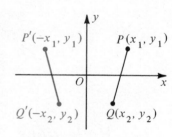

It can also be shown that a triangle is reflected to a congruent triangle, and an angle is reflected to an equal angle.

Classroom Exercises

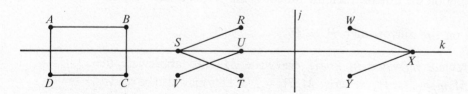

Complete the following.

1. M_k stands for ___?___.

2. $M_k : A \to$ __?__

3. $M_k : \overline{AB} \to$ __?__

4. $M_k : \overline{AD} \to$ __?__

5. $M_k : S \to$ __?__

6. $M_k : \angle RST \to$ __?__

7. $M_j : R \to$ __?__

8. $M_j : \overline{RS} \to$ __?__

9. $M_j : \underline{\ ?\ } \to \overline{XY}$

Points *A–D* are reflected in the *x*-axis. Points *E–H* are reflected in the *y*-axis.
State the coordinates of the images.

10.

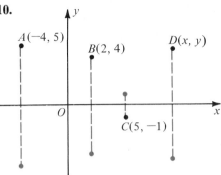

11.

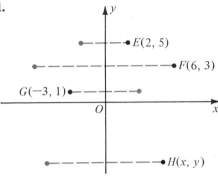

Sketch each figure on the chalkboard. With a different color, sketch its image,
using the dashed line as the mirror of reflection.

12.

13.

14.

15.

Written Exercises

Copy each figure on squared paper. Then draw the image by reflection in line *k*.

A **1.**

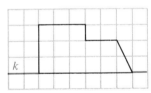

2.

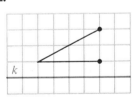

3.

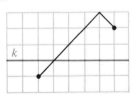

4.

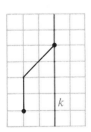

5.

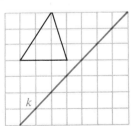

6.

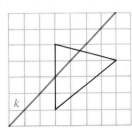

Write the coordinates of the image of each point by reflection in
(a) the *x*-axis and in (b) the *y*-axis.

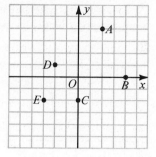

   7. *A*   **8.** *B*   **9.** *C*

10. *D*   **11.** *E*   **12.** *O*

B **13–18.** Write the coordinates of the image of each point in
Exercises 7–12 by reflection in the line $y = x$.

In Exercises 19–21, refer to the diagrams on page 428. Given the reflection
$M_j : \overline{PQ} \rightarrow \overline{P'Q'}$, write the key steps of a proof that $PQ = P'Q'$ for each case.
19. Case 2   **20.** Case 3   **21.** Case 4

Copy the figures shown. Use a straightedge and compass.
22. Find the image of *A* by M_t.   **23.** Find the mirror line *t* so that
                                         $M_t : B \rightarrow B'$.

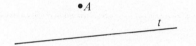

24. a. Reflections can be used to solve problems in miniature
golf. A ball which does not have much spin will roll off a
wall so that $\angle 1 = \angle 2$. To roll a ball off a wall and into
hole *H*, aim for *H'*, the mirror image of *H* by reflection in
the wall. Why does $\angle 1 = \angle 3$? Why does $\angle 3 = \angle 2$?
You can conclude that $\angle 1 = \angle 2$ and that the ball will
roll to *H*.

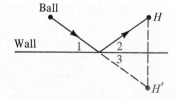

   b. Show how to score a hole-in-one on the fifth hole of the
golf course shown by rolling the ball off one wall.

   c. Repeat (b) but roll the ball off two walls.

   d. Repeat (b) but roll the ball off three walls.

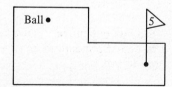

25. The line with equation $y = 2x + 3$ is reflected in the *y*-axis. What is the
equation of the image line?

26. The line with equation $y = x + 5$ is reflected in the *x*-axis. What is the
equation of the image line?

C **27.** Draw the *x*- and *y*-axes and the line *l* with equation $y = -x$. Plot several points and their images by M_l. What is the image of (x, y)?

28. Draw the *x*- and *y*-axes and the vertical line *j* through $(5, 0)$. Find the images under M_j of the following points.

    a. $(4, 3)$ **b.** $(0, -2)$ **c.** $(-3, 1)$ **d.** (x, y)

29. Repeat Exercise 28 letting *j* be the horizontal line through $(0, 6)$.

For each exercise make a sketch showing A' as the mirror image of point A in a line *k*. Then find the equation of line *k*.

	30.	**31.**	**32.**	**33.**	**34.**	**35.**	**36.**
A	$(2, 3)$	$(5, 0)$	$(1, 4)$	$(4, 0)$	$(5, 1)$	$(0, 2)$	$(-1, 2)$
A'	$(-2, 3)$	$(9, 0)$	$(3, 4)$	$(4, 6)$	$(1, 5)$	$(4, 6)$	$(4, 5)$

37. Use the distance formula to prove Theorem 12-2 by coordinate geometry methods.

Application

Mirrors

If a ray of light strikes a mirror at an angle of 30°, it will be reflected off the mirror at an angle of 30° also. The angle between the mirror and the reflected ray always equals the angle between the mirror and the initial light ray. In the diagram at the left below, $\angle 2 = \angle 1$.

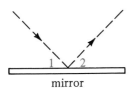

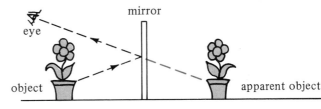

We see objects in a mirror when the reflected light ray reaches the eye. The object appears to lie behind the mirror as shown in the diagram at the right above.

You can see all of yourself in a mirror which is only half as tall as you are if the mirror is in a position as shown. You see the top of your head at the top of the mirror and your feet at the bottom of the mirror. If the mirror is too high or too low, you will not see your entire body.

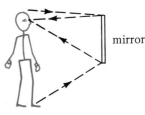

A periscope uses mirrors to enable a viewer to see above the line of sight. The diagram at the right is a simple illustration of the principle used in a periscope. It has two mirrors, parallel to each other, at the top and at the bottom. The mirrors are placed at an angle of 45° with the horizontal. Horizontal light rays from an object entering at the top are reflected down to the mirror at the bottom. They are then reflected to the eye of the viewer.

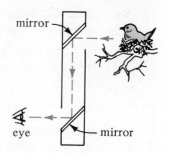

Exercises

1. What are the measures of the angles which the initial light ray and the reflected light rays make with the mirrors in the diagram of the periscope shown above?

2. If you can see the eyes of someone when you look into a mirror, is it *always* true that the other person can see your eyes in that same mirror?

3. A person with eyes at *A*, 150 cm above the floor, faces a mirror 1 m away. The mirror extends 30 cm above eye level. How high can the person see on a wall 2 m behind point *A*?

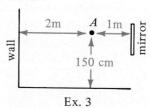

Ex. 3

4. Prove that the height of your image in a mirror is one-half your height when you are just able to see the top of your head at the top of a mirror and your feet at the bottom.

5. Prove that the point *D* which is as far behind the mirror as the object *A* is in front of the mirror lies in $\overleftrightarrow{BC}$. (*Hint:* Show that ∠*CBE* and ∠*EBD* are supplementary.)

6. Show that the light ray follows the shortest possible path by proving that for any point *E* on the mirror (other than *B*) *AE* + *EC* > *AB* + *BC*. (*Hint:* *AE* + *EC* = ? + *EC*)

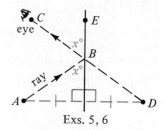

Exs. 5, 6

12-3 *Rotations and Half-Turns*

A *rotation* is a transformation suggested by a rotating Ferris wheel. When the Ferris wheel moves, each chair rotates to a new position. When the wheel stops, the position of a chair (P') can be referred to mathematically as the image of the initial position of the chair (P).

For the counterclockwise rotation shown about point *O* through 90°, we write $R_{O,90}$. A counterclockwise rotation is considered positive and a clockwise rotation is considered negative. If the red chair is rotated about *O* clockwise until it moves into the position of the black chair, the rotation is denoted by $R_{O,-90}$.

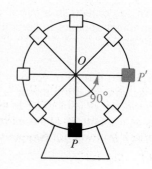

A 90° counterclockwise rotation about center O is considered the same as a 270° clockwise rotation about O. (See the left diagram below.) A 390° rotation about O is the same as a 30° rotation about O. (See the right diagram below.) What matters is not the actual rotating motion, but the final correspondence $P \rightarrow P'$.

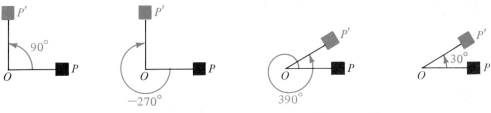

$$R_{0,90} = R_{0,-270}$$
Notice: $90 - 360 = -270$

$$R_{0,390} = R_{0,30}$$
Notice: $390 - 360 = 30$

In the following definition of a rotation, we shall allow angle measures to be positive or negative and more than 180° in absolute value.

A **rotation** about point O through $x°$ is a transformation such that:
(1) If point P is different from O, then $OP' = OP$ and $\angle POP' = x°$.
(2) If point P is the same as O, then $P' = O$.

Theorem 12-3 A rotation is an isometry.

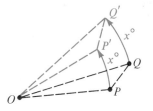

Given: $R_{0,x}$ maps P to P' and Q to Q'.

Prove: $PQ = P'Q'$

Key steps of proof:
1. $OP = OP'$; $OQ = OQ'$ (Definition of rotation)
2. $\angle POP' = \angle QOQ' = x°$ (Definition of rotation)
3. $\angle POQ = \angle P'OQ'$ (Subtraction Postulate – subtract $\angle P'OQ$)
4. $\triangle POQ \cong \triangle P'OQ'$ (SAS)
5. $PQ = P'Q'$ (Corr. parts of $\cong$ ⧍ are $=$.)

A rotation about point O through 180° is called a **half-turn** about O and is usually denoted by H_0 instead of $R_{0,180}$. The diagram shows $\triangle PQR$ and its image $\triangle P'Q'R'$ by H_0. Notice that O is the midpoint of $\overline{PP'}$, $\overline{QQ'}$, and $\overline{RR'}$.

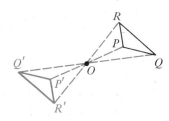

Classroom Exercises

State another name for each rotation.

Example: $R_{O,100}$

Solution: For example, $R_{O,-260}$ and $R_{O,460}$

1. $R_{O,10}$ **2.** $R_{O,50}$ **3.** $R_{O,-40}$ **4.** $R_{O,-90}$

5. $R_{O,380}$ **6.** $R_{O,400}$ **7.** $R_{O,-180}$ **8.** $R_{O,360}$

In the diagram for Exercises 9–13, O is the center of equilateral $\triangle PST$. State the images of points P, S, and T for each rotation.

9. $R_{O,120}$ **10.** $R_{O,-120}$ **11.** $R_{O,360}$

12. What is the image of S by $R_{T,60}$?

13. What is the image of P by $R_{T,-60}$?

Exs. 9-13

Draw a coordinate grid on the chalkboard and plot the following points: $A(2, 0)$, $B(2, 1)$, $C(4, 1)$, $D(4, -2)$.

14. Plot the images of these points by a half-turn about the origin O.

15. Plot the images of these points by $R_{O,90}$.

16. Plot the images of these points by $R_{O,-90}$.

Written Exercises

State another name for each rotation.

A **1.** $R_{O,120}$ **2.** $R_{O,-5}$ **3.** $R_{O,-150}$ **4.** $R_{O,800}$

In the diagram for Exercises 5–9, $ABCD$ is a square with center O. Give the images of points A, B, C, and D for each rotation.

5. $R_{O,90}$ **6.** $R_{O,-90}$ **7.** H_O

8. What is the image of C by $R_{D,90}$?

9. What is the image of D by $R_{C,-90}$?

Complete each statement below by using one of the words *reflection, translation, rotation,* or *half-turn.*

10. A ___?___ maps Triangle (1) to (2).

11. A ___?___ maps Triangle (1) to (3).

12. A ___?___ maps Triangle (1) to (4).

13. A ___?___ maps Triangle (1) to (5).

14. A ___?___ maps Triangle (2) to (4).

15. A ___?___ maps Triangle (2) to (7).

16. A ___?___ maps Triangle (4) to (6).

17. A ___?___ maps Triangle (4) to (8).

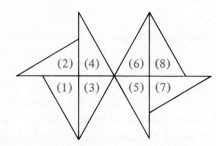

Copy the figure on squared paper. Draw the image of the figure by the rotation specified.

B 18. $R_{O,90}$ **19.** $R_{O,-90}$ **20.** H_O

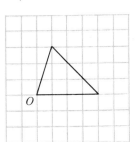 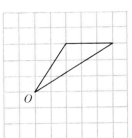 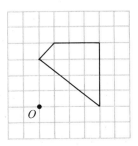

21. a. Draw a coordinate grid with origin O and plot the points $A(5,0)$ and $B(2,4)$.

    b. Plot A' and B', the images of A and B by H_O.

    c. Use the distance formula to verify that $AB = A'B'$.

    d. Use the midpoint formula to verify that O is the midpoint of $\overline{AA'}$ and $\overline{BB'}$.

    e. What special kind of quadrilateral is $ABA'B'$? Why?

22. a. Draw a coordinate grid with origin O and plot the points $A(0,3)$ and $B(4,1)$.

    b. Plot A' and B', the images of A and B by $R_{O,90}$.

    c. Compare the slopes of $\overleftrightarrow{AB}$ and $\overleftrightarrow{A'B'}$. What does this tell you about these lines?

    d. Without using the distance formula, you know that $A'B' = AB$. What theorem tells you this?

23. A half-turn about $(3, 2)$ maps P to P'. Where does this half-turn map the following points?

    a. P' **b.** $(0, 0)$ **c.** $(3, 0)$

    d. $(1, 4)$ **e.** $(-2, 1)$ **f.** (x, y)

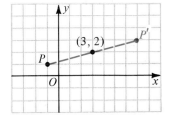

24. $\triangle ABC$ and $\triangle DCE$ are equilateral.

    a. $R_{C,-60}$ maps A to _?_ and D to _?_ .

    b. $AD = BE$ because a rotation is an _?_ .

C 25. Point P, line l, and $\odot O$ are given. You are to construct an
equilateral triangle with one vertex at point P, one vertex on
line l, and one vertex on $\odot O$. Carry out the construction
step-by-step.

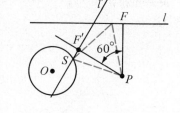

Construct $\overline{PF} \perp l$.

Construct a ray so that a 60° angle is formed at P. On this
ray take $PF' = PF$.

Construct a line $l' \perp$ to $\overline{PF'}$ at F'.

Just as $R_{P,60}$ maps F into F', it maps every other point of l
into a point of l'. In particular, there is some point of l that
is mapped into point S. $\overline{PS}$ is one side of the desired
triangle. Complete the construction.

26. Given three parallel lines, construct an equilateral triangle with one vertex on
each line.

SELF-TEST

Transformation S maps $\triangle PQR$ to $\triangle P'Q'R'$.

1. The image of Q is __?__.

2. The preimage of R' is __?__.

3. If S is an isometry, then $PQ = $ __?__.

4. What are the coordinates of the image of $(2, 3)$ by the translation
$T:(x, y) \rightarrow (x + 1, y - 2)$?

5. A reflection in line k maps A to A'. If A is not on line k, then k is the
perpendicular bisector of __?__.

6. Give two other names for the rotation $R_{O,-30}$.

Complete.

7. $M_y : A \rightarrow$ __?__

8. $M_x : B \rightarrow$ __?__

9. $M_x : \overline{DC} \rightarrow$ __?__

10. $M_y : \overline{OA} \rightarrow$ __?__

11. $H_O : K \rightarrow$ __?__

12. $H_O : \overline{CO} \rightarrow$ __?__

13. $R_{O,90}$ maps M to __?__.

14. $R_{O,-90}$ maps $\triangle MCO$ to $\triangle$ __?__.

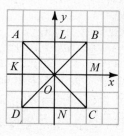

Mappings and Products of Mappings

Objectives

1. Recognize and use the terms *symmetry, product, inverse,* and *identity* in relation to mappings.
2. Locate images of figures by products of mappings and by dilations.

12-4 *Symmetry*

Each of the figures below has **line symmetry.** This means that for each figure there is a *symmetry line k* such that the reflection M_k maps the figure into itself.

 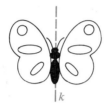 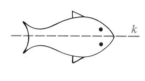

Each of the figures below has **point symmetry.** This means that for each figure there is a *symmetry point O* such that the half-turn H_O maps the figure into itself.

Besides having a symmetry point, the middle figure above has a vertical symmetry line and a horizontal symmetry line.

A third kind of symmetry is **rotational symmetry.** A figure has rotational symmetry if there is a rotation which maps the figure into itself. The figure shown has 90° rotational symmetry because $R_{O,90}$ maps the figure into itself. It also has 270° rotational symmetry because $R_{O,270}$ maps the figure into itself. Finally, it has 180° rotational symmetry, which is just another name for point symmetry.

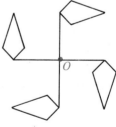

One final example of symmetry occurs when the entire plane is filled with a repeating pattern. For example, see the zigzag pattern in the Extra on page 454. Imagine that this pattern continues indefinitely to fill the whole plane. Then you can translate the whole pattern two rows up or down and it will map into itself. You can also translate the pattern an even number of columns left or right. Do you see that you can also translate the pattern along slanted lines? We say that this pattern has **translational symmetry.**

Classroom Exercises

Tell how many symmetry lines each figure has.

1.   **2.**   **3.**   **4.**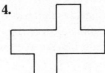

5. Which figures above have point symmetry?

6. Describe all of the rotational symmetries of Figure 2 above.

7. Describe all of the rotational symmetries of Figure 3 above.

Draw each figure on the chalkboard and describe all of its symmetries.

8. isosceles triangle   **9.** parallelogram

10. rectangle   **11.** rhombus

12. Imagine that the pattern shown fills the entire plane. Does the pattern have the symmetry named?

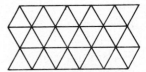

    a. translational symmetry   **b.** line symmetry

    c. point symmetry   **d.** rotational symmetry

Written Exercises

For each figure in Exercises 1–4:

a. State how many symmetry lines each figure has.

b. State whether or not the figure has a symmetry point.

c. List all the rotational symmetries of each figure between 0° and 360°. If none, say so.

A 1.    **2.**   **3.**   **4.**

5. Which capital letters of the alphabet have just one line of symmetry? (One answer is "D".)

6. Which capital letters of the alphabet have two lines of symmetry?

7. Which capital letters of the alphabet have a point of symmetry?

Copy the line *k* and the figure shown on squared paper. Then complete the figure so that *k* is a symmetry line.

8.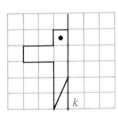

9.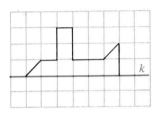

10.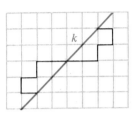

Copy the point *O* and the figure shown on squared paper. Then complete the figure in such a way that *O* is a point of symmetry.

B 11.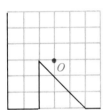

12.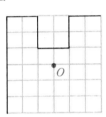

13.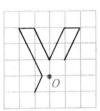

Copy the figure shown. Then complete the figure so that it has the symmetries specified.

14.

60°, 120°, and 180° rotational symmetry

15.

90°, 180°, and 270° rotational symmetry

16.

2 symmetry lines and 1 symmetry point

Draw a figure if there is one that meets the conditions. Otherwise, write *Not possible.*

17. A trapezoid with no symmetry lines

18. A trapezoid with exactly one symmetry line

19. A trapezoid with a symmetry point

20. A parallelogram with four symmetry lines

21. A parallelogram with two symmetry points

22. A parallelogram with exactly one symmetry line

23. A polygon with exactly five symmetry lines

24. A polygon with exactly six symmetry lines

25. If you use tape to hinge together two pocket mirrors as shown and place the mirrors at a 120° angle, then a penny placed between the mirrors will be reflected giving a pattern with 120° and 240° rotational symmetry.

   a. What kinds of symmetries occur when the mirrors are at a right angle?

   b. Experiment by forming various angles with two mirrors. Be sure to try 60°, 45°, and 30° angles. Record the number of pennies you see, including the actual penny.

26. You can make a mosaic by tracing around any quadrilateral, placing copies of the quadrilateral systematically as shown.

   a. The mosaic shown has many symmetry points but none of these are at vertices of the quadrilateral. Where are they?

   b. What other kind of symmetry does this mosaic have?

The coordinates of the vertices of a quadrilateral are given.

a. State the equations of all symmetry lines, if there are any.

b. State the coordinates of a symmetry point, if there is one.

C **27.** (3, 2); (−3, 2); (−3, −2); (3, −2)   **28.** (1, 1); (−5, 1); (−5, −3); (1, −3)

29. (2, 5); (−2, 2); (−2, −3); (2, 0)   **30.** (2, 3); (0, 0); (4, 1); (6, 4)

12-5 *Products of Mappings*

The diagram below shows figure F reflected in line j to figure F', and F' then reflected in line k to figure F''.

 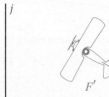

F j F' k F''

   When you study the diagram, you should think not only of the separate mappings M_j and M_k which were combined, but also of the single mapping from F to F'' that the combination produced. This single mapping is called the **product** of M_j and M_k and is denoted by $M_j \circ M_k$. The symbol $M_j \circ M_k$ may also be read as M_j *circle* M_k, or M_j *followed by* M_k. We write $M_j \circ M_k : F \rightarrow F''$ to abbreviate $M_j \circ M_k$ *maps* F *to* F''.

Did you notice that the product of M_j and M_k shown appears to be a translation mapping F to F''? The next theorem proves that this is so.

Theorem 12-4   A product of reflections in parallel lines is a translation.

Given: $j \parallel k$

Prove: $M_j \circ M_k$ is a translation.

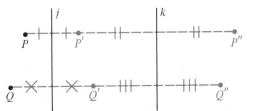

Analysis:

To prove $M_j \circ M_k$ is a translation, we must use the definition of a translation and show (1) $PP'' = QQ''$ and (2) $PQ = P''Q''$.

1. By the definition of reflection, equal segments can be marked as shown. Thus $PP'' = QQ'' =$ twice the distance between j and k.

2. Since M_j and M_k are isometries, $PQ = P'Q'$ and $P'Q' = P''Q''$. Thus $PQ = P''Q''$.

Theorem 12-5   A product of reflections in intersecting lines is a rotation.

Given: j intersects k forming an angle of $y°$ at O.

Prove: $M_j \circ M_k = R_{O,2y}$

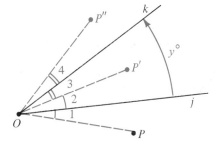

Analysis:

To prove $M_j \circ M_k = R_{O,2y}$, we must use the definition of a rotation and show (1) $OP'' = OP$ and (2) $\angle POP'' = 2y°$.

1. Since M_j is a reflection, $OP = OP'$ and $\angle 1 = \angle 2$.

2. Since M_k is a reflection, $OP' = OP''$ and $\angle 3 = \angle 4$.

3. Thus $OP = OP''$ and $\angle POP'' = \angle 1 + \angle 2 + \angle 3 + \angle 4$
$$= 2\angle 2 + 2\angle 3$$
$$= 2(\angle 2 + \angle 3) = 2y°$$

As you work with transformations, you will see that their products behave very much like products of numbers. There is one important exception, however, where the product of transformations is not at all like the product of numbers. For numbers, the products *ab* and *ba* are equal. But for transformations, the products $S \circ T$ and $T \circ S$ are usually not equal. The following example shows this.

Example: Show that $M_j \circ H_O \neq H_O \circ M_j$.

Solution: Study the two diagrams below and notice that the two products map P to different points.

$$M_j \circ H_O \qquad\qquad\qquad H_O \circ M_j$$

The previous example shows that the order in a product of transformations can be very important, but this is not always the case. For example, if S and T are two translations, then order is not important, since $S \circ T = T \circ S$. See Exercises 22 and 24.

Classroom Exercises

Copy the figure on the chalkboard and find its image by $M_j \circ M_k$. Then copy the figure again and find its image by $M_k \circ M_j$.

1.
2.

Complete the following:

3. $M_x \circ M_y : A \to$?
4. $M_x \circ M_y : D \to$?
5. $H_O \circ M_y : B \to$?
6. $M_y \circ H_O : B \to$?
7. $H_O \circ H_O : A \to$?
8. $M_y \circ M_y : C \to$?

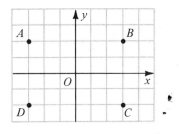

Written Exercises

X, Y, and Z are midpoints of the sides of $\triangle ABC$. Complete each statement.

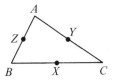

A **1.** $H_Z \circ H_X : A \to$?   **2.** $H_X \circ H_Y : B \to$?
    3. $H_Y \circ H_X : A \to$?   **4.** $H_X \circ H_X : C \to$?
    5. $H_X \circ H_Z : C \to$?   **6.** $H_B \circ H_X : B \to$?

7. Copy the figure above and locate point D, the image of B by the half-turn H_Y. What kind of figure is quadrilateral $ABCD$?

For each exercise copy the figure twice. Then show the image of the red flag by each of the products given.

8. a. $H_A \circ H_B$ **9. a.** $M_j \circ H_O$ **10. a.** $M_j \circ M_k$
   b. $H_B \circ H_A$ **b.** $H_O \circ M_j$ **b.** $M_k \circ M_j$

For each exercise draw a grid, show $A(3, 1)$, and find the coordinates of the image point.

B **11.** $M_x \circ M_y : (3, 1) \to ($? , ? $)$ **12.** $M_x \circ M_y : (4, -2) \to ($? , ? $)$
    13. $H_O \circ M_y : (1, 2) \to ($? , ? $)$ **14.** $H_A \circ H_O : (3, 0) \to ($? , ? $)$
    15. $H_O \circ H_A : (3, 0) \to ($? , ? $)$ **16.** $H_A \circ M_x : (0, 1) \to ($? , ? $)$
    17. $H_O \circ H_A : (-1, -1) \to ($? , ? $)$ **18.** $H_O \circ H_A : (-3, -1) \to ($? , ? $)$
    19. $M_y \circ H_A : (-3, 2) \to ($? , ? $)$ **20.** $H_A \circ M_y : (1, 0) \to ($? , ? $)$

21. A *glide reflection* is the product of a translation and a reflection in a line parallel to the direction of the translation.

a. Consider the translation T which maps all points 3 cm upward. Now copy the footprint F and sketch its image F'' by $T \circ M_j$.

b. Also sketch the image of F'' by $T \circ M_j$.

If two figures are congruent, it is possible to prove that one is the image of the other by either a reflection, translation, rotation, or glide reflection. There are only these four isometries of the plane.

Use the following mappings in Exercises 22–29.

$T:(x, y) \to (x + 5, y + 3)$ $M_x:(x, y) \to (x, -y)$ $H_O:(x, y) \to (-x, -y)$

$S:(x, y) \to (x + 1, y + 4)$ $M_y:(x, y) \to (-x, y)$

For Exercises 22–24, let $A = (4, 1)$, $B = (1, 5)$, and $C = (0, 1)$. Draw $\triangle ABC$ and its images by the given products.

22. a. T followed by S **23. a.** $M_x \circ H_O$ **24. a.** $M_x \circ M_y$

  b. S followed by T **b.** $H_O \circ M_x$ **b.** $M_y \circ M_x$

Complete.

25. $T \circ S:(x, y) \to (\underline{?}, \underline{?})$

26. $M_x \circ H_O:(x, y) \to (\underline{?}, \underline{?})$

27. $M_x \circ M_y:(x, y) \to (\underline{?}, \underline{?})$

28. Your answer to Exercise 27 shows that $M_x \circ M_y$ is equivalent to $\underline{?}$.

29. What special result does Theorem 12-5 give when $j \perp k$?

A theorem from advanced geometry states that if two figures are congruent, then one can be mapped to the other by a single reflection or by a product of two or three reflections.

C 30. Copy the figure and construct lines j and k so that

$M_j \circ M_k:\triangle ABC \to \triangle A''B''C''$

31. Copy the figure and construct lines j, k, and l so that

$M_j \circ M_k \circ M_l:\triangle ABC \to \triangle A'''B'''C'''$

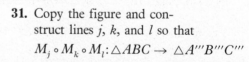

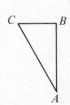

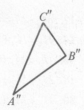

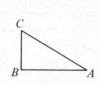

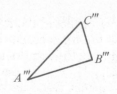

12-6 *Inverses and Identities*

Suppose that the pattern below continues indefinitely far to both the left and the right. Then the translation T that slides each runner one place to the right is a symmetry. The translation which slides every runner one place to the *left* is called the **inverse** of T, and is denoted T^{-1}. Notice that T followed by T^{-1} keeps all points fixed:

$$T \circ T^{-1} : P \to P$$

The product $T \circ T$, usually denoted T^2, slides each runner two places to the right.

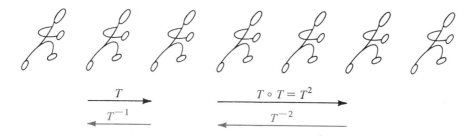

Any mapping $S \circ S^{-1}$ which maps every point to itself is called the **identity transformation *I***. The words "identity" and "inverse" are used for mappings in much the same way that they are used for numbers.

Relating Geometry and Algebra

For products of numbers, 1 is the identity. $a \cdot 1 = a$
For products of mappings, I is the identity. $S \circ I = S$

For numbers, the inverse of a is written a^{-1} $\left(\text{or } \dfrac{1}{a}\right)$. $a \cdot a^{-1} = 1$

For mappings, the inverse of S is written S^{-1}. $S \circ S^{-1} = I$

For numbers, $a^{-1} \cdot a^{-1} = a^{-2}$, and a^{-2} is the inverse of a^2.
For mappings, $S^{-1} \circ S^{-1} = S^{-2}$, and S^{-2} is the inverse of S^2.

In general, the inverse of a transformation T is defined as the transformation S such that $T \circ S = I$. If we apply this definition to the figure at the right, we see that $M_j \circ M_j = I$ so that the inverse of M_j is M_j itself. In symbols, $M_j^{-1} = M_j$.

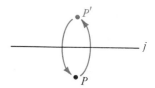

Example 1: Given the transformation $T:(x, y) \rightarrow (2x, 2y)$, find T^{-1}.

Solution: $T^{-1}:(x, y) \rightarrow (\frac{1}{2}x, \frac{1}{2}y)$

Example 2: Given the transformation $S:(x, y) \rightarrow (x + 5, y - 2)$, find S^{-1}.

Solution: $S^{-1}:(x, y) \rightarrow (x - 5, y + 2)$

Classroom Exercises

The symbol 2^{-1} stands for the multiplicative inverse of 2, or $\frac{1}{2}$. Likewise, give the value of each of the following.

1. 3^{-1} **2.** 7^{-1} **3.** $(\frac{4}{5})^{-1}$ **4.** $(2^{-1})^{-1}$

The translation T maps all points five units right. Describe each of the following transformations.

5. T^2 **6.** T^3 **7.** T^{-1}

8. T^{-2} **9.** $T \circ T^{-1}$ **10.** $(T^{-1})^{-1}$

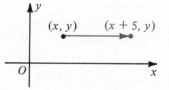

The rotation R maps all points $120°$ about G, the center of equilateral $\triangle ABC$. Give the image of A by each of the following.

11. R **12.** R^2 **13.** R^3

14. R^6 **15.** R^{-1} **16.** R^{-2}

17. $R^2 \circ R^{-2}$ **18.** $R^2 \circ R^{-3}$ **19.** R^{100}

20. What number is the identity for multiplication?

21. The product of any number t and the identity for multiplication is ___?___.

22. The product of any transformation T and the identity transformation is ___?___.

Written Exercises

Give the value of each of the following.

A **1.** 4^{-1} **2.** 9^{-1} **3.** $(\frac{2}{3})^{-1}$ **4.** $(5^{-1})^{-1}$

The rotation R maps all points $90°$ about O, the center of square $ABCD$. Give the image of A by each of the following.

5. R^2 **6.** R^3 **7.** R^4

8. R^{-1} **9.** R^{-2} **10.** R^{-3}

11. $R^{-3} \circ R^3$ **12.** R^5 **13.** R^{50}

Complete.

14. By definition, the identity mapping *I* maps every point *P* to ___?___.

15. $H_0{}^2$ is the same as the mapping ___?___.

16. The inverse of H_O is ___?___.

17. $H_0{}^3$ is the same as the mapping ___?___.

18. If $T:(x, y) \rightarrow (x + 2, y)$, then $T^2:(x, y) \rightarrow (\underline{}, \underline{})$.

19. If $T:(x, y) \rightarrow (x + 3, y + 4)$, then $T^2:(x, y) \rightarrow (\underline{}, \underline{})$.

In each exercise, a rule is given for a mapping *S*. Write the rule for S^{-1}.

B **20.** $S:(x, y) \rightarrow (x + 5, y + 2)$ **21.** $S:(x, y) \rightarrow (x - 3, y - 1)$

22. $S:(x, y) \rightarrow (3x, 3y)$ **23.** $S:(x, y) \rightarrow (\tfrac{1}{4}x, \tfrac{1}{4}y)$

24. $S:(x, y) \rightarrow (x + 4, 4y)$ **25.** $S:(x, y) \rightarrow (y, x)$

C **26. a.** *j* and *k* are vertical lines 1 unit apart. According to Theorem 12-4, $M_j \circ M_k$ and $M_k \circ M_j$ are both translations. Describe in words the distance and direction of each translation.

b. Show that $M_j \circ M_k$ and $M_k \circ M_j$ are inverses by showing their product is *I*.
Hint: $(M_j \circ M_k) \circ (M_k \circ M_j) = M_j \circ (M_k \circ M_k) \circ M_j$

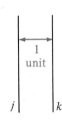

27. The figure shows that $H_A \circ H_B:P \rightarrow P''$.

a. By measuring, verify that $PP'' = 2 \cdot AB$. What theorem about the midpoints of the sides of a triangle does this suggest?

b. Copy the figure, choose another point *Q*, and carefully locate *Q''*, the image of *Q* by $H_A \circ H_B$.

c. By measuring, verify that the distance and direction from *Q* to *Q''* are the same as the distance and direction from *P* to *P''*.

d. What kind of transformation does $H_A \circ H_B$ appear to be?

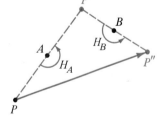

28. The black lines in the diagram illustrate the statement $H_A \circ H_B = $ translation *T*. The red lines show that $H_B \circ H_A = $ translation *S*.

a. How is translation *S* related to translation *T*?

b. Prove your answer correct by showing that $(H_A \circ H_B) \circ (H_B \circ H_A) = I$. (See the hint in Exercise 26.)

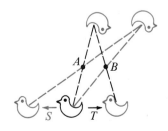

29. The diagram shows a figure and its images after successive half-turns about A, B, and C.

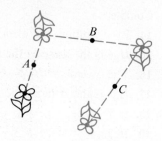

    a. Does there appear to be a single half-turn mapping the original figure to the final figure?

    b. Mentally locate the point D about which such a half-turn could be made. How is D related to A, B, and C?

Solve each equation for the unknown transformation X. Then tell what kind of transformation X is.

Example: $H_A \circ X = H_B$

Solution: "Multiply" both sides of the equation by H_A, which is the inverse of H_A.

$$H_A \circ H_A \circ X = H_A \circ H_B$$
$$I \circ X = H_A \circ H_B$$
$$X = H_A \circ H_B$$

    X is a translation (see Exercise 28).

30. $M_k \circ X = M_j$ where $k \perp j$.   **31.** $M_k \circ X = M_j$ where $k \parallel j$.

32. $M_k \circ X = M_j$ where k is the x-axis and j is the line $y = x$.

12-7 *Dilations*

Thus far the only kind of mappings we have studied are isometries, or congruence mappings. In this section, we shall study a particular transformation that is related to similarity.

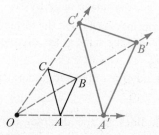

Begin with $\triangle ABC$ and point O.
Let A' lie on $\overrightarrow{OA}$ so that $OA' = 2 \cdot OA$.
Let B' lie on $\overrightarrow{OB}$ so that $OB' = 2 \cdot OB$.
Let C' lie on $\overrightarrow{OC}$ so that $OC' = 2 \cdot OC$.

    The transformation, $ABC \to A'B'C'$, described above is called a *dilation* with center O and scale factor 2. It is abbreviated $D_{O,2}$.

    A **dilation** with center O and scale factor k ($k > 0$) is abbreviated $D_{O,k}$. It is a mapping such that:

(1) If P is different from O, then P' lies on $\overrightarrow{OP}$ and $OP' = k \cdot OP$.

(2) If P is the point O, then P' is the same point as P.

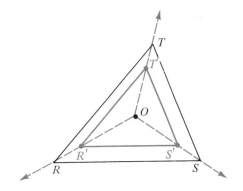

Example: Given point O and $\triangle RST$, find the image of $\triangle RST$ under the dilation with center O and the scale factor $\frac{2}{3}$.

Solution: $\triangle R'S'T'$ shown

A dilation can also have a negative scale factor. In this case, the image of point P is on the ray *opposite* $\overrightarrow{OP}$. The diagram illustrates the dilation $D_{O,-\frac{1}{2}}:P \rightarrow P'$. $\overrightarrow{OP'}$ is opposite $\overrightarrow{OP}$ and $OP' = \frac{1}{2}OP$.

The dilation shown in the example is called a **contraction**. In general, $D_{O,k}$ is a contraction if $0 < |k| < 1$. If $|k| > 1$, the dilation is called an **expansion**.

Theorem 12-6   The dilation $D_{O,k}$ maps every line segment to a parallel line segment which is $|k|$ times as long.

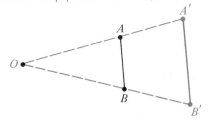

Given: $D_{O,k}:\overline{AB} \rightarrow \overline{A'B'}$

Prove: $\overline{A'B'} \| \overline{AB}$;
$A'B' = |k| \cdot AB$

Proof:

Statements	Reasons
1. $OA' = |k| \cdot OA,\ OB' = |k| \cdot OB$	1. Definition of dilation
2. $\triangle OA'B' \sim \triangle OAB$	2. SAS Similarity Theorem
3. $\dfrac{A'B'}{AB} = \dfrac{OA'}{OA} = |k|$	3. Corr. sides of $\sim \triangle$ are proportional.
(so that $A'B' = |k| \cdot AB$)	
4. $\angle OA'B' = \angle OAB$	4. Corr. $\angle$ of $\sim \triangle$ are equal.
5. $\overline{A'B'} \| \overline{AB}$	5. ___?___

The diagram for the proof of Theorem 12-6 shows the case when $k > 0$. You should draw the diagram for $k < 0$ and convince yourself that the proof is the same.

Classroom Exercises

Sketch each triangle on the chalkboard. Then sketch its image by the given dilation.

1. $D_{O,3}$

2. $D_{H,\frac{1}{2}}$

3. $D_{E,-2}$

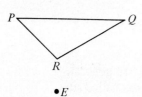

4. $D_{N,-\frac{1}{3}}$

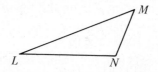

5. Find the coordinates of the images of points *A*, *B*, and *C* by the dilation $D_{O,2}$.

6. Find the image of (x, y) by $D_{O,2}$.

7. What are the images of points *A*, *B*, and *C* by the inverse of $D_{O,2}$?

8. Explain the meaning of the statement $(D_{O,2})^{-1} = D_{O,\frac{1}{2}}$.

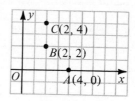

Complete each statement with one of the words *expansion, contraction, identity,* or *half-turn*.

9. If its scale factor is $\frac{2}{5}$, a dilation is a(n) ___?___.

10. If its scale factor is 4, a dilation is a(n) ___?___.

11. If its scale factor is -1, a dilation is a(n) ___?___.

12. If its scale factor is 1, a dilation is the ___?___.

Written Exercises

Find the coordinates of the images of *A*, *B*, and *C* by the given dilation.

 A

1. $D_{O,2}$

2. $D_{O,3}$

3. $D_{O,\frac{1}{2}}$

4. $D_{O,-\frac{1}{2}}$

5. $D_{O,-2}$

6. $D_{O,1}$

7. $D_{A,-\frac{1}{2}}$

8. $D_{A,2}$

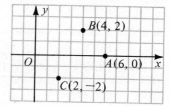

A dilation with the origin, O, as center maps the given point to the image point named. Find the scale factor of the dilation. Is the dilation an expansion or a contraction?

9. $(2, 0) \rightarrow (8, 0)$ **10.** $(2, 3) \rightarrow (4, 6)$ **11.** $(3, 9) \rightarrow (1, 3)$

12. $(4, 10) \rightarrow (-2, -5)$ **13.** $(0, \frac{1}{6}) \rightarrow (0, \frac{2}{3})$ **14.** $(-6, 2) \rightarrow (18, -6)$

Draw quadrilateral $PQRS$ in black. Then draw in color the image, quadrilateral $P'Q'R'S'$, by the dilation given.

B 15. $P(-1, 1)$ $Q(0, -1)$ $R(4, 0)$ $S(2, 2)$; $D_{O,3}$

16. $P(12, 0)$ $Q(0, 15)$ $R(-9, 6)$ $S(3, -9)$; $D_{O,\frac{2}{3}}$

17. $P(3, 0)$ $Q(3, 4)$ $R(6, 6)$ $S(5, -1)$; $D_{O,-2}$

18. $P(-2, -2)$ $Q(0, 0)$ $R(4, 0)$ $S(6, -2)$; $D_{O,-\frac{1}{2}}$

19. $D_{O,3}$ maps $\triangle ABC$ to $\triangle A'B'C'$.

   a. What is the ratio of the perimeters of $\triangle A'B'C'$ and $\triangle ABC$?

   b. What is the ratio of the areas of $\triangle A'B'C'$ and $\triangle ABC$?

20. The diagram illustrates a dilation of three-dimensional space. $D_{O,2}$ maps the smaller cube to the larger cube.

   a. What is the ratio of the surface areas of these cubes?

   b. What is the ratio of the volumes of these cubes?

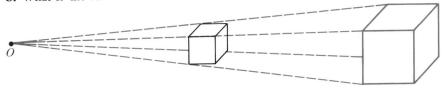

21. G is the intersection of the medians of $\triangle XYZ$. Complete the following statements.

   a. $\dfrac{XG}{XM} = ?$ **b.** $\dfrac{GM}{GX} = ?$

   c. What dilation maps X to M?

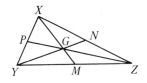

22. $D_{O,k}$ maps $\overline{PQ}$ to $\overline{P'Q'}$.

   a. Show that the slopes of $\overline{PQ}$ and $\overline{P'Q'}$ are equal.

   b. Part (a) proves that $\overline{PQ}$ and $\overline{P'Q'}$ are ___?___.

C 23. Use the distance formula to show that

$$P'Q' = |k| \sqrt{(a - c)^2 + (b - d)^2} = |k|\, PQ.$$

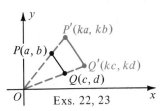

Exs. 22, 23

24. $D_{A,2}{:}\overline{PQ} \rightarrow \overline{P'Q'}$ and $D_{B,\frac{1}{2}}{:}\overline{P'Q'} \rightarrow \overline{P''Q''}$. What kind of special transformation is the product $D_{A,2} \circ D_{B,\frac{1}{2}}$? Explain.

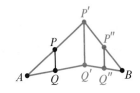

SELF-TEST

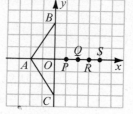

1. What is the symmetry line of $\triangle ABC$?
2. Does $\triangle ABC$ have point symmetry?

3. $R_{0,90} \circ M_y : B \to \underline{\ ?\ }$
4. $M_x \circ H_0 : C \to \underline{\ ?\ }$
5. $D_{0,4} : P \to \underline{\ ?\ }$
6. $D_{0,-\frac{1}{2}} : S \to \underline{\ ?\ }$

7. For any transformation T, $T \circ T^{-1} : P \to \underline{\ ?\ }$.
8. The product of any transformation T and the identity is $\underline{\ ?\ }$.

COMPUTER KEY-IN

How can you find the area of the region bounded by the x-axis, the line $x = 1$, and the graph of the equation $y = x^2$? The second diagram suggests one way to approximate this area. Since each of the rectangles has one vertex on the graph of the equation $y = x^2$, you can find the height of each rectangle.

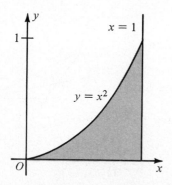
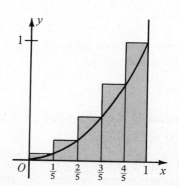

Rectangle:	1st	2nd	3rd	4th	5th
Height (h):	$\frac{1}{25}$	$\frac{4}{25}$	$\frac{9}{25}$	$\frac{16}{25}$	1

The base (b) of each rectangle is $\frac{1}{5}$. Using the formula $A = bh$, you can find the sum of the areas of the five rectangles:

$$\frac{1}{5}\left(\frac{1}{25}\right) + \frac{1}{5}\left(\frac{4}{25}\right) + \frac{1}{5}\left(\frac{9}{25}\right) + \frac{1}{5}\left(\frac{16}{25}\right) + \frac{1}{5}(1) =$$

$$\frac{1}{125} + \frac{4}{125} + \frac{9}{125} + \frac{16}{125} + \frac{25}{125} = \frac{55}{125} = \frac{11}{25} = 0.44$$

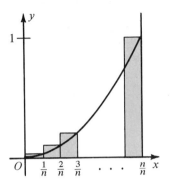

You can improve this estimate by increasing the number of rectangles. For n rectangles each with width $\frac{1}{n}$, the total area of the rectangles is

$$\frac{1}{n}\left(\frac{1}{n^2}\right) + \frac{1}{n}\left(\frac{4}{n^2}\right) + \frac{1}{n}\left(\frac{9}{n^2}\right) + \cdots + \frac{1}{n}\left(\frac{n^2}{n^2}\right), \text{ or}$$

$$\frac{1 + 4 + 9 + \cdots + n^2}{n^3}.$$

The program below in BASIC calculates the total area of the rectangles for as many rectangles as we please.

```
10   PRINT "HOW MANY RECTANGLES DO YOU WANT";
20   INPUT N
30   LET S=0
40   FOR I=1 TO N
50   LET A=I*I
60   LET S=S+A
70   NEXT I
80   LET S=S/N↑3
90   PRINT "AREA OF";N;" RECTANGLES IS";S;"."
100  END
```

Exercises

1. If your computer uses a language other than BASIC, write a similar program in that language.

2. Run the program for $n = 50$, $n = 100$, $n = 300$, and $n = 500$. Give an estimate of the area bounded by the x-axis, the curve $y = x^2$, and the line $x = 1$.

3. **a.** How would you change the program to estimate the area bounded by the x-axis, the curve $y = x^2$, and the line $x = 2$?

   b. Revise and run the program to give an estimate of the area of the region described in (a).

4. **a.** How would you change the program to estimate the area bounded by the x-axis, the curve $y = x^3$, and the line $x = 1$?

   b. Revise and run the program to give an estimate of the area of the region described in (a).

Extra Tessellations _____

A tessellation is a design in which congruent copies of a figure completely fill the plane without overlapping.

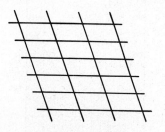

A tessellation of parallelograms

A tessellation of the letter H

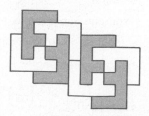

A tessellation of the letter F

You can make an original tessellation as shown below by making a design on one side of a square and then drawing a congruent copy of the design on the opposite side of the square. Your altered square will tessellate the plane.

 →   →

It is not necessary to start with a square. Any parallelogram will work. And you can alter *both* pairs of opposite sides.

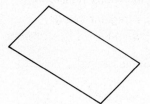

Start with a parallelogram.

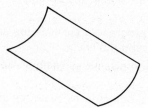

Alter one pair of opposite sides.

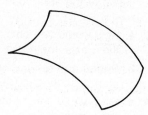

Then alter the other pair of opposite sides.

The altered parallelogram may remind you of a sting ray or of a bird.

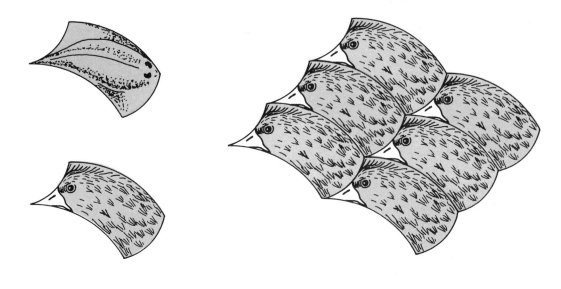

Exercises

Make a tesselation of the given figure on graph paper.

1. **2.** **3.** **4.**

 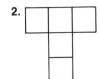 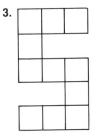 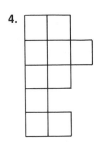

Tell whether or not a tessellation can be made with the given figure.

5. A rhombus

6. A regular hexagon

7. A regular pentagon

8. An isosceles triangle

9. A scalene triangle

10. A regular octagon

11. A circle

12. A trapezoid

13. An arbitrary quadrilateral

14. Starting with a square, create a tessellation of your own.

15. Starting with a parallelogram, create a tessellation of your own.

16. Make a tessellation using congruent copies of the figure shown.

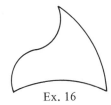

Ex. 16

Chapter Summary

1. A transformation of the plane is a mapping such that (1) every point of the plane has exactly one image, and (2) every point of the plane has exactly one preimage.

2. Some basic isometries are:

   Translation

   Reflection in line j (M_j)

   Rotation about O through $x°$ ($R_{O,x}$)

   Half-turn about O (H_O)

3. A figure may have line symmetry, point symmetry, rotational symmetry, or translational symmetry.

4. The combination of one mapping followed by another is called a *product* of mappings. A product of reflections in parallel lines is a translation. A product of reflections in intersecting lines is a rotation.

5. A transformation followed by its inverse keeps all points fixed. The mapping $S \circ S^{-1}$ is the identity transformation.

6. Dilations are transformations, but most dilations are not isometries. A dilation with center O and scale factor k is denoted by $D_{O,k}$.

Chapter Review

12-1 *Isometries*

The transformation $S:(x, y) \rightarrow (x + 3, y + 4)$ is given.

1. The image of (2, 5) is (_?_, _?_).
2. The preimage of (3, 4) is (_?_, _?_).
3. Is S a translation? an isometry?

12-2 *Reflections*

Copy the figure on squared paper. Draw the image by reflection in line k.

4.

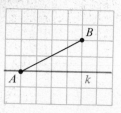

5.

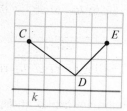

6.

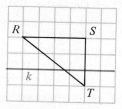

12-3 *Rotations and Half-Turns*

Plot the following points on a coordinate grid: $A(1, 0), B(2, 1), C(4, 1), D(3, -1)$. Label the origin O. Draw the images of quad. $ABCD$ by the transformations specified.

7. H_O **8.** $R_{O, 90}$ **9.** $R_{O, -90}$

10. Give another name for $R_{O, 405}$.

12-4 *Symmetry*

11. Does a scalene triangle have line symmetry?

12. Does a rectangle have point symmetry?

13. Does an equilateral triangle have 90° rotational symmetry?

14. Does a regular hexagon have 60° rotational symmetry?

12-5 *Products of Mappings*

The transformations $H_O: \triangle ABC \to \triangle DEF$ and $M_j: \triangle DEF \to \triangle GHI$ are given.

15. $H_O \circ M_j : A \to \underline{\ ?\ }$ **16.** $H_O \circ M_j : \underline{\ ?\ } \to \overline{GH}$ **17.** $H_O \circ H_O : \angle ABC \to \angle \underline{\ ?\ }$

12-6 *Inverses and Identities*

The translation $T: (x, y) \to (x + 3, y)$ is given.

18. a. $T: (5, 3) \to (\underline{\ ?\ }, \underline{\ ?\ })$ **b.** $T^{-1}: (8, 3) \to (\underline{\ ?\ }, \underline{\ ?\ })$

19. a. $T \circ T^{-1}: (x, y) \to (\underline{\ ?\ }, \underline{\ ?\ })$ **b.** $T \circ \underline{\ ?\ } = I$

20. $T^{-2}: (10, 2) \to (\underline{\ ?\ }, \underline{\ ?\ })$

12-7 *Dilations*

Give the coordinates of the image of A by the dilation specified.

21. $D_{O, 2}$ **22.** $D_{B, 2}$

Give the coordinates of the image of B by the dilation specified.

23. $D_{O, \frac{1}{3}}$ **24.** $D_{O, 6} \circ D_{O, \frac{1}{2}}$

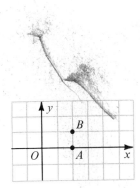

Chapter Test _____

Transformation Z maps $\triangle RST$ to $\triangle R'S'T'$.

1. The image of T is _?_.

2. The preimage of $\overline{R'S'}$ is _?_.

3. If the mapping is an isometry, then $ST =$ _?_.

State whether the mapping $\angle ABC \rightarrow \angle DEF$ can be described as a translation, a reflection, or a rotation.

4.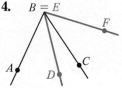

5.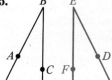

6.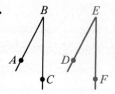

Points E, F, G, and K are the midpoints of the sides of rectangle $ABCD$. Classify each statement as true or false.

7. $\overleftrightarrow{KF}$ is a line of symmetry.

8. $\overleftrightarrow{BD}$ is a line of symmetry.

9. G is a point of symmetry.

10. O is a point of symmetry.

11. $R_{O,90} : E \rightarrow F$ **12.** $H_O : \triangle KOD \rightarrow \triangle FOB$

13. $R_{O,135} : K \rightarrow B$ **14.** $M_j : \triangle KOD \rightarrow \triangle FOC$

15. $M_j \circ H_O : \triangle AOE \rightarrow \triangle DOG$ **16.** $R_{O,-180} \circ H_O : E \rightarrow E$

Point O is the origin. Give the coordinates of the image of $(2, 4)$ by the transformation specified.

17. $D_{O,5}$ **18.** $D_{O,\frac{1}{2}}$ **19.** $D_{O,-\frac{1}{2}}$ **20.** $(D_{O,2})^{-1}$

Cumulative Review: Chapters 7–12 ———

True–False Exercises

Write T or F to indicate your answer.

A   **1.** The total area of a solid figure is measured in cubic units.

   2. Two parallelograms which have equal bases and equal altitudes must have equal areas.

   3. An equilateral polygon inscribed in a circle is regular.

   4. $x^2 + y^2 = 4$ is the equation of a circle of radius 4.

   5. If two rectangles have equal perimeters, then they must have equal areas.

   6. If a circle has circumference C and diameter d, then $\dfrac{C}{d} = \pi$.

   7. If the radius of $\odot O$ is twice that of $\odot P$, then the area of $\odot O$ is twice that of $\odot P$.

   8. In a plane, the locus of points equidistant from two given points is a point.

   9. If a polygon is inscribed in a circle, then it is a regular polygon.

   10. If two circles have common external tangents which are parallel, then the diameters of the circles are equal.

   11. If two squares have equal areas, then they have equal perimeters.

   12. A reflection in a line is an isometry.

   13. If an angle is inscribed in a semicircle, then it is a right angle.

   14. The slope of the line containing $(0, 5)$ and $(5, 0)$ is negative.

   15. If each side of a polygon is tangent to a circle, then the circle is inscribed in the polygon.

   16. If a cone and a pyramid have equal base areas and equal heights, then they have equal volumes.

B   **17.** If the circumference of $\odot O$ is twice the circumference of $\odot P$, then the radius of $\odot O$ is twice the radius of $\odot P$.

   18. The circle $x^2 + y^2 = 25$ is tangent to the line $y = 5$.

   19. If the area and perimeter of one rhombus are equal respectively to the area and perimeter of another rhombus, then the rhombuses are congruent.

   20. If a triangular prism has a base of 300 cm^2 and a height of 42 cm, then the volume is 4200 cm^3.

   21. $H_A{}^2 = I$

   22. If the area of $\odot O$ is twice the area of $\odot P$, then the radius of $\odot O$ is twice the radius of $\odot P$.

   23. Every figure which has rotational symmetry has line symmetry.

   24. For any rhombus, a circle can be inscribed in the rhombus.

25. If one square pyramid has height 20 and base edge 10, while another square pyramid has height 30 and base edge 15, then the pyramids are similar.

26. The circle $(x - 1)^2 + (y - 2)^2 = 9$ intersects the line $x = 3$.

27. If a sector has radius r and arc length s, then it has area $\frac{1}{2}rs$.

28. If two triangles are inscribed in the same circle, then they have equal areas.

29. If $D_{O,k}$ is an isometry, then $|k| = 1$.

Multiple Choice Exercises

Write the letter that indicates the best answer.

A **1.** If point A lies inside $\odot O$ and point B lies outside $\odot O$, what is the intersection of $\overleftrightarrow{AB}$ with $\odot O$?

    a. one point   **b.** two points   **c.** no points   **d.** can't tell

2. Which transformation is not always an isometry?

    a. reflection   **b.** half-turn   **c.** rotation   **d.** dilation

3. What is the volume of a sphere with radius 6 cm?

    a. 288π cm^3 **b.** 72π cm^3 **c.** 144π cm^3 **d.** none of these

4. What is the perimeter of a regular hexagon which is inscribed in a circle of diameter 12?

    a. 72   **b.** 36   **c.** $36\sqrt{3}$ **d.** $108\sqrt{3}$

5. At 2:00 P.M. what is the angle formed by the hands of a clock?

    a. $60°$ **b.** $70°$ **c.** $75°$ **d.** $72\frac{1}{2}°$

6. If a box has length 8 cm, width 10 cm, and height 10 cm, what is the total area?

    a. 800 cm^2 **b.** 360 cm^2 **c.** 520 cm^2 **d.** none of these

7. If two intersecting lines have slopes m_1 and m_2, then:

    a. $m_1 > m_2$ **b.** $m_1 < m_2$ **c.** $m_1 = m_2$ **d.** $m_1 \neq m_2$

8. What is the midpoint of the line segment which joins (x_1, y_1) and (x_2, y_2)?

    a. $\left(\dfrac{x_1 + y_1}{2}, \dfrac{x_2 + y_2}{2}\right)$ **b.** $\left(\dfrac{x_1 + y_2}{2}, \dfrac{x_2 + y_1}{2}\right)$

    c. $\left(\dfrac{x_1 + x_2}{2}, \dfrac{y_1 + y_2}{2}\right)$ **d.** $\left(\dfrac{x_1}{2}, \dfrac{y_2}{2}\right)$

9. Which of the points does not lie on the line $3x - 4y = -16$?

    a. $(0, 4)$ **b.** $(8, -2)$ **c.** $(-4, 1)$ **d.** $(4, 7)$

10. What is the total area of a cylinder with radius 10 cm and height 30 cm?

    a. 3000π cm^2 **b.** 800π cm^2 **c.** 700π cm^2 **d.** 600π cm^2

11. Two intersecting chords intercept arcs which measure 30° and 100°. What is the measure of an angle formed?

    a. 15°   **b.** 35°   **c.** 50°   **d.** 65°

12. For which line is slope not defined?

    a. a horizontal line   **b.** a line containing $(0, 0)$

    c. a vertical line   **d.** a line perpendicular to the y-axis

B 13. A rectangle with height 24 has area 696. What is the perimeter of the rectangle?

    a. 122   **b.** 120   **c.** 144   **d.** none of these

14. M_x is a reflection in the x-axis and O is the origin. What is the image of $(-3, 3)$ by $M_x \circ R_{0,90}$?

    a. $(-3, 3)$ **b.** $(-3, -3)$ **c.** $(3, -3)$ **d.** $(3, 3)$

15. If a circle has the equation $(x + 1)^2 + (y - 3)^2 = 16$, which of the points does not lie on the circle?

    a. $(-1, -1)$ **b.** $(-2, 1)$ **c.** $(3, 3)$ **d.** $(-1, 7)$

C 16. If a triangle has sides of 10 cm, 13 cm, and 13 cm, what is the radius of the circumscribed circle?

    a. $\sqrt{13}$ **b.** $\dfrac{169}{24}$ **c.** $\dfrac{181}{24}$ **d.** none of these

17. If point P is 6 cm from point Q, what is the locus of points 4 cm from Q and equidistant from P and Q?

    a. a sphere   **b.** a line   **c.** a plane   **d.** a circle

Completion Exercises

Write the correct expression on your paper.

A 1. If the slopes of two lines are $\dfrac{1}{3}$ and -3, then the lines are ___?___ to each other.

2. If a parallelogram is inscribed in a circle, then the parallelogram must be a(n) ___?___ .

3. If $\triangle ABC \sim \triangle DEF$, the perimeter of $\triangle ABC$ is 36 cm, and the perimeter of $\triangle DEF$ is 54 cm, then the ratio of the area of $\triangle ABC$ to $\triangle DEF$ is ___?___ .

4. The slope of a horizontal line is ___?___ .

5. If P is $(-6, 4)$ and Q is $(2, 7)$, then the midpoint of $\overline{PQ}$ is ___?___ .

6. If an inscribed angle measures 82°, then its intercepted arc measures ___?___°.

7. A sphere of radius 7 has a volume of ___?___ .

8. If $M_k: \triangle ABC \rightarrow \triangle A'B'C'$, then the preimage of B' is ? .

9. The ratio of the sides of two cubes is $3:4$. The ratio of the volumes is ? .

10. A chord of a circle which passes through the center is a ___?___ .

11. The image of $(4, -3)$ by reflection in the y-axis is (? , ?).

12. If the ratio of the radii of two spheres is $8:13$, then the ratio of their surface areas is ? .

13. The distance between $(-1, 2)$ and $(2, 4)$ is ? .

14. In the plane of a rectangle, the locus of points equidistant from the vertices of the rectangle is a ___?___ .

15. If the area of a parallelogram is 91 cm² and the base is 13 cm, then the height is ? cm.

16. If a cone has slant height 17 cm and lateral area 153π cm², then the radius of the base is ? cm.

17. If a circle has area 64π, then its circumference is ? .

18. If line l_1 has slope $\dfrac{2}{3}$ and l_2 is perpendicular to l_1, then the slope of l_2 is ? .

19. If $S:(x, y) \rightarrow (2x + 1, y - 3)$, then the image of $(-3, 2)$ is (? , ?).

20. A prism with a square base 15 cm on each side and a height of 42 cm has lateral area ? cm².

21. If three vertices of a rectangle are $(7, 0)$, $(7, 17)$, and $(15, 0)$, then the area is ? .

22. Two circles can have at most ? common tangents.

23. If a line passes through $(6, 1)$ and $(1, -3)$, then its slope is ___?___ .

24. If a right triangle has legs 19 cm and 28 cm, then the area is ? cm².

25. If a pyramid has a square base, 10 cm on each side, and has height 12 cm, then the volume of the pyramid is ? cm³.

B 26. Use the pyramid described in Exercise 25. The lateral area is ? cm².

27. If the total area of a cylinder is 228π cm² and the radius is 6 cm, then the height is ? cm.

28. If two cylinders have equal heights but the radius of one is 3 times the radius of the other, the ratio of their volumes is ? .

29. If $T:(x, y) \rightarrow (\frac{1}{2}x, y - 3)$, then $T^{-1}:(x, y) \rightarrow$ (? , ?).

30. The locus of points in space equidistant from three non-collinear points is ___?___ .

31. A chord is 8 cm from the center of a circle of radius 10 cm. The length of the chord is ? cm.

32. If the perimeter of a rectangle is 64 cm and its base is 23 cm, then the area is ___?___ cm².

33. If a circle has diameter 12 and passes through $(5, 4)$ and $(-7, 4)$, then the equation of the circle is ___?___ .

34. If a sphere has volume 36π, then the radius of the sphere is ___?___ .

35. If two parallel chords, each 8 cm long, are 6 cm apart, then the diameter of the circle is ___?___ cm.

36. If the area of a trapezoid is 98 cm² and the bases are 12 cm and 16 cm, then the height is ___?___ cm.

37. If the base of a prism is a right triangle having legs $3\sqrt{2}$ and $4\sqrt{2}$, and the height of the prism is 12, the total area of the prism is ___?___ .

38. If the number of square centimeters in the area of a sphere equals the number of cubic centimeters in the volume, then the radius of the sphere is ___?___ cm.

Construction Exercises

A

1. Construct an equilateral triangle.

2. Construct a 30° angle.

3. Draw a scalene triangle. Inscribe a circle in the triangle.

4. Construct a triangle whose sides are in the ratio of $2:3:4$.

5. Draw an acute angle. Then construct an angle whose measure is $\frac{3}{2}$ times the measure of the angle you drew.

B

6. Draw a circle and a point P outside the circle. Construct an isosceles triangle with vertex at P which circumscribes the circle.

7. Draw a line segment AB. By construction, find a point C on $\overrightarrow{AB}$ so that $AB:AC = 3:5$.

8. Draw two segments and label the lengths j and k. Construct a segment whose length equals $\sqrt{3jk}$.

C

9. Draw a segment and label its length t. Then construct a rhombus with perimeter $6t$ and with one diagonal t units long.

10. Draw a scalene triangle. Then construct a square whose area is equal to the area of the triangle.

11. In an equilateral triangle, construct three circles of equal radius, each tangent to two sides of the triangle and externally tangent to the other two circles.

Examinations

Chapter 1

Indicate the best answer by writing the appropriate letter.

1. Which of the following sets of points are collinear?
   - **a.** G, O, J
   - **b.** H, O, E
   - **c.** E, O, F
   - **d.** K, O, G

2. Which of the following sets of points are contained in *more* than one plane?
   - **a.** G, O, J
   - **b.** E, O, G
   - **c.** H, E, G
   - **d.** G, O, H

3. If $\angle KOH = 30°$, which of the following *must* be true?
   - **a.** $\angle EOK = 60°$
   - **b.** $\angle GOJ = 30°$
   - **c.** $\angle EOG = 90°$
   - **d.** $\angle KOF = 150°$

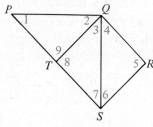

Exs. 1-3

4. Which of the following *cannot* be used as a reason in a proof?
   - **a.** a definition
   - **b.** a postulate
   - **c.** yesterday's theorem
   - **d.** tomorrow's theorem

5. On a number line, point K has coordinate 5, and point Q has coordinate 13. Point S is on $\overrightarrow{KQ}$, and $QS = 7$. What is the coordinate of S?
   - **a.** either 6 or 20
   - **b.** neither 6 nor 20
   - **c.** 6
   - **d.** 20

6. If $\angle 1$ and $\angle 2$ are complements and $\angle 2$ and $\angle 4$ are complements, what are $\angle 1$ and $\angle 4$?
   - **a.** supplements
   - **b.** complements
   - **c.** equal angles
   - **d.** adjacent angles

7. The statement "If $\angle A = \angle B$ and $\angle D = \angle A + \angle C$, then $\angle D = \angle B + \angle C$" is justified by which property from algebra?
   - **a.** Reflexive
   - **b.** Symmetric
   - **c.** Transitive
   - **d.** Substitution

8. If $\overline{TQ} \perp \overline{QR}$, which angles *must* be complementary angles?
   - **a.** $\angle 2$ and $\angle 3$
   - **b.** $\angle 3$ and $\angle 4$
   - **c.** $\angle 5$ and $\angle 8$
   - **d.** $\angle 3$ and $\angle 7$

9. If $\angle 8 = (x + 80)°$, what is the measure of $\angle 9$?
   - **a.** $(100 - x)°$
   - **b.** $(100 + x)°$
   - **c.** $(x - 80)°$
   - **d.** $(x - 180)°$

10. Which angle is obtuse?
    - **a.** $\angle 7$
    - **b.** $\angle PQR$
    - **c.** $\angle 9$
    - **d.** $\angle R$

11. If $\overrightarrow{SQ}$ bisects $\angle RST$, which angles *must* be equal?
    - **a.** $\angle 6$ and $\angle 7$
    - **b.** $\angle 3$ and $\angle 4$
    - **c.** $\angle 2, \angle 3,$ and $\angle 4$
    - **d.** $\angle 4$ and $\angle 6$

Exs. 8-11

464

Chapter 2

Indicate the best answer by writing the appropriate letter.

1. If $\overrightarrow{BE}$ bisects $\angle ABC$, what does $\angle AEB$ equal?

    a. 30°   **b.** 35°   **c.** 40°   **d.** 45°

2. If $\angle ABE = 40°$, what does $\angle BED$ equal?

    a. 140°   **b.** 40°   **c.** 75°   **d.** 135°

3. If $\overline{AB} \parallel \overline{DC}$, what does $\angle D$ equal?

    a. 70°   **b.** 80°   **c.** 90°   **d.** 100°

4. Which of the following would allow you to conclude that $\overline{AD} \parallel \overline{BC}$?

    a. $\angle BEC = \angle BCE$ **b.** $\angle ABE = \angle BEC$

    c. $\angle DEC = \angle BCE$ **d.** $\angle A + \angle AEC = 180°$

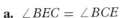

Exs. 1–4

5. Given: (1) If A is white, then B is red.
             (2) B is not red.

  Which of the following *must* be true?

    a. B is white.   **b.** B is not white.   **c.** A is red.   **d.** A is not white.

6. If a statement is known to be true, then which of the following *must* also be true?

    a. its converse   **b.** its inverse

    c. its contrapositive   **d.** none of these

7. What is the measure of each angle of a regular octagon?

    a. 120°   **b.** 135°   **c.** 144°   **d.** 150°

8. The plane containing S, A, C, K appears to be parallel to the plane containing which points?

    a. Q, E, K, S **b.** Q, U, R, E

    c. A, S, Q, U **d.** U, R, C, A

9. Which of the following appear to be skew lines?

    a. $\overleftrightarrow{QE}$ and $\overleftrightarrow{AC}$ **b.** $\overleftrightarrow{QU}$ and $\overleftrightarrow{KC}$

    c. $\overleftrightarrow{QS}$ and $\overleftrightarrow{AC}$ **d.** $\overleftrightarrow{AC}$ and $\overleftrightarrow{UR}$

10. $\overleftrightarrow{EK}$ does *not* appear to be parallel to the plane containing which points?

    a. U, A, C **b.** Q, U, A **c.** Q, S, C **d.** Q, U, R

Exs. 8–10

11. The sum of the interior angles of a certain polygon is the same as the sum of its exterior angles. How many sides does the polygon have?

    a. four   **b.** six   **c.** eight   **d.** ten

12. What is the sum of the interior angles of a pentagon?

    a. 180°   **b.** 360°   **c.** 540°   **d.** 900°

Chapter 3

In Exercises 1–8, write a method (SSS, SAS, ASA, AAS, or HL) that can be used to prove the two triangles congruent.

1.   **2.**

3.   **4.**

5.   **6.**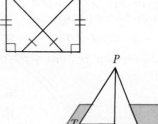

7. Given: $\overline{PO} \perp$ plane X; $OT = OS$

8. Given: $\overline{PO} \perp$ plane X; $PT = PS$

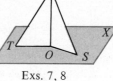

Exs. 7, 8

In Exercises 9–13, indicate the best answer by writing the appropriate letter.

9. $\overline{RT}$ is the base of $\triangle RXT$. $\angle R = \angle T$, $RT = 2x + 5$, $RX = 5x - 7$, and $TX = 2x + 8$. What is the perimeter of $\triangle RXT$?

    a. 5   **b.** 15   **c.** 18   **d.** 51

10. If $\triangle DEF \cong \triangle PRS$, which of these equations must be true?

    a. $DF = PS$ **b.** $EF = PR$ **c.** $\angle E = \angle S$ **d.** $\angle F = \angle R$

11. What is the *principal* basis for inductive reasoning?

    a. definitions   **b.** previously proved theorems

    c. postulates   **d.** past observations

12. $\overline{AC}$ is a diagonal of regular pentagon $ABCDE$. What does $\angle ACD$ equal?

    a. $36°$ **b.** $54°$ **c.** $72°$ **d.** $108°$

13. In $\triangle ABC$, $AB = AC$, $\angle A = 46°$, and $\overline{BD}$ is an altitude. What does $\angle CBD$ equal?

    a. $23°$ **b.** $44°$ **c.** $67°$ **d.** $134°$

Chapter 4

Indicate the best answer by writing the appropriate letter.

1. Both pairs of opposite sides of a quadrilateral are parallel. Which special kind of quadrilateral *must* it be?

    a. parallelogram   **b.** rectangle   **c.** rhombus   **d.** trapezoid

2. The perpendicular bisectors of the sides of a triangle intersect in a point. Where *must* the point be located?

    a. equidistant from the vertices   **b.** equidistant from the sides

    c. inside the triangle   **d.** outside the triangle

3. The diagonals of a certain quadrilateral are equal. Which term could *not* be used to describe the quadrilateral?

    a. isosceles trapezoid   **b.** rectangle

    c. rhombus   **d.** parallelogram with a 60° angle

4. You don't need a figure to do this exercise. Given that $\angle A = \angle B$, $AC = BD$, and $l \parallel m$, you want to prove that $\angle 3 = \angle 4$. To write an indirect proof, you should begin by supposing what?

    a. $\angle A \neq \angle B$ **b.** $AC \neq BD$ **c.** $l \perp m$ **d.** $\angle 3 \neq \angle 4$

5. A diagonal of a parallelogram bisects one of its angles. Which special kind of parallelogram *must* it be?

    a. rectangle   **b.** rhombus

    c. square   **d.** parallelogram with a 60° angle

6. Three segments related to $\triangle ABC$ intersect at a point that is two-thirds of the distance from each vertex to the midpoint of the opposite side. What are these segments called?

    a. altitudes   **b.** angle bisectors   **c.** medians   **d.** $\perp$ bisectors

7. The lengths of the bases of a trapezoid are 18 and 26. What is the length of the median?

    a. 8   **b.** 22   **c.** 44   **d.** none of these

8. In $\triangle RST$, $RS = 8$ and $ST = 10$. Which of these *must* be true?

    a. $RT > 2$ **b.** $RT < 2$ **c.** $RT > 10$ **d.** $RT < 10$

9. In $\triangle ABC$, $AB = 8$, $BC = 10$, and $AC = 12$. M is the midpoint of $\overline{AB}$, and N is the midpoint of $\overline{BC}$. What is the length of $\overline{MN}$?

    a. 4   **b.** 5   **c.** 6   **d.** 9

10. If $EFGH$ is a parallelogram, which of the following *must* be true?

    a. $\angle E = \angle F$ **b.** $\angle F = \angle H$ **c.** $\overline{FG} \parallel \overline{GH}$ **d.** $\angle E + \angle G = 180°$

11. Which information does *not* prove that quad. $ABCD$ is a parallelogram?

    a. $\overline{AC}$ and $\overline{BD}$ bisect each other.   **b.** $\overline{AD} \parallel \overline{BC}$; $AD = BC$

    c. $\overline{AB} \parallel \overline{CD}$; $AD = BC$ **d.** $\angle A = \angle C$; $\angle B = \angle D$

Chapter 5

Indicate the best answer by writing the appropriate letter.

1. $\triangle ABC \sim \triangle DEF$, $AB = 8$, $BC = 12$, $AC = 16$, and $DE = 12$. What is the perimeter of $\triangle DEF$?

   a. 36   **b.** 40   **c.** 48   **d.** 54

2. Which of the following pairs of polygons *must* be similar?

   a. two rectangles   **b.** two regular hexagons

   c. two isosceles triangles   **d.** two parallelograms with a 60° angle

3. What is the value of u?

   a. 8   **b.** 10   **c.** 16   **d.** 25

4. What is the value of z?

   a. 25   **b.** 28   **c.** $\dfrac{28}{3}$ **d.** $\dfrac{70}{3}$

5. Quad. $GHJK \sim$ quad. $RSTU$, $GH = JK = 10$, $HJ = KG = 14$, and $RS = TU = 16$. What is the scale factor of quad. $GHJK$ to quad. $RSTU$?

   a. $\dfrac{5}{7}$ **b.** $\dfrac{5}{8}$ **c.** $\dfrac{7}{8}$ **d.** $\dfrac{16}{10}$

6. If $\dfrac{a}{b} = \dfrac{x}{y}$, what does $\dfrac{y}{b}$ equal?

   a. $\dfrac{x}{a}$ **b.** $\dfrac{a}{x}$ **c.** $\dfrac{y}{x}$ **d.** $\dfrac{b}{y}$

7. In $\triangle APC$, the bisector of $\angle P$ meets $\overline{AC}$ at B. $PA = 30$, $PC = 50$, and $AB = 12$. What is the length of $\overline{BC}$?

   a. $\dfrac{36}{5}$ **b.** 12   **c.** 20   **d.** 32

8. Which of the following can you use to prove that the two triangles are similar?

   a. SAS Similarity Theorem   **b.** AA Postulate

   c. SSS Similarity Theorem   **d.** def. of similar triangles

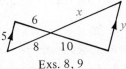

Exs. 8, 9

9. Which statement is correct?

   a. $\dfrac{6}{10} = \dfrac{8}{x}$ **b.** $\dfrac{6}{8} = \dfrac{x}{10}$ **c.** $6 \cdot 10 = 8x$ **d.** $\dfrac{5}{y} = \dfrac{8}{10}$

10. If the angles of a triangle are in the ratio $3:3:4$, what is the largest angle of the triangle?

   a. 40°   **b.** 54°   **c.** 72°   **d.** 90°

11. If $\triangle ABC \sim \triangle JOT$, which of these is a correct proportion?

   a. $\dfrac{BC}{AC} = \dfrac{JT}{OT}$ **b.** $\dfrac{AB}{JT} = \dfrac{AC}{JO}$ **c.** $\dfrac{AB}{BC} = \dfrac{OT}{JT}$ **d.** $\dfrac{AC}{JT} = \dfrac{BC}{OT}$

Chapter 6

Indicate the best answer by writing the appropriate letter.

1. The shorter leg of a 30-60-90 triangle is 7. Find the hypotenuse.

    a. 14   **b.** $7\sqrt{2}$ **c.** $7\sqrt{3}$ **d.** $\sqrt{14}$

2. What is the simplified form of the number $3\sqrt{60}$?

    a. $5\sqrt{15}$ **b.** $6\sqrt{15}$ **c.** $7\sqrt{15}$ **d.** $12\sqrt{15}$

3. A certain right triangle has an $n°$ acute angle. If $\sin n° = \dfrac{3}{5}$, what does $\tan n°$ equal?

    a. $\dfrac{4}{3}$ **b.** $\dfrac{4}{5}$ **c.** $\dfrac{3}{4}$ **d.** none of these

4. The altitude to the 55 cm hypotenuse of a right triangle divides the hypotenuse into segments 25 cm and 30 cm long. How long is the altitude?

    a. $15\sqrt{3}$ cm   **b.** $15\sqrt{5}$ cm   **c.** $5\sqrt{30}$ cm   **d.** $5\sqrt{55}$ cm

5. The hypotenuse and one leg of a right triangle are 61 and 11. Find the other leg.

    a. 36   **b.** $5\sqrt{2}$ **c.** 60   **d.** $\sqrt{3842}$

6. Each side of an equilateral triangle is 10. Find an altitude.

    a. 5   **b.** 10   **c.** $5\sqrt{2}$ **d.** $5\sqrt{3}$

7. In $\triangle RST$, $\angle S = 90°$. What does $\sin T$ equal?

    a. $\dfrac{ST}{RT}$ **b.** $\dfrac{RS}{ST}$ **c.** $\dfrac{RS}{RT}$ **d.** $\dfrac{RT}{RS}$

8. Which equation could be used to find the value of x?

    a. $\cos 58° = \dfrac{x}{18.9}$ **b.** $\sin 32° = \dfrac{x}{16}$

    c. $\cos 44° = \dfrac{x}{10.4}$ **d.** $\tan 46° = \dfrac{x}{10.4}$

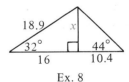

Ex. 8

9. One side of a square is s. Find a diagonal.

    a. $2\sqrt{s}$ **b.** $s\sqrt{2}$ **c.** $\dfrac{s}{2}\sqrt{3}$ **d.** $s\sqrt{3}$

10. What is the geometric mean between $\dfrac{5}{7}$ and 2?

    a. $\dfrac{19}{14}$ **b.** $2\sqrt{\dfrac{5}{7}}$ **c.** $\dfrac{1}{10}\sqrt{70}$ **d.** $\dfrac{1}{7}\sqrt{70}$

11. The legs of a right triangle are 4 and 7. Find the hypotenuse.

    a. $2\sqrt{7}$ **b.** $\sqrt{28}$ **c.** $\sqrt{33}$ **d.** $\sqrt{65}$

12. What is a triangle whose sides are 12, 13, and 18?

    a. an obtuse triangle   **b.** a right triangle

    c. an acute triangle   **d.** an impossibility

Chapter 7

Indicate the best answer by writing the appropriate letter.

1. $\overline{ZX}$ and $\overline{ZY}$ are tangent to a circle at X and Y. If $\angle XZY = 50°$, what does $\overset{\frown}{XY}$ equal?

    a. 50°   **b.** 130°   **c.** 140°   **d.** none of these

In Exercises 2 and 3, $\overline{PT}$ is tangent to $\odot M$ at T.

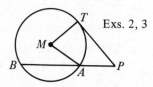

Exs. 2, 3

2. If $\angle TMA = 70°$, what is the measure of $\overset{\frown}{TBA}$?

    a. 35°   **b.** 70°   **c.** 290°   **d.** 145°

3. If $PA = 9$ and $AB = 16$, what does PT equal?

    a. 12   **b.** $\dfrac{25}{2}$ **c.** 15   **d.** 20

4. In a plane, two circles intersect in two points. What is the number of common tangents that can be drawn to the circles?

    a. zero   **b.** one   **c.** two   **d.** none of these

5. Points A, B, and C lie on a circle. $\overline{AB}$ is a diameter, and $\overset{\frown}{AC} = 110°$. What is the measure of $\angle BAC$?

    a. 35°   **b.** 55°   **c.** 90°   **d.** 110°

6. In Exercise 5, point D is in such a position that $ABCD$ is an inscribed quadrilateral. What is the sum of $\angle ABC$ and $\angle ADC$?

    a. 90°   **b.** 110°   **c.** 180°   **d.** 145°

7. If $\overset{\frown}{BC} = 120°$ and $\overset{\frown}{AD} = 50°$, what does $\angle X$ equal?

    a. 25°   **b.** 35°   **c.** 60°   **d.** 70°

Exs. 7-9

8. If $\overset{\frown}{BC} = 120°$ and $\overset{\frown}{AD} = 50°$, what does $\angle 1$ equal?

    a. 60°   **b.** 85°   **c.** 90°   **d.** 95°

9. If $AY = j$, $YC = k$, and $YD = 7$, what does BY equal?

    a. $\dfrac{jk}{7}$ **b.** $\dfrac{7j}{k}$ **c.** $\dfrac{7k}{j}$ **d.** $\dfrac{k}{7j}$

10. R and S are points on a circle. $\overline{RS}$ could be which of these?

    a. radius   **b.** diameter   **c.** secant   **d.** tangent

In Exercises 11-13, $\overleftrightarrow{XA}$ is tangent to $\odot O$ at X.

11. Which of these equals $\angle AXZ$?

Exs.11-13

    a. $\overset{\frown}{XYZ}$ **b.** $\angle OXM$ **c.** $\tfrac{1}{2}\overset{\frown}{XY}$ **d.** $\angle XOY$

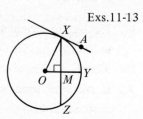

12. If the radius is 13 and $XZ = 24$, what is the distance from O to any chord that is equal to XZ?

    a. 5   **b.** 8   **c.** 11   **d.** $\sqrt{407}$

13. If $OM = 8$ and $MY = 9$, what does XZ equal?

    a. $6\sqrt{2}$ **b.** $2\sqrt{17}$ **c.** $\sqrt{145}$ **d.** 30

Chapter 8

Indicate the best answer by writing the appropriate letter.

1. To inscribe a circle in a triangle, what should you construct first?

a. two medians

b. two angle bisectors

c. two altitudes

d. two perpendicular bisectors

2. In a plane, what is the locus of points equidistant from two given points?

a. a point **b.** a circle **c.** a line **d.** a pair of lines

3. The lengths of two segments are r and s, with $r > s$. It is *not* possible to construct a segment with which of these lengths?

a. $\frac{1}{3}(r+s)$ **b.** $s-r$ **c.** $\sqrt{rs}$ **d.** $\sqrt{r^2+s^2}$

4. Point P lies on line l in a plane. What is the locus of points in that plane that lie 8 cm from P and 2 cm from l?

a. no points **b.** two points **c.** three points **d.** four points

5. You are to construct a tangent to a given $\odot O$ from a point P outside the circle. In the process, it would be *useless* to construct which of these?

a. $\overline{OP}$

b. the perpendicular bisector of $\overline{OP}$

c. a circle with O and P on it

d. a line parallel to $\overline{OP}$

6. It *is* possible to construct an angle with which of these measures?

a. $10°$ **b.** $20°$ **c.** $30°$ **d.** $40°$

7. Which of these could *not* be the intersection of a plane and a sphere?

a. a pair of circles **b.** a circle **c.** a point **d.** the empty set

8. Where *must* the perpendicular bisectors of the sides of a triangle meet?

a. inside the triangle

b. on the triangle

c. outside the triangle

d. none of these

9. What is the locus of points 3 cm from a given point A?

a. a line

b. a plane

c. a cylindrical surface

d. a sphere

10. In a plane, what is the locus of points equidistant from the sides of a square?

a. a square

b. a pair of squares

c. a circle

d. a point

11. You are to construct a perpendicular to a line l at a given point X on l. In how many places on l will you need to position the point of your compass in order to do this construction?

a. one **b.** two **c.** three **d.** four

12. In which polygon is it *not* always possible to inscribe a circle?

a. a rectangle

b. a right triangle

c. an obtuse triangle

d. an equilateral triangle

Chapter 9

Indicate the best answer by writing the appropriate letter.

1. A parallelogram and a triangle have equal areas. The base and height of the parallelogram are 12 and 9. If the base of the triangle is 36, find its height.

 a. 3   **b.** 6   **c.** 9   **d.** 12

2. One side of a rectangle is 14 and the perimeter is 44. What is the area?

 a. 112   **b.** 210   **c.** 224   **d.** 420

3. A square is inscribed in a circle with radius 8. What is the area of the square?

 a. 32   **b.** 64   **c.** $64\sqrt{2}$ **d.** 128

4. The area of a circle is 25π. What is its circumference?

 a. 5π **b.** 10π **c.** 12.5π **d.** 50π

5. What is the area of a trapezoid with bases 7 and 8 and height 6?

 a. 90   **b.** 336   **c.** 45   **d.** 168

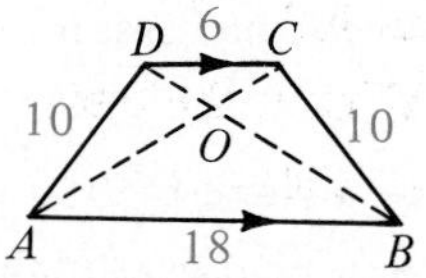

6. What is the area of trapezoid $ABCD$?

 a. 96   **b.** 120   **c.** 144   **d.** 192

7. What is the ratio of the areas of $\triangle AOB$ and $\triangle DOC$?

 a. $\sqrt{3}:1$ **b.** $\sqrt{3}:3$ **c.** $3:1$ **d.** $9:1$

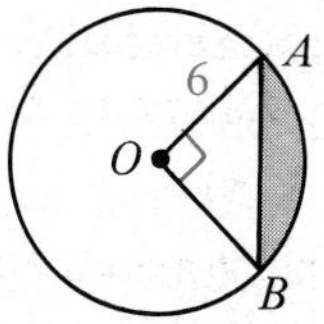

Exs. 6, 7

8. In the diagram, what is the length of $\overset{\frown}{AB}$?

 a. $6\sqrt{2}$ **b.** 6π **c.** 3π **d.** 36π

9. In the diagram, what is the area of the shaded region?

 a. $9\pi - 36$ **b.** $12\pi - 36$ **c.** $9\pi - 18$ **d.** $12\pi - 18$

Exs. 8, 9

10. What is the area of a circle with diameter 12?

 a. $24\pi^2$ **b.** 12π **c.** 144π **d.** 36π

11. What is the area of an equilateral triangle with perimeter 24?

 a. $64\sqrt{3}$ **b.** $32\sqrt{3}$ **c.** $\dfrac{32\sqrt{3}}{3}$ **d.** $16\sqrt{3}$

12. What is the area of a triangle with sides 15, 15, and 24?

 a. 54   **b.** 108   **c.** 180   **d.** 216

13. A rhombus has diagonals 6 and 8. What is its area?

 a. 12   **b.** 24   **c.** 36   **d.** 48

Chapter 10

Indicate the best answer by writing the appropriate letter.

1. Two similar cones have heights 4 and 16. What is the ratio of their volumes?

   a. $1:64$ **b.** $1:4$ **c.** $1:16$ **d.** $4:16$

2. What is the volume of a regular square pyramid with base edge 16 and height 6?

   a. 128   **b.** 256   **c.** 512   **d.** 1536

3. What is the lateral area of the pyramid in Exercise 2?

   a. 256   **b.** 320   **c.** 576   **d.** 640

4. A sphere has area 16π. What is its volume?

   a. $\dfrac{8\pi}{3}$ **b.** $\dfrac{32\pi}{3}$ **c.** $\dfrac{64\pi}{3}$ **d.** $\dfrac{256\pi}{3}$

5. What is the volume of a rectangular solid with dimensions 12, 9, and 6?

   a. 108   **b.** 216   **c.** 432   **d.** 648

6. What is the total surface area of the solid in Exercise 5?

   a. 234   **b.** 468   **c.** 252   **d.** 360

7. A cone has radius 5 and height 12. A cylinder with radius 10 has the same volume as the cone. What is the cylinder's height?

   a. 1   **b.** 2   **c.** 3   **d.** 4

8. A square prism is inscribed in a cylinder with radius 6 and height 10. What is the volume of the prism?

   a. 1080   **b.** 720   **c.** 360   **d.** 240

9. A plane passes 2 cm from the center of a sphere with radius 4 cm. What is the area of the circle of intersection?

   a. $12\pi \text{ cm}^2$ **b.** $16\pi \text{ cm}^2$ **c.** $18\pi \text{ cm}^2$ **d.** $20\pi \text{ cm}^2$

10. The base of a right prism is a regular hexagon with side 4. The height of the prism is 6. What is the volume of the prism?

    a. $144\sqrt{3}$ **b.** $72\sqrt{3}$ **c.** $48\sqrt{3}$ **d.** $36\sqrt{3}$

11. What is the lateral area of the prism in Exercise 10?

    a. 24   **b.** 36   **c.** 72   **d.** 144

12. Find the total surface area of a cylinder with radius 4 and height 6.

    a. 16π **b.** 32π **c.** 48π **d.** 80π

13. Two similar pyramids have volumes 27 and 125. If the smaller has lateral area 18, what is the lateral area of the larger?

    a. 30   **b.** $83\frac{1}{3}$ **c.** 50   **d.** 25

Chapter 11

Indicate the best answer by writing the appropriate letter.

1. A line with slope $\frac{2}{5}$ passes through point $(1, 4)$. What is an equation of the line?

   a. $y - 4 = \frac{2}{5}(x - 1)$ **b.** $y - 4 = \frac{5}{2}(x - 1)$

   c. $y + 4 = \frac{2}{5}(x + 1)$ **d.** $y - 1 = \frac{5}{2}(x - 4)$

2. The midpoint of $\overline{AB}$ is $(3, 4)$. If the coordinates of B are $(6, 6)$, what are the coordinates of A?

   a. $(9, 10)$ **b.** $(4.5, 5)$ **c.** $(0, 2)$ **d.** $(9, 10)$

3. The slope of line l is $\frac{2}{7}$. What is the slope of any line perpendicular to l?

   a. $\frac{2}{7}$ **b.** $-\frac{2}{7}$ **c.** $\frac{7}{2}$ **d.** $-\frac{7}{2}$

4. What is the distance between points $(-2, 0)$ and $(2, 5)$?

   a. 5 **b.** 3 **c.** $\sqrt{29}$ **d.** $\sqrt{41}$

5. Which point lies on the line $3x + 2y = 12$?

   a. $(0, 4)$ **b.** $(\frac{14}{3}, 1)$ **c.** $(\frac{10}{3}, -1)$ **d.** $(\frac{10}{3}, 1)$

6. What is an equation of the circle with center $(3, 0)$ and radius 8?

   a. $x^2 + y^2 = 64$ **b.** $(x - 3)^2 + y^2 = 64$

   c. $(x + 3)^2 + y^2 = 8$ **d.** $(x - 3)^2 + y^2 = 8$

7. What is the *best* term for a quadrilateral with vertices $(-5, 0)$, $(3, 6)$, $(6, 2)$, and $(-2, -4)$?

   a. trapezoid   **b.** parallelogram   **c.** rectangle   **d.** rhombus

8. Three consecutive vertices of a parallelogram are $(j, 7)$, $(0, 0)$, and $(7, 0)$. Which is the fourth vertex?

   a. $(7, 7)$ **b.** $(j - 7, 7)$ **c.** $(j + 7, 7)$ **d.** $(7, j)$

9. Points $(2, 2)$ and $(8, v)$ lie on a line whose slope is $\frac{1}{2}$. What is the value of v?

   a. -10 **b.** -1 **c.** 5 **d.** 14

10. What is the *best* term for a triangle with vertices $(1, -3)$, $(6, 2)$, and $(0, 4)$?

    a. isosceles triangle   **b.** equilateral triangle

    c. right triangle   **d.** none of these

11. Which point is the intersection of lines $3x + 2y = 17$ and $x - 4y = 1$?

    a. $(1, 5)$ **b.** $(5, 1)$ **c.** $(-1, 5)$ **d.** $(\frac{33}{5}, \frac{7}{5})$

12. What is the slope of the segment that joins points $(-3, 5)$ and $(2, 8)$?

    a. -1 **b.** $-\frac{3}{5}$ **c.** $\frac{5}{3}$ **d.** $\frac{3}{5}$

Chapter 12

Indicate the best answer by writing the appropriate letter.

1. What is the image of the point $(2, 3)$ by reflection in the x-axis?

   a. $(3, 2)$ **b.** $(-2, 3)$ **c.** $(2, -3)$ **d.** $(-2, -3)$

2. $T: (x, y) \rightarrow (x, y - 2)$. What is the preimage of $(3, 5)$?

   a. $(5, 7)$ **b.** $(3, 7)$ **c.** $(3, 3)$ **d.** $(5, 3)$

3. A regular hexagon does *not* have which symmetry?

   a. line   **b.** point   **c.** $30°$ rotational   **d.** $120°$ rotational

4. If O is the point $(0, 0)$, what is the image of $(3, 6)$ by $D_{0, \frac{1}{3}}$?

   a. $(9, 18)$ **b.** $(2, 4)$ **c.** $(1, 2)$ **d.** $(-1, -2)$

5. What is the image of $(-1, 3)$ by a half-turn about $(1, 2)$?

   a. $(3, 1)$ **b.** $(1, -3)$ **c.** $(-1, -2)$ **d.** $(3, -1)$

6. What is the image of J by $M_y \circ M_x$?

   a. J **b.** K **c.** L **d.** M

7. T is a translation that maps K to N. What is the image of J by T?

   a. K **b.** O **c.** N **d.** L

8. What is the image of J by $M_x \circ H_O$?

   a. J **b.** K **c.** L **d.** M

9. What is the image of $\triangle LMJ$ by $R_{0, 90}$?

   a. $\triangle JKL$ **b.** $\triangle KLM$ **c.** $\triangle LMJ$ **d.** $\triangle MJK$

Exs. 6-9

10. $T: (x, y) \rightarrow (x, y - 2)$. What is the image of $(5, 3)$ by T^{-1}?

   a. $(3, 3)$ **b.** $(5, 1)$ **c.** $(3, 5)$ **d.** $(5, 5)$

11. Isometry $S: \square ABCD \rightarrow \square JKLM$. Which statement *must* be true?

   a. $\angle DAB = \angle JKL$ **b.** $AC = JL$ **c.** $S: C \rightarrow M$ **d.** $CD = MJ$

12. What is the mirror of reflection for a transformation that maps $(-2, 1)$ to $(2, 1)$?

   a. the x-axis   **b.** the line $y = x$ **c.** the y-axis   **d.** the origin

13. How many lines of symmetry does a rhombus have?

   a. none   **b.** one   **c.** two   **d.** four

Logic

Statements and Truth Tables

In algebra, you have used letters to represent numbers. In logic, letters are used to represent simple statements which are either true or false. For example, p might represent the statement "Paris is the capital city of France," and q might represent the statement "The moon is made of green cheese."

Simple statements can be joined to form **compound statements.** Two important compound statements are defined below.

A **conjunction** is a compound statement composed of two simple statements joined by the word "and." The symbol $\wedge$ is used to represent the word "and."

A **disjunction** is a compound statement composed of two simple statements joined by the word "or." The symbol $\vee$ is used to represent the word "or."

Example:

Simple statements:	p	Mom plays the guitar.
	q	Dad plays the piano.
Conjunction:	$p \wedge q$	Mom plays the guitar and Dad plays the piano.
Disjunction:	$p \vee q$	Mom plays the guitar or Dad plays the piano.

The table at the right is called a **truth table.** It tells you the conditions under which a conjunction is a true statement. "T" stands for "true" and "F" for "false." The first row of the table shows that when statement p is true and statement q is true, the conjunction $p \wedge q$ is true. The other rows of the table show that $p \wedge q$ is false when either of its component statements is false.

Truth table for conjunction

p	q	$p \wedge q$
T	T	T
T	F	F
F	T	F
F	F	F

In everyday speech, the word "or" is used in two different ways.

Inclusive use of "or": Rosa will go or Mary will go. This statement is true if Rosa goes, or if Mary goes, or if both Rosa and Mary go.

Exclusive use of "or": Jake will be elected class president or Carol will be elected class president. This statement is true if either Jake or Carol is elected class president. Obviously they can't both be elected. The statement means that Jake will be chosen or else Carol will be.

We shall deal only with the inclusive use of "or" in this course. The first row of the truth table for disjunction shows that when both p and q are true, $p \lor q$ is true. The next two rows show that the compound statement $p \lor q$ is true when either of its components is true. The last row shows that a disjunction is false when both of its components are false.

Truth table for disjunction

p	q	$p \lor q$
T	T	T
T	F	T
F	T	T
F	F	F

In addition to the words "and" and "or," the word "not" is an important word in logic. If p is a statement, then the statement "p is not true," usually shortened to "not p" and written $\sim p$, is called the **negation** of p.

Example:

Statement:	p	Sam is sleeping in class.
Negation:	$\sim p$	It is not true that Sam is sleeping in class.
(Two forms)	$\sim p$	Sam is not sleeping in class.

The truth table for negation shows that when p is true, $\sim p$ is false. When p is false, $\sim p$ is true.

Truth table for negation

p	$\sim p$
T	F
F	T

An example will show how to make truth tables for some other compound statements.

Example: Make a truth table for the statement $\sim p \lor \sim q$.

Solution:

1. Make a column for p and a column for q. Write all possible combinations of T and F in the standard pattern shown.

2. Since $\sim p$ is a part of the given statement, add a column for $\sim p$. To fill out this column, use the first column and refer to the truth table for negation above. Similarly, add a column for $\sim q$.

p	q	$\sim p$	$\sim q$	$\sim p \lor \sim q$
T	T	F	F	F
T	F	F	T	T
F	T	T	F	T
F	F	T	T	T

3. Using the columns for $\sim p$ and $\sim q$, refer to the truth table for disjunction above in order to fill out the column for $\sim p \lor \sim q$. Remember that a disjunction is false only when both of its components are false.

To make a truth table for a compound statement involving three simple statements p, q, and r, you would need an eight-row table to show all possible combinations of T and F. The standard pattern across the three columns headed p, q, and r is as follows: TTT, TTF, TFT, TFF, FTT, FTF, FFT, FFF.

Exercises

Suppose *p* stands for "I like the city," and *q* stands for "You like the country." Express in words each of the following statements.

1. $p \land q$ **2.** $\sim p$ **3.** $\sim q$ **4.** $p \lor q$

5. $p \lor \sim q$ **6.** $\sim(p \land q)$ **7.** $\sim p \land q$ **8.** $\sim(p \lor q)$

Suppose *p* stands for "Hawks swoop," and *q* stands for "Gulls glide." Express in symbolic form each of the following statements.

9. Hawks swoop or gulls glide.

10. Gulls do not glide.

11. Hawks do not swoop and gulls do not glide.

12. "Hawks swoop or gulls glide" is not true.

13. Hawks do not swoop or gulls do not glide.

14. "Hawks swoop and gulls glide" is not true.

15. Do the statements in Exercises 11 and 12 mean the same thing?

16. Do the statements in Exercises 13 and 14 mean the same thing?

Make a truth table for each of the following statements.

17. $p \lor \sim q$ **18.** $\sim p \lor q$ **19.** $\sim(\sim p)$ **20.** $\sim(p \lor q)$

21. $p \lor \sim p$ **22.** $p \land \sim p$ **23.** $p \land (q \lor r)$ **24.** $(p \land q) \lor (p \land r)$

Truth Tables for Conditionals

The conditional statement "If *p* then *q*," which is discussed in Section 2-6, is symbolized as $p \to q$. This is also read as "*p* implies *q*" and as "*q* follows from *p*." The truth table for $p \to q$ is shown at the right. Notice that the only time a conditional is false is when the hypothesis *p* is true and the conclusion *q* is false. The example below will show why this is a reasonable way to make out the truth table.

Truth table for conditionals

p	q	$p \to q$
T	T	T
T	F	F
F	T	T
F	F	T

Example:

Mom promises, "If I catch the early train home, I'll take you swimming." Consider the four possibilities of the truth table.

1. Mom catches the early train home and takes you swimming. She kept her promise; her statement was *true*.

2. Mom catches the early train home but does not take you swimming. She broke her promise; her statement was *false*.

3. Mom does not catch the early train home but still takes you swimming. She has not broken her promise; her statement was *true*.

4. Mom does not catch the early train home and does not take you swimming. She has not broken her promise; her statement was *true*.

The tables below show the converse and contrapositive of $p \rightarrow q$. Make sure that you understand how these tables were made. Notice that the last column of the table for the contrapositive $\sim q \rightarrow \sim p$ is identical with the last column of the above table for $p \rightarrow q$. In other words, the contrapositive of a statement is true (or false) exactly when the statement itself is true (or false). This is what we mean when we say that a statement and its contrapositive are logically equivalent. On the other hand, a statement and its converse are not logically equivalent. The last columns in their truth tables are different.

Converse of $p \rightarrow q$

p	q	$q \rightarrow p$
T	T	T
T	F	T
F	T	F
F	F	T

Contrapositive of $p \rightarrow q$

p	q	$\sim q$	$\sim p$	$\sim q \rightarrow \sim p$
T	T	F	F	T
T	F	T	F	F
F	T	F	T	T
F	F	T	T	T

Exercises

Suppose p represents "You love bananas," q represents "You are a monkey's uncle," and r represents "Your nephew is a chimp." Express in words each of the following statements.

1. $p \rightarrow q$ **2.** $q \rightarrow r$ **3.** $\sim q \rightarrow \sim r$ **4.** $\sim(p \rightarrow q)$

5. $(p \wedge q) \rightarrow r$ **6.** $p \wedge (q \rightarrow r)$ **7.** $(r \vee q) \rightarrow p$ **8.** $r \vee (q \rightarrow p)$

Let b, s, and w represent the following statements.
b: Bonnie bellows. s: Sheila shouts. w: Wilbur whispers.
Express in symbolic form each of the following statements.

9. If Bonnie bellows, then Wilbur whispers.

10. If Wilbur whispers, then Sheila does not shout.

11. If Bonnie does not bellow or Wilbur does not whisper, then Sheila shouts.

12. Sheila shouts, and if Bonnie bellows, then Wilbur whispers.

13. It is not true that Sheila shouts if Bonnie bellows.

14. If Bonnie does not bellow, then Wilbur whispers and Sheila shouts.

15. **a.** Make a truth table for $\sim p \to \sim q$ (the inverse of $p \to q$). Your first two columns should be the same as the first two columns of the table for $p \to q$. The last columns of the two tables should be different. Are they? Is $\sim p \to \sim q$ logically equivalent to $p \to q$?

    b. Compare the truth table for $\sim p \to \sim q$ (the inverse of $p \to q$) with the truth table for $q \to p$ (the converse of $p \to q$). Are the last columns the same? Are the inverse and the converse logically equivalent?

Make truth tables for the following statements.

16. $p \to \sim q$ 17. $\sim(p \to q)$ 18. $p \wedge \sim q$

19. By comparing the truth tables in Exercises 16–18, you should find that two of the three statements are logically equivalent. Which two?

20. The statement "p if and only if q" is defined as $(p \to q) \wedge (q \to p)$. Make a truth table for this statement.

Some Rules of Inference

Four rules for making logical inferences are symbolized below. A horizontal line separates the given information, or premises, from the conclusion. If you accept the given statement or statements as true, then you must accept as true the conclusion shown.

1. Modus Ponens

   $p \to q$
   $\underline{p}$
   Therefore, q

2. Modus Tollens

   $p \to q$
   $\underline{\sim q}$
   Therefore, $\sim p$

3. Simplification

   $\underline{p \wedge q}$
   Therefore, p

4. Disjunctive Syllogism

   $p \vee q$
   $\underline{\sim p}$
   Therefore, q

You should convince yourself that these rules make good sense. For example, Rule 4 says that if you know that "p or q" is true and then you find out that p is not true, you must conclude that q is true.

Example 1:   If today is Tuesday, then tomorrow is Wednesday.
               Today is Tuesday.
               _____
               Therefore, tomorrow is Wednesday. (Rule 1)

Example 2:   If a figure is a triangle, then it is a polygon.
This figure is not a polygon.
Therefore, this figure is not a triangle. (Rule 2)

Example 3:   It is Tuesday and it is April.
Therefore, it is Tuesday. (Rule 3)

Example 4:   It is a square or it is a trapezoid.
It is not a square.
Therefore, it is a trapezoid. (Rule 4)

Rules 1-4 can be used to prove more complicated arguments as in the following example.

Example 5:

Given: $p \rightarrow q$; $p \vee r$; $\sim q$

Prove: r

Proof:

Statements	Reasons
1. $p \rightarrow q$	1. Given
2. $\sim q$	2. Given
3. $\sim p$	3. Steps 1 and 2 and Modus Tollens
4. $p \vee r$	4. Given
5. r	5. Steps 3 and 4 and Disjunctive Syllogism

Exercises

Supply the reasons to complete each proof.

1. Given: $p \wedge q$; $p \rightarrow s$
  Prove: s

Statements
1. $p \wedge q$
2. p
3. $p \rightarrow s$
4. s

2. Given: $r \rightarrow s$; r; $s \rightarrow t$
  Prove: t

Statements
1. $r \rightarrow s$
2. r
3. s
4. $s \rightarrow t$
5. t

Write two-column proofs for the following.

3. Given: $p \lor q$; $\sim p$; $q \to s$
   Prove: s

4. Given: $a \to b$; $a \lor c$; $\sim b$
   Prove: c

5. Given: $a \land b$; $a \to \sim c$; $c \lor d$
   Prove: d

6. Given: $p \land q$; $p \to \sim s$; $r \to s$
   Prove: $\sim r$

Symbolize the statements, accept them as true, and write two-column proofs.

7. If Jim jogs, then the dog barks.
   If the dog barks, then the cat scats.
   Jim jogs and Willa runs.

   Prove that the cat scats. (Use the letters j, d, c, and w.)

8. Alice is on the team and Rachel is also.
   Barb is on the team or Carol is.
   If Alice is on the team, then Barb is not.

   Prove that Carol is on the team. (Use the letters a, r, b, and c.)

9. Turner will play fullback or Rizzo will play fullback.
   If Turner plays fullback, then Packard will play halfback.
   If Rizzo plays fullback, then Sullivan will be quarterback.
   Packard will not play halfback.

   Prove that Sullivan will be quarterback. (Use the letters t, r, p, and s.)

Some Rules of Replacement

The symbol $\equiv$ means "is logically equivalent to." Thus Rule 5 below states that the conditional statement $p \to q$ is logically equivalent to its contrapositive, $\sim q \to \sim p$. Rules 6–10 give other logical equivalences. These can be verified by comparing the truth tables of the statements on both sides of the $\equiv$ sign.

5. Contrapositive Rule

$$p \to q \equiv \sim q \to \sim p$$

6. Double Negation

$$\sim(\sim p) \equiv p$$

7. Commutative Rules

$$p \land q \equiv q \land p$$
$$p \lor q \equiv q \lor p$$

8. Associative Rules

$$(p \land q) \land r \equiv p \land (q \land r)$$
$$(p \lor q) \lor r \equiv p \lor (q \lor r)$$

9. Distributive Rules

$$p \land (q \lor r) \equiv (p \land q) \lor (p \land r)$$
$$p \lor (q \land r) \equiv (p \lor q) \land (p \lor r)$$

10. DeMorgan's Rules

$$\sim(p \land q) \equiv \sim p \lor \sim q$$
$$\sim(p \lor q) \equiv \sim p \land \sim q$$

Any logically equivalent expressions can replace each other wherever they occur in a proof.

Example:

Given: $p \wedge q$; $q \to \sim(r \vee s)$

Prove: $\sim r \wedge \sim s$

Proof:

Statements	Reasons
1. $p \wedge q$	1. Given
2. $q \wedge p$	2. Step 1 and Commutative Rule
3. q	3. Step 2 and Simplification
4. $q \to \sim(r \vee s)$	4. Given
5. $\sim(r \vee s)$	5. Steps 3 and 4 and Modus Ponens
6. $\sim r \wedge \sim s$	6. Step 5 and DeMorgan's Rule

Exercises

Supply the reasons to complete each proof.

1. Given: $a \to \sim b$; b

Prove: $\sim a$

Statements

1. b
2. $\sim(\sim b)$
3. $a \to \sim b$
4. $\sim a$

2. Given: $a \vee (b \wedge c)$; $\sim b$

Prove: a

Statements

1. $a \vee (b \wedge c)$
2. $(a \vee b) \wedge (a \vee c)$
3. $a \vee b$
4. $b \vee a$
5. $\sim b$
6. a

Write two-column proofs for the following.

3. Given: $a \wedge (b \wedge c)$

Prove: c

5. Given: $p \vee (\sim q)$; q

Prove: p

7. Given: $p \vee (q \wedge s)$

Prove: $p \vee s$

4. Given: $(p \wedge q) \to s$; $\sim s$

Prove: $\sim p \vee \sim q$

6. Given: $\sim q \to \sim p$; $q \to r$; p

Prove: r

8. Given: $t \vee (r \vee s)$; $\sim r \wedge \sim s$

Prove: t

Symbolize the statements and write two-column proofs.

9. Observers agree on three things:

If the strike is settled by Monday, then the workers will return to work on Tuesday.

If the strike is not settled by Monday, then the picket lines will be crossed. The picket lines will never be crossed.

Prove that the workers will return to work on Tuesday. (Let s = strike is settled by Monday, p = picket lines will be crossed, and w = workers will return to work on Tuesday.)

10. A detective has established these facts:

Lady Eastwick stole the jewels or Sir Castleton stole them.

The butler is telling the truth or the maid is lying.

If the butler is telling the truth, then the safe was left open.

If the maid is lying, then Lady Eastwick did not steal the jewels.

The safe was not left open.

Prove that Sir Castleton stole the jewels. (Let e = Lady Eastwick stole the jewels, c = Sir Castleton stole the jewels, b = butler is telling truth, m = maid is lying, and s = safe was left open.)

Application of Logic to Electrical Circuits

The diagram at the right represents part of an electrical circuit. When switch p is open, the electricity which is flowing from A will not reach B. When switch p is closed, as in the second diagram, the electricity flows through the switch to B.

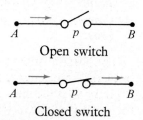

Open switch

Closed switch

The left diagram below represents two switches p and q which are *connected in series*. Notice that current will flow if and only if both switches are closed. The right diagram below represents the switches p and q *connected in parallel*. Notice that current will flow if either switch is closed or if both switches are closed. If switches p and q are both open, the current cannot flow.

Series circuit

Parallel circuit

In order to understand how circuits are related to truth tables, let us do the following:

1. If a switch is closed, label it T. If it is open, label it F.
2. If current will flow in a circuit, label the circuit T. If the current will not flow, label the circuit F.

With these agreements, we can use truth tables to show what happens in the series circuit and the parallel circuit on page 484.

Series circuit

p	q	circuit
T	T	T
T	F	F
F	T	F
F	F	F

Parallel circuit

p	q	circuit
T	T	T
T	F	T
F	T	T
F	F	F

Notice that the truth table for the series circuit is just like the truth table for $p \wedge q$. Also, the truth table for the parallel circuit is just like the table for $p \vee q$.

Now study the circuit shown at the right. Notice that one of the switches is labeled $\sim q$. This means that this switch is open if switch q is closed, and vice versa. (In many electrical circuits, switch $\sim q$ will open automatically when switch q is closed.)

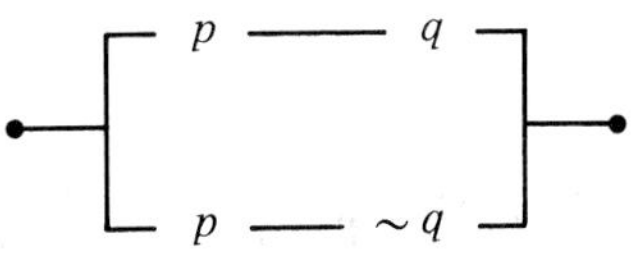

$(p \wedge q) \vee (p \wedge \sim q)$

The circuit shown above is basically a parallel circuit, but in each branch of the circuit there are two switches connected in series. This explains why the circuit is labeled $(p \wedge q) \vee (p \wedge \sim q)$. A truth table for this circuit is given below.

p	q	$\sim q$	$p \wedge q$	$p \wedge \sim q$	$(p \wedge q) \vee (p \wedge \sim q)$
T	T	F	T	F	T
T	F	T	F	T	T
F	T	F	F	F	F
F	F	T	F	F	F

Notice that the first and last columns of the truth table are identical. This means that the complicated circuit shown above can be replaced by a simpler circuit which contains just switch p! In other words, logic can be used to replace a complex electrical circuit by a simpler one.

Exercises

Symbolize each circuit, using ∧, ∨, ~, and the letters given for the switches in each diagram.

1.

2.

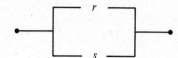

3.

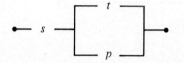

4.

5.

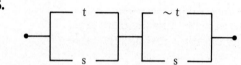

6.

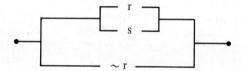

7. Draw a diagram for the circuit $p \land \sim p$; also for the circuit $p \lor \sim p$. Electricity can always pass through one of these circuits and can never pass through the other. Which is which?

8. According to the commutative rule, $p \land q \equiv q \land p$. This means that the circuit $p \land q$ does the same thing as the circuit $q \land p$. Make a diagram of each circuit.

9. According to the associative rule, $(p \lor q) \lor r \equiv p \lor (q \lor r)$. Draw diagrams for each circuit.

10. According to the distributive rule, $p \land (q \lor r) \equiv (p \land q) \lor (p \land r)$. Draw diagrams for each circuit.

11. Make both a diagram and a truth table for the circuit $(p \lor q) \lor \sim q$. Notice that the last column of your table is always T so that current always flows. This means that all of the switches could be eliminated.

12. Make both a diagram and a truth table for the circuit $(p \lor q) \land (p \lor \sim q)$. Describe a simpler circuit equivalent to this circuit.

Projections; Dihedral Angles

Projections

The purpose of this section is to help you develop skill in working with three-dimensional figures. As a start, you should read the first paragraph on page 423 carefully. On that page, point P' is called the image of point P. You can also say that P' is the **projection** of P into the plane.

In the figure below, each point of $\triangle ABC$ is mapped into a point on plane M by dropping a perpendicular from the point of $\triangle ABC$ to plane M.

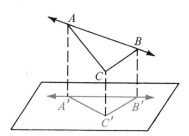

Original figure	**Projection into plane M**
Point A	Point A'
$\overline{AB}$	$\overline{A'B'}$
$\overleftrightarrow{AB}$	$\overleftrightarrow{A'B'}$
$\angle CAB$	$\angle C'A'B'$
$\triangle CAB$	$\triangle C'A'B'$

Exercises

Exercises 1–10 deal with a square $ABCD$, whose sides are 20 units long, and the projection of that square into a plane F. You can improve your thinking by cutting a square out of cardboard and holding your model in various positions with respect to the floor. Tell whether it is possible for the square to be in such position that its projection into F becomes the figure named.

A 1. A square

    3. A 1 by 20 rectangle

    5. A 30-unit segment

    7. A $20\sqrt{2}$-unit segment

2. A 10 by 20 rectangle

4. A 20-unit segment

6. A 10-unit segment

8. A trapezoid

B 9. A nonsquare rhombus

10. A parallelogram that is neither a rectangle nor a rhombus

For each exercise, draw a diagram to show that the projection of an acute angle RUZ into a plane Q is the figure named.

11. A ray

13. An angle equal to $\angle RUZ$

12. A line

14. An obtuse angle

The figure named is to be projected into some plane. Describe the various projections that can be formed.

C 15. Two parallel lines

16. A circle

17. An equilateral triangle

18. A regular hexagon

19. An ordinary parallelogram

20. An obtuse angle

Dihedral Angles

Your answers to Exercises 4 and 7 on page 487 should have been yes. You might naturally say, in each case, that the plane of square *ABCD* is perpendicular to plane *F.* To describe the ways in which two planes can intersect, we use the terms *half-plane* and *dihedral angle.* Any line in a plane separates the plane into two **half-planes,** and all the points on one side of the line lie in the same half-plane. The diagrams below suggest that a **dihedral angle** is the figure formed by two half-planes that have the same edge.

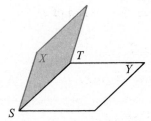

Dihedral angle *X-ST-Y*

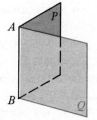

Dihedral angle *P-AB-Q*

The diagram at the left above suggests an *acute dihedral angle.* But just what does this phrase mean? You measure a dihedral angle as shown below.

In X, $\overrightarrow{EJ} \perp \overleftrightarrow{ST}$.

In Y, $\overrightarrow{EK} \perp \overleftrightarrow{ST}$.

$\angle JEK$ is a **plane angle** of dihedral angle *X-ST-Y.*

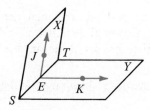

Dihedral angle *X-ST-Y* is measured in terms of one of its plane angles. In Exercise 13 you will prove that any two plane angles of a dihedral angle are equal. Suppose $\angle JEK = 80°$. Then dihedral angle *X-ST-Y* is an 80° dihedral angle, and we say that planes *X* and *Y* meet at an 80° angle.

The diagram below shows how to measure the angle that a line makes with a plane.

$\overrightarrow{AL}$ is the projection of $\overrightarrow{AP}$ into plane M.

$\angle PAL = 30°$

Noting that $\overleftrightarrow{AL}$ is a very special line, namely the projection of $\overleftrightarrow{AP}$ into the plane, we say that $\overleftrightarrow{AP}$ makes a 30° angle with the plane and that $\overleftrightarrow{AP}$ is inclined at a 30° angle to the plane.

Exercises

Draw a diagram that represents the figure named.

A **1.** An acute dihedral angle

2. A right dihedral angle

3. An obtuse dihedral angle

4. Adjacent dihedral angles

B **5.** Write your own definition of perpendicular planes. (*Hint:* Use the idea of a plane angle of a dihedral angle.)

6. a. Draw a plane X and a line k that is perpendicular to X.
   b. Draw a plane Y that contains k.
   c. What appears to be true about plane Y with respect to plane X?
   d. Complete this statement suggested by parts (a)-(c):
      If a line is perpendicular to a plane, then a second plane . . .

7. State theorems dealing with planes and dihedral angles suggested by the three theorems found on pages 46 and 47.

In Exercises 8-10 use a regular square pyramid each of whose edges is 6 units long.

C **8. a.** Estimate the angle a lateral edge makes with the base of the pyramid.
   b. Compute to the nearest degree the angle named in part (a). (*Hint:* Draw your own figure, including the altitude and a segment joining the foot of the altitude to a vertex of the base.)

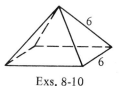

Exs. 8-10

9. a. Estimate the measure of the dihedral angle formed by a lateral face and the base.
   b. Compute to the nearest degree the angle described in part (a). You will need to use a trigonometry table.

10. Repeat Exercise 9, but use the dihedral angle formed by two intersecting lateral faces.

11, 12. Using a regular triangular pyramid, repeat Exercises 8 and 9.

13. To show that it is reasonable to speak of *the* plane angle of a dihedral angle, you can show that any two plane angles of a dihedral angle are equal. Explain how you can show this. You may use, for this exercise, a theorem we haven't proved: If two lines are parallel to a third line, the two lines are parallel to each other.

14. *Challenge problem:* In the diagram, $\overrightarrow{AX}$ is oblique to plane ABC, $\angle CAB = 50°$, $\angle XAB = 65°$, $\angle XAC = 70°$, $AX = 10$, $\overline{XP} \perp$ plane ABC, and all right angles are as shown. Find:
a. AQ, AR, and XR
b. The angle that AX makes with plane ABC

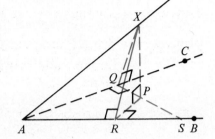

Postulates

Postulate 1 A line contains at least two points; a plane contains at least three points not all in one line; space contains at least four points not all in one plane. (p. 31)

Postulate 2 Through any two points there is exactly one line. (p. 31)

Postulate 3 Through any three points there is at least one plane, and through any three noncollinear points there is exactly one plane. (p. 31)

Postulate 4 If two points are in a plane, then the line joining the points is in that plane. (p. 31)

Postulate 5 If two planes intersect, then their intersection is a line. (p. 31)

Postulate 6 (Ruler Postulate) The points on a line can be paired with the real numbers in such a way that:
a. Any two desired points can have coordinates 0 and 1.
b. The distance between any two points equals the absolute value of the difference of their coordinates. (p. 32)

Postulate 7 (Protractor Postulate) Select a line $\overleftrightarrow{AB}$ and choose any point O between A and B. Consider the rays $\overrightarrow{OA}$ and $\overrightarrow{OB}$ and all the rays that can be drawn from O on one side of $\overleftrightarrow{AB}$. These rays can be paired with the real numbers from 0 to 180 in such a way that:
a. $\overrightarrow{OA}$ is paired with 0, and $\overrightarrow{OB}$ with 180.
b. If $\overrightarrow{OP}$ is paired with x, and $\overrightarrow{OQ}$ with y,
then $\angle POQ = |x - y|°$. (p. 32)

Postulate 8 If two parallel lines are cut by a transversal, then corresponding angles are equal. (p. 46)

Postulate 9 If two lines are cut by a transversal and corresponding angles are equal, then the lines are parallel. (p. 50)

Postulate 10 (SSS Postulate) If three sides of one triangle are equal to three sides of another triangle, then the triangles are congruent. (p. 85)

Postulate 11 (SAS Postulate) If two sides and the included angle of one triangle are equal to two sides and the included angle of another triangle, then the triangles are congruent. (p. 85)

Postulate 12   (ASA Postulate)   If two angles and the included side of one triangle are equal to two angles and the included side of another triangle, then the triangles are congruent.   (p. 86)

Postulate 13   (AA Similarity Postulate)   If two angles of one triangle are equal to two angles of another triangle, then the triangles are similar. (p. 180)

Postulate 14   The area of a square is the square of the length of a side. $(A = s^2)$ (p. 322)

Postulate 15   (Area Congruence Postulate)   If two figures are congruent, then they have the same area.   (p. 322)

Postulate 16   (Area Addition Postulate)   The area of a polygonal region is the sum of the areas of its non-overlapping parts.   (p. 322)

Theorems

Points, Lines, Planes, and Angles

1-1 If two angles are complements of equal angles (or of the same angle), then the two angles are equal. (p. 22)

1-2 If two angles are supplements of equal angles (or of the same angle), then the two angles are equal. (p. 22)

1-3 Vertical angles are equal. (p. 23)

1-4 If two lines intersect, then they intersect in exactly one point. (p. 33)

1-5 If there is a line and a point not in the line, then exactly one plane contains them. (p. 33)

1-6 If two lines intersect, then exactly one plane contains them. (p. 33)

1-7 Every segment has exactly one midpoint. (p. 33)

1-8 Every angle has exactly one bisector. (p. 33)

Parallel Lines and Planes

2-1 If two parallel planes are cut by a third plane, then the lines of intersection are parallel. (p. 42)

2-2 If two parallel lines are cut by a transversal, then alternate interior angles are equal. (p. 46)

2-3 If two parallel lines are cut by a transversal, then same-side interior angles are supplementary. (p. 46)

2-4 If a transversal is perpendicular to one of two parallel lines, then it is perpendicular to the other one also. (p. 47)

2-5 If two lines are cut by a transversal and alternate interior angles are equal, then the lines are parallel. (p. 50)

2-6 If two lines are cut by a transversal and same-side interior angles are supplementary, then the lines are parallel. (p. 50)

2-7 If a transversal is perpendicular to two lines, then the two lines are parallel. (p. 51)

2-8 Through a point outside a line, exactly one parallel can be drawn to the line. (p. 51)

2-9 Through a point outside a line, exactly one perpendicular can be drawn to the line. (p. 51)

2-10 The sum of the angles of a triangle is 180°. (p. 57)

    Corollary 1 If two angles of one triangle are equal to two angles of another triangle, then the third angles are equal.

    Corollary 2 Each angle of an equiangular triangle has measure 60°.

    Corollary 3 In a triangle, there can be at most one right angle or obtuse angle.

    Corollary 4 The acute angles of a right triangle are complementary.

2-11 If one side of a triangle is extended, then the exterior angle formed equals the sum of the two remote interior angles. (p. 58)

2-12 The sum of the angles of a convex polygon with n sides is $(n - 2)180°$. (p. 62)

2-13 The sum of the exterior angles of any convex polygon, one angle at each vertex, is 360°. (p. 62)

Congruent Triangles

3-1 If two sides of a triangle are equal, then the angles opposite those sides are equal. (p. 94)

    Corollary 1 An equilateral triangle is also equiangular. (p. 94)

    Corollary 2 An equilateral triangle has three 60° angles. (p. 94)

    Corollary 3 The bisector of the vertex angle of an isosceles triangle is perpendicular to the base at its midpoint. (p. 94)

3-2 If two angles of a triangle are equal, then the sides opposite those angles are equal. (p. 95)

    Corollary An equiangular triangle is also equilateral. (p. 95)

3-3 (AAS Theorem) If two angles and a non-included side of one triangle are equal to the corresponding parts of another triangle, then the triangles are congruent. (p. 98)

3-4 (HL Theorem) If the hypotenuse and a leg of one right triangle are equal to the corresponding parts of another right triangle, then the triangles are congruent. (p. 98)

3-5 If a point lies on the bisector of an angle, then the point is equidistant from the sides of the angle. (p. 99)

3-6 If a point is equidistant from the sides of an angle, then the point lies on the bisector of the angle. (p. 99)

3-7   If a point lies on the perpendicular bisector of a segment, then the point is equidistant from the endpoints of the segment.   (p. 104)

3-8   If a point is equidistant from the endpoints of a segment, then the point lies on the perpendicular bisector of the segment.   (p. 104)

Using Congruent Triangles

4-1   Opposite sides of a parallelogram are equal.   (p. 124)

   Corollary   If two lines are parallel, then all points on each line are equidistant from the other line.   (p. 124)

4-2   Opposite angles of a parallelogram are equal.   (p. 124)

4-3   The diagonals of a parallelogram bisect each other.   (p. 124)

4-4   If one pair of opposite sides of a quadrilateral are equal and parallel, then the quadrilateral is a parallelogram.   (p. 127)

4-5   If both pairs of opposite sides of a quadrilateral are equal, then the quadrilateral is a parallelogram.   (p. 127)

4-6   If both pairs of opposite angles of a quadrilateral are equal, then the quadrilateral is a parallelogram.   (p. 128)

4-7   If the diagonals of a quadrilateral bisect each other, then the quadrilateral is a parallelogram.   (p. 128)

4-8   If three parallel lines cut off equal segments on one transversal, then they cut off equal segments on every transversal.   (p. 128)

4-9   The diagonals of a rectangle are equal.   (p. 132)

4-10   The diagonals of a rhombus are perpendicular.   (p. 132)

4-11   Each diagonal of a rhombus bisects a pair of opposite angles.   (p. 132)

4-12   The midpoint of the hypotenuse of a right triangle is equidistant from the three vertices.   (p. 133)

4-13   Base angles of an isosceles trapezoid are equal.   (p. 137)

4-14   The median of a trapezoid has two properties:
   (1) It is parallel to the bases.
   (2) Its length equals half the sum of the lengths of the bases.   (p. 138)

4-15   The segment that joins the midpoints of two sides of a triangle has two properties:
   (1) It is parallel to the third side.
   (2) Its length equals half the length of the third side.   (p. 138)

4-16 The bisectors of the angles of a triangle intersect in a point which is equidistant from the three sides of the triangle. (p. 142)

4-17 The perpendicular bisectors of the sides of a triangle intersect in a point which is equidistant from the three vertices of the triangle. (p. 143)

4-18 The lines that contain the altitudes of a triangle intersect in a point. (p. 143)

4-19 The medians of a triangle intersect in a point that is two thirds of the distance from each vertex to the midpoint of the opposite side. (p. 143)

4-20 If two sides of a triangle are unequal, then the angles opposite those sides are unequal in the same order. (p. 152)

4-21 If two angles of a triangle are unequal, then the sides opposite those angles are unequal in the same order. (p. 152)

Corollary 1 The perpendicular segment from a point to a line is the shortest segment from the point to the line. (p. 152)

Corollary 2 The perpendicular segment from a point to a plane is the shortest segment from the point to the plane. (p. 152)

4-22 The sum of the lengths of any two sides of a triangle is greater than the length of the third side. (p. 153)

4-23 (SAS Inequality) If two sides of one triangle are equal to two sides of another triangle, but the included angle of the first triangle is larger than the included angle of the second, then the third side of the first triangle is longer than the third side of the second. (p. 156)

4-24 (SSS Inequality) If two sides of one triangle are equal to two sides of another triangle, but the third side of the first triangle is longer than the third side of the second, then the included angle of the first triangle is larger than the included angle of the second. (p. 157)

Similar Polygons

5-1 (SAS Similarity Theorem) If an angle of one triangle is equal to an angle of another triangle and the sides including those angles are in proportion, then the triangles are similar. (p. 186)

5-2 (SSS Similarity Theorem) If the sides of two triangles are in proportion, then the triangles are similar. (p. 187)

5-3 (Triangle Proportionality Theorem) If a line is parallel to one side of a triangle and intersects the other two sides, then it divides them proportionally. (p. 190)

Corollary If three parallel lines intersect two transversals, then they divide the transversals proportionally. (p. 191)

5-4 If a ray bisects an angle of a triangle, it divides the opposite side into segments proportional to the other two sides. (p. 191)

Right Triangles

6-1 If the altitude is drawn to the hypotenuse of a right triangle, then the two triangles formed are similar to the original triangle and to each other. (p. 204)

  Corollary 1 When the altitude is drawn to the hypotenuse of a right triangle, the altitude is the geometric mean between the segments of the hypotenuse. (p. 204)

  Corollary 2 When the altitude is drawn to the hypotenuse of a right triangle, either leg is the geometric mean between the hypotenuse and the segment of the hypotenuse that is adjacent to that leg. (p. 204)

6-2 (Pythagorean Theorem) In a right triangle, the square of the hypotenuse is equal to the sum of the squares of the legs. (p. 207)

6-3 If the square of one side of a triangle is equal to the sum of the squares of the other two sides, then the triangle is a right triangle. (p. 212)

6-4 If the square of the longest side of a triangle is greater than the sum of the squares of the other two sides, then the triangle is an obtuse triangle. (p. 213)

6-5 If the square of the longest side of a triangle is less than the sum of the squares of the other two sides, then the triangle is an acute triangle. (p. 213)

6-6 (45-45-90 Theorem) In a 45-45-90 triangle, the hypotenuse is $\sqrt{2}$ times as long as a leg. (p. 216)

6-7 (30-60-90 Theorem) In a 30-60-90 triangle, the hypotenuse is twice as long as the shorter leg, and the longer leg is $\sqrt{3}$ times as long as the shorter leg. (p. 216)

Circles

7-1 If a line is tangent to a circle, then the line is perpendicular to the radius drawn to the point of tangency. (p. 249)

  Corollary Tangents to a circle from a point are equal. (p. 249)

7-2 If a line in the plane of a circle is perpendicular to a radius at its outer endpoint, then the line is tangent to the circle. (p. 250)

7-3   In the same circle or in equal circles,
(1) Equal chords have equal arcs;
(2) Equal arcs have equal chords. (p. 257)

7-4   A diameter that is perpendicular to a chord bisects the chord and its arc. (p. 258)

7-5   In the same circle or in equal circles,
(1) Equal chords are equally distant from the center;
(2) Chords equally distant from the center are equal. (p. 258)

7-6   An inscribed angle is equal to half its intercepted arc. (p. 264)

Corollary 1   If two inscribed angles intercept the same arc, then the angles are equal. (p. 265)

Corollary 2   If a quadrilateral is inscribed in a circle, then opposite angles are supplementary. (p. 265)

Corollary 3   An angle inscribed in a semicircle is a right angle. (p. 265)

7-7   An angle formed by a chord and a tangent is equal to half the intercepted arc. (p. 265)

7-8   An angle formed by two chords intersecting inside a circle is equal to half the sum of the intercepted arcs. (p. 269)

7-9   An angle formed by two secants, two tangents, or by a secant and a tangent drawn from a point outside a circle is equal to half the difference of the intercepted arcs. (p. 269)

7-10   When two chords intersect inside a circle, the product of the segments of one chord equals the product of the segments of the other. (p. 272)

7-11   When two secants are drawn to a circle from an outside point, the product of one secant and its external segment equals the product of the other secant and its external segment. (p. 273)

7-12   When a tangent and a secant are drawn to a circle from an outside point, the square of the tangent is equal to the product of the secant and its external segment. (p. 273)

Areas of Plane Figures

9-1   The area of a rectangle equals the product of its base length and height. ($A = bh$) (p. 323)

9-2   The area of a parallelogram equals the product of its base length and height. ($A = bh$) (p. 326)

9-3 The area of a triangle equals half the product of its base length and height. $(A = \frac{1}{2}bh)$ (p. 326)

> **Corollary** The area of a rhombus equals half the product of its diagonals. $(A = \frac{1}{2}d_1d_2)$ (p. 326)

9-4 The area of a trapezoid equals half the product of the height and the sum of the bases. $(A = \frac{1}{2}h(b_1 + b_2))$ (p. 330)

9-5 The area of a regular polygon is equal to half the product of the apothem and the perimeter. $(A = \frac{1}{2}ap)$ (p. 335)

> **Related formulas** In a circle: $C = 2\pi r$ $A = \pi r^2$

9-6 If the scale factor of two similar figures is $a:b$, then
(1) the ratio of the perimeters is $a:b$, and
(2) the ratio of the areas is $a^2:b^2$. (p. 347)

Areas and Volumes of Solids

10-1 The lateral area of a right prism equals the perimeter of a base times the height of the prism. (L.A. $= ph$) (p. 360)

10-2 The volume of a right prism equals the area of a base times the height of the prism. $(V = Bh)$ (p. 361)

> **Related formulas** In a pyramid: L.A. $= \frac{1}{2}pl$ $V = \frac{1}{3}Bh$
>
> In a cylinder: L.A. $= 2\pi rh$ $V = \pi r^2 h$
>
> In a cone: L.A. $= \pi rl$ $V = \frac{1}{3}\pi r^2 h$
>
> In a sphere: $A = 4\pi r^2$ $V = \frac{4}{3}\pi r^3$

10-3 If the scale factor of two similar solids is $a:b$, then
(1) the ratio of corresponding perimeters is $a:b$;
(2) the ratios of the base areas, of the lateral areas, and of the total areas are $a^2:b^2$; and
(3) the ratio of the volumes is $a^3:b^3$. (p. 377)

Coordinate Geometry

11-1 The distance d between points (x_1, y_1) and (x_2, y_2) is given by $d = \sqrt{(x_2 - x_1)^2 + (y_2 - y_1)^2}$. (p. 388)

11-2 The circle with center (a, b) and radius r has the equation $(x - a)^2 + (y - b)^2 = r^2$. (p. 391)

11-3 The midpoint of the segment that joins points (x_1, y_1) and (x_2, y_2) is the point $\left(\dfrac{x_1 + x_2}{2}, \dfrac{y_1 + y_2}{2}\right)$. (p. 395)

11-4 Two nonvertical lines are parallel if and only if their slopes are equal. (p. 401)

11-5 Two nonvertical lines are perpendicular if and only if the product of their slopes is -1. (p. 401)

$$m_1 \cdot m_2 = -1, \text{ or } m_1 = -\frac{1}{m_2}$$

11-6 The graph of any equation that can be written in the form $ax + by = c$, with a and b not both zero, is a line. (p. 406)

11-7 The equation of the line that passes through point (x_1, y_1) and has slope m is $y - y_1 = m(x - x_1)$. (p. 407)

Transformations

12-1 A translation is an isometry. (p. 425)

12-2 A reflection in a line is an isometry. (p. 427)

12-3 A rotation is an isometry. (p. 433)

12-4 A product of reflections in parallel lines is a translation. (p. 441)

12-5 A product of reflections in intersecting lines is a rotation. (p. 441)

12-6 The dilation $D_{O,k}$ maps every line segment to a parallel line segment which is $|k|$ times as long. (p. 449)

Constructions

Answers for Self-Tests

Chapter 1

PAGE 16
1. T **2.** F **3.** F **4.** T **5.** T **6.** T **7.** Point, line, plane **8.** 4 **9.** B **10.** 90°
11. ∠1 **12.** 180° **13.** obtuse **14.** Addition Postulate **15.** Subtraction Postulate **16.** Transitive
Property *or* Substitution Property

PAGE 26
1. $\overline{AD}$ and $\overline{AC}$ **2.** $\overrightarrow{CA}$ **3.** ∠DAB, ∠B **4.** ∠D, ∠BCD **5.** ∠BCA and ∠ACD; *or* ∠BAC and
∠CAD **6.** If two angles are complements of equal angles, then the two angles are equal. (Theorem 1-1)
7. Vertical angles are equal. (Theorem 1-3) **8.** If two angles are supplements of the same angle, then the two
angles are equal. (Theorem 1-2)

PAGE 36
1. 1. $\overrightarrow{OB}$ bisects ∠AOC. (Given) 2. ∠1 = ∠2 (Def. of ∠ bisector) 3. ∠2 = ∠3 (Vertical ∡ are
equal.) 4. ∠1 = ∠3 (Transitive Property *or* Substitution Property) **2.** Exactly one **3.** Exactly one
4. No **5.** Yes **6.** No **7.** No

Chapter 2

PAGE 55
1. T **2.** T **3.** F **4.** T **5.** T **6.** F **7.** F **8.** sometimes **9.** never **10.** always
11. sometimes **12.** always

PAGE 65
1. 90° **2.** equilateral **3.** scalene **4.** $x = 120; y = 30$ **5.** 1440°, 360° **6.** A regular polygon is a
polygon which is both equiangular and equilateral. **7.** 144° **8.** 18

PAGE 74
1. Ned can't go. **2.** Valerie can go. **3.** If Valerie can go, then Ned can't go. **4.** If Ned can go, then
Valerie can't go. **5.** If Valerie can't go, then Ned can go. **6. a.** Valerie can go. **b.** No conclusion
c. No conclusion **d.** Ned can go. **7. a.** If a figure is a square, then it is a rectangle.
b.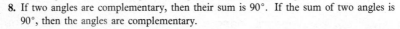
8. If two angles are complementary, then their sum is 90°. If the sum of two angles is
90°, then the angles are complementary.

Chapter 3

PAGE 93
1. △AKN and △OKL; ASA Postulate **2.** △BNO and △SON; SSS Postulate **3.** △IKT and △EKT;
SAS Postulate **4.** 1. $QS = RS$; $\overline{PS} \perp \overline{QR}$ (Given) 2. ∠QSP = ∠RSP (Def. of ⊥ lines) 3. $SP = SP$
(Reflexive Property) 4. △QSP ≅ △RSP (SAS Postulate) 5. ∠QPS = ∠RPS (Corr. parts of ≅ ▵
are =.) 6. $\overrightarrow{PS}$ bisects ∠QPR. (Def. of ∠ bisector) **5.** 1. $LA = SA$; $\overline{BL} \| \overline{SH}$ (Given) 2. ∠L = ∠S (If ∥
lines are cut by a transversal, then alt. int. ∡ are =.) 3. ∠BAL = ∠HAS (Vertical ∡ are =.)
4. △LBA ≅ △SHA (ASA Postulate) 5. $BA = HA$ (Corr. parts of ≅ ▵ are =.) 6. A is the midpoint of $\overline{BH}$.
(Def. of midpoint)

502

PAGE 102

1. 42   **2.** 5   **3.** $\triangle ADY$ and $\triangle CDY$ **4.** *DZ*   **5.** $\overrightarrow{CD}$; $\angle BCA$ **6.** $\angle DCA$ **7.** *AD*   **8.** No; Theorem 3-2 does not apply because $\angle Q$ and $\angle R$ are not in the same $\triangle$, nor is there enough information to prove that $\triangle PQS \cong \triangle PRS$.   **9.** 1. $\angle Q = \angle R$; $\overline{PS} \perp$ plane *X* (Given)   2. $\overline{PS} \perp \overline{QS}$; $\overline{PS} \perp \overline{RS}$ (Def. of line $\perp$ plane)   3. $\angle QSP = \angle RSP = 90°$ (Def. of $\perp$ lines)   4. $SP = SP$ (Reflexive Property)   5. $\triangle PQS \cong \triangle PRS$ (AAS Theorem)

PAGE 115

1. *Q* is the midpoint of $\overline{RS}$ **2.** $\overleftrightarrow{PQ} \perp \overleftrightarrow{RS}$ **3.** $\overleftrightarrow{PQ} \perp \overline{RS}$; $\overleftrightarrow{PQ}$ bisects $\overline{RS}$ **4.** *P* is equidistant from *R* and *S*; *or PR = PS*   **5.** No   **6.** No   **7. a.** 1. $\angle 1 = \angle 2$; $\angle D = \angle C$; $AD = BC$ (Given)   2. $\triangle ADX \cong \triangle BCY$ (AAS Theorem)   **b.** See Part **(a)** for Steps 1 and 2.   3. $AX = BY$ (Corr. parts of $\cong$ $\triangle$ are =.)   4. $XY = YX$ (Reflexive Property)   5. $\angle 3$ is supp. to $\angle 1$; $\angle 4$ is supp. to $\angle 2$ (Def. of supp. $\triangle$)   6. $\angle 3 = \angle 4$ (Supps. of = $\triangle$ are =.)   7. $\triangle AXY \cong \triangle BYX$ (SAS Postulate)   **8.** 625   **9.** 13

Chapter 4

PAGE 136

1. Always   **2.** Sometimes   **3.** Always   **4.** Always   **5.** Never   **6.** Always   **7.** Yes   **8.** No   **9.** No   **10.** Yes   **11.** 1. Quad. *ABCD* is a rectangle. (Given)   2. $AC = BD$ (Diagonals of a rectangle are =.)   3. $CD = DC$ (Reflexive Property)   4. $DA = CB$ (Opp. sides of a rectangle are =.)   5. $\triangle ACD \cong \triangle BDC$ (SSS Postulate)   6. $\angle 1 = \angle 2$ (Corr. parts of $\cong$ $\triangle$ are =.)

PAGE 150

1. No   **2.** $\angle A$ and $\angle B$; $\angle C$ and $\angle D$ **3.** $\frac{1}{2}(DC + AB)$, *or* 8   **4.** It is parallel to the third side; its length is half the length of the third side.   **5.** Angle bisectors; lines containing the altitudes; medians   **6. (b)**   **7.** contradiction

PAGE 160

1. $\angle B > \angle A$ **2.** 9   **3.** 95   **4.** $<$ **5.** $>$ **6.** $=$

Chapter 5

PAGE 178

1. $\frac{3}{4}$, *or* 3:4   **2.** $\frac{5}{3}$, *or* 5:3   **3.** 36° and 54°   **4.** No   **5.** Yes   **6.** Yes   **7.** Yes   **8.** $\angle U$ **9.** *VR*   **10.** 6

PAGE 195

1. AA Similarity Postulate   **2.** SAS Similarity Theorem   **3.** SSS Similarity Theorem   **4.** *x*   **5.** *z*   **6.** *z*   **7.** *w*   **8.** 4   **9.** 12

Chapter 6

PAGE 211

1. $\frac{\sqrt{6}}{4}$ **2.** 10   **3.** $x = 6$ **4.** $y = 2\sqrt{10}$ **5.** $z = 6\sqrt{10}$ **6.** $t = 10$ **7.** $r = 5$ **8.** $t = 3\sqrt{2}$ **9.** $r = 7$ **10.** $t = 10$

PAGE 219
1. Acute   **2.** Obtuse   **3.** Right   **4.** $b = 9$; $c = 9\sqrt{2}$ **5.** $a = 12$; $b = 12$ **6.** $t = 14$; $s = 7\sqrt{3}$
7. $r = 10$; $s = 10\sqrt{3}$ **8.** $r = \frac{5}{2}$; $t = 5$

PAGE 232
1. $\frac{15}{17}$ **2.** $\frac{8}{17}$ **3.** $\frac{15}{8}$ **4.** $\frac{8}{17}$ **5.** 62°   **6.** 6.4   **7.** 13.0   **8.** 12.6   **9.** 8.1

Chapter 7

PAGE 263
1. a. $\overleftrightarrow{OX}$, $\overleftrightarrow{OY}$ **b.** $\overrightarrow{AB}$ **c.** $\overline{XY}$ **d.** $\overline{AB}$, $\overline{XY}$ **2. a.**   **b.**
3. circle   **4.** 12   **5.** three   **6.** $\overline{QX} \perp \overline{TX}$
7. $TX = TY$ **8.** 58°   **9.** 302°   **10.** In the
same circle, equal arcs have equal chords. (Theorem 7-3)   **11.** 8   **12.** 119°   **13.** 5 units

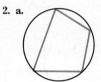
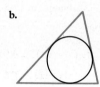

PAGE 276
1. 22°   **2.** 55°   **3.** 160°   **4.** 8   **5.** 90°   **6.** 20°   **7.** 4   **8.** 1. $TZ = TY$ (Given)   2. $\angle 3 = \angle 4$
(If two sides of a △ are =, the △ opp. the sides are =.)   3. $\angle 2 = \angle 3$ (If two inscribed △ intercept the same
arc, the △ are =.)   4. $\angle 2 = \angle 4$ (Transitive Property *or* Substitution Property)

Chapter 8

PAGE 302
1. Use Construction 3.   **2.** Use Construction 5.   **3.** Use Construction 7.   **4.** Use Construction 8.
5. Use Construction 9.   **6.** Use Construction 12.   **7.** Use Construction 10.

PAGE 314
1. The bisector of the angle   **2.** In the plane, the circle with center at the point and radius 5 cm   **3.** The
point of intersection of the diagonals of the rectangle   **4.** In the plane, the two points of intersection of (1) a
line parallel to and 4 cm from $\overleftrightarrow{BC}$, and (2) a circle with center at A and radius 4 cm   **5.** The two planes parallel to and 5 cm from the given plane   **6.** The circle of intersection of two spheres, each with radius 5 cm and
center at one of the given points   **7.** 1. Draw any line and take a point A on it.   2. Determine B so that
$AB = 2n$.   3. At A and B construct angles equal to $\angle 1$.   4. Construct the perpendicular to $\overleftrightarrow{AB}$ at A.   5. On the
perpendicular constructed in Step 4, determine H so that $AH = n$.   6. Construct the perpendicular to $\overleftrightarrow{AH}$ at
H.   7. Extend the sides of $\angle A$ and $\angle B$ to determine D and C.

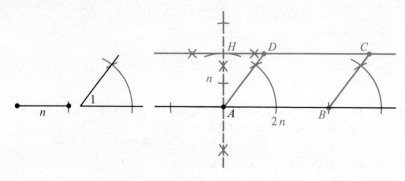

Chapter 9

PAGE 337

1. 25 **2.** 108 **3.** 70 **4.** 35 **5.** $3\sqrt{7}$ **6.** 24 **7.** 8 **8.** 48 **9.** 40 **10.** $48\sqrt{3}$ **11.** Radius $= 4$; apothem $= 2\sqrt{3}$; area $= 24\sqrt{3}$

PAGE 350

1. Circumference $= 14\pi$; area $= 49\pi$ **2.** 12π **3.** 42 **4. a.** 6π **b.** 27π **5. a.** 4π **b.** $16\pi - 32$
6. $9:25$ **7.** $4:9$ **8.** $5:7$

Chapter 10

PAGE 370

1. 150 cm² **2.** 6 **3.** $V = 60$ cm³; T.A. $= 94$ cm² **4.** $V = 54\pi$; T.A. $= 54\pi$ **5.** $V = 400$;
T.A. $= 360$ **6.** $V = 128\pi$; T.A. $= 144\pi$ **7.** $V = \dfrac{27}{2}\sqrt{3}$; T.A. $= \dfrac{108 + 9\sqrt{3}}{2}$, or $54 + \dfrac{9}{2}\sqrt{3}$

PAGE 380

1. 36π **2.** 5 **3.** 9π **4.** $1:4$; $1:8$ **5.** $1:3$

Chapter 11

PAGE 397

1. a. 6 **b.** $(5, 3)$ **2. a.** 12 **b.** $(1, 1)$ **3. a.** 10 **b.** $(-3, 4)$ **4. a.** 17 **b.** $\left(-1, -\dfrac{3}{2}\right)$

5. $x^2 + y^2 = 49$ **6.** $(x - 3)^2 + (y + 2)^2 = 4$ **7.** $(0, -1)$, 5

PAGE 412

1. $\dfrac{5}{2}$ **2.** $-\dfrac{4}{5}$ **3.** A vertical line **4. a.** $-\dfrac{7}{2}$ **b.** $\dfrac{2}{7}$ **5. a.** $\dfrac{1}{2}$ **b.** $(4, -3)$ **6.**
7. $(4, -1)$

PAGE 418

1. $(2e, 0)$ **2.** $(g + j, h)$ **3.** $(j - g, h)$ **4.** Let $A = (2, 0)$, $B = (19, 0)$, and $C = (10, 15)$. Then
$AC = \sqrt{(10 - 2)^2 + (15 - 0)^2} = 17$, and $AB = 17$. Since $AB = AC$, $\triangle ABC$ is isosceles. **5.** The slope of
$\overline{ON} = \dfrac{0 - 0}{9 - 0} = 0$, and the slope of $\overline{RD} = \dfrac{7 - 7}{12 - 3} = 0$. Since their slopes are equal, $\overline{ON} \| \overline{RD}$. Since $ON = 9$ and
$RD = 9$, $ON = RD$. Since opp. sides $\overline{ON}$ and $\overline{RD}$ of quad. $ROND$ are $=$ and $\|$, quad. $ROND$ is a $\square$.

Chapter 12

PAGE 436

1. Q' **2.** R **3.** $P'Q'$ **4.** $(3, 1)$ **5.** $\overline{AA'}$ **6.** $R_{O, 330}$; $R_{O, -390}$ **7.** B **8.** C **9.** $\overline{AB}$
10. $\overline{OB}$ **11.** M **12.** $\overline{AO}$ **13.** L **14.** $\triangle NDO$

PAGE 452

1. The x-axis **2.** No **3.** R **4.** C **5.** S **6.** A **7.** P **8.** T

Glossary

acute angle: An angle with measure between 0° and 90°. (p. 10)

acute triangle: A triangle with three acute angles. (p. 56)

adjacent angles: Two angles in a plane that have a common vertex and a common side but no common interior points. (p. 17)

alternate interior angles: There are two pairs in the diagram, ∠3 and ∠6, and ∠4 and ∠5. (p. 43)

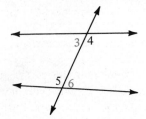

altitude to a base of a rectangle or other parallelogram: Any segment perpendicular to the base line from any point on the opposite side. (p. 322)

altitude of a prism: A segment joining the two base planes and perpendicular to both. (p. 359)

altitude of a pyramid: *See* pyramid.

altitude of a right cone: *See* cone.

altitude of a right cylinder: *See* cylinder.

altitude of a trapezoid: Any segment perpendicular to a base line from a point on the opposite base line. (p. 330)

altitude of a triangle: A segment from a vertex perpendicular to the line that contains the opposite side. (p. 103)

angle: A figure formed by two rays that have the same endpoint. (p. 8)

angle of depression: The angle through which a telescope must be lowered from the horizontal to sight an object. (p. 229)

angle of elevation: The angle through which a telescope must be raised from the horizontal to sight an object. (p. 229)

angle of a triangle: *See* triangle.

apothem of a regular polygon: The (perpendicular) distance from the center of the polygon to a side. (p. 334)

arc of a circle: A part of a circle. There are three kinds of arcs, a *semicircle* ("half" a circle), a *minor arc* (less than a semicircle), and a *major arc* (greater than a semicircle). In the diagrams, a semicircle ($\widehat{ABC}$) and a minor arc ($\widehat{DE}$) are shown in color, and a major arc ($\widehat{EFD}$) is shown in black. Three letters must be used to name a semicircle or a major arc. (p. 253)

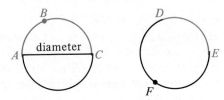

arc of a chord: The minor arc cut off by the chord. (p. 257)

area of a circle: The limit of the areas of the inscribed regular polygons. (p. 338)

auxiliary line: A line, ray, or segment added to a diagram to help in a proof. (p. 94)

axiom: A statement that is accepted without proof. (p. 22)

base angles of an isosceles triangle: The angles at the base of the isosceles triangle. (p. 94)

base of a cone: *See* cone.

base of a cylinder: *See* cylinder.

base of an isosceles triangle: The side of the isosceles triangle other than the legs. (p. 94)

base of a prism: *See* prism.

base of a pyramid: *See* pyramid.

base of a rectangle or other parallelogram: Any side of a rectangle or other parallelogram can be considered to be a base. (p. 322)

bases of a trapezoid: The parallel sides of a trapezoid. (p. 137)

bisector of an angle: A ray that divides the angle into two equal adjacent angles. (p. 17)

bisector of an arc: Any line, segment, or ray that contains the midpoint of the arc. (p. 258)

bisector of a segment: A line, segment, ray, or plane that intersects the segment at its midpoint. (p. 6)

center of a circle: *See* circle.

center of a regular polygon: The center of the circumscribed circle. (p. 334)

central angle of a circle: An angle whose vertex is the center of the circle. (p. 253)

central angle of a regular polygon: An angle formed by two radii drawn to consecutive vertices. (p. 334)

chord of a circle: A segment that joins two points on a circle. (p. 245)

circle: The set of all points in a plane that are a given distance from a given point in the plane. The given point is the *center*. The given distance is the *radius*. (p. 245)

circumference of a circle: The limit of the perimeters of the inscribed regular polygons. (p. 338)

circumscribed circle: *See* inscribed polygon.

circumscribed polygon: A polygon is said to be *circumscribed about a circle* and the circle is said to be *inscribed in the polygon* when each side of the polygon is tangent to the circle. (p. 250)

collinear points: Points all in one line. (p. 4)

common tangent: A line that is tangent to each of two coplanar circles. Common *internal* tangents intersect the segment joining the centers of the circles. Common *external* segments do *not* intersect the segment joining the centers. (p. 250)

complementary angles: Two angles whose measures have the sum 90°. Each angle is a *complement* of the other. (p. 18)

concentric circles: Circles that lie in the same plane and have the same center. (p. 246)

conclusion: *See* conditional.

concurrent lines: Two or more lines that intersect in one point. (p. 142)

conditional: A statement that can be expressed in the basic form *If p, then q,* where *p* and *q* are statements. *p* is the *hypothesis*. *q* is the *conclusion*. (p. 66)

cone: In the diagrams, the *right cone* A and the *oblique cone* B have circular *bases*. In a right cone, the segment joining the vertex and the center of the base is the *altitude* of the cone. The length of the altitude is the *height h* of the cone. A radius of the base is the *radius r* of the cone. The length of a segment joining the vertex and a point on the circular base is the *slant height l* of the cone. In the text the word "cone" always refers to a right cone. (p. 367)

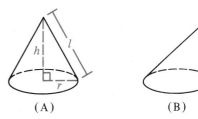

(A)   (B)

congruent figures: Figures having the same size and shape. (p. 81)

congruent triangles: Two triangles are *congruent* if their vertices can be matched up so that the corresponding parts of the triangles are equal. (p. 82)

construction: In the *construction* of a geometric figure, a *straightedge* is used only to draw lines, segments, or rays; and a *compass* is used only to draw circles or arcs of circles. No other instruments may be used. (p. 283)

contraction: *See* dilation.

contrapositive of a conditional: The *contrapositive* of the statement *If p, then q* is *If not q, then not p*. (p. 70)

converse of a conditional: The *converse* of the statement *If p, then q* is the statement *If q, then p*. (p. 66)

convex polygon: A polygon such that no line containing a side of the polygon contains a point in the interior of the polygon. (p. 61)

coordinate plane: The plane of the *x*-axis and *y*-axis. (p. 387)

coordinate(s) of a point: On the number line, the number paired with a point. In the coordinate plane, the ordered pair associated with a point. The first number in the ordered pair is the *x-coordinate* of the point; the second number is the *y-coordinate* of the point. (pp. 5, 387)

coplanar points: Points all in one plane. (p. 5)

corollary: A statement that can be proved easily by applying a theorem. (p. 57)

corresponding angles: There are four pairs in the diagram, $\angle 1$ and $\angle 5$, $\angle 2$ and $\angle 6$, $\angle 3$ and $\angle 7$, and $\angle 4$ and $\angle 8$. (p. 43)

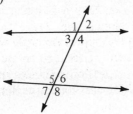

cosine ratio: The *cosine* of an acute angle of a right triangle is the ratio of the length of the leg *adjacent* to the angle to the length of the hypotenuse. (p. 226)

cylinder: In the diagrams, the *right cylinder* A and the *oblique cylinder* B have circular *bases*. In a right cylinder, the segment joining the centers of the bases is an *altitude*. The length of an altitude is the *height h* of the cylinder. A radius of a base is a *radius r* of the cylinder. In the text, the word "cylinder" always refers to a right cylinder. (p. 367)

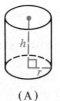   

(A)   (B)

deductive reasoning: The process of reasoning from accepted statements to conclusions. (p. 27)

diagonal of a polygon: A segment whose endpoints are two nonconsecutive vertices of a polygon. (p. 61)

diagonal of a rectangular solid: *See* rectangular solid.

diameter of a circle: A chord that passes through the center of the circle. (p. 245)

dilation: A *dilation* with center *O* and scale factor *k* ($k > 0$) is a mapping such that:

(1) If *P* is different from *O*, then *P'* lies on $\overrightarrow{OP}$ and $OP' = k \cdot OP$.

(2) If *P* is the point *O*, then *P'* is the same point as *P*.

If $0 < |k| < 1$, the dilation is called a *contraction*. If $|k| > 1$, the dilation is called an *expansion*. (pp. 448, 449)

distance from a point to a line (or plane): The length of the perpendicular segment from the point to the line (or plane). (p. 99)

edge of a rectangular solid: *See* rectangular solid.

endpoint of a ray: *See* ray.

endpoints of a segment: *See* segment.

equal angles: Angles that have equal measures. (p. 9)

equal arcs: Arcs that have equal measures. (p. 254)

equal circles: Circles that have equal radii. (p. 246)

equal segments: Segments that have equal lengths. (p. 6)

equiangular triangle: A triangle with all angles equal. (p. 56)

equilateral triangle: A triangle with all sides. equal. (p. 56)

expansion: *See* dilation.

extended proportion: When you want to state that three or more ratios are equal, you can write an *extended proportion:*

$$\frac{a}{b} = \frac{c}{d} = \frac{e}{f}.\text{ (p. 168)}$$

exterior angle of a triangle: An angle formed by extending one side of the triangle. In the diagram, $\angle 1$ is an *exterior angle;* and $\angle 2$ and $\angle 3$ are *remote interior angles.* (p. 58)

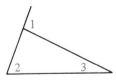

extremes of a proportion: *See* proportion.

face: *See* prism, pyramid, and rectangular solid.

geometric mean: Let a, x, and d be positive numbers. If $\dfrac{a}{x} = \dfrac{x}{d}$ then x is the *geometric mean* between a and d. (p. 203)

graph of an ordered pair: The point in the coordinate plane whose x-coordinate is the first number in the ordered pair and whose y-coordinate is the second number in the ordered pair. (p. 387)

half-turn: A *half-turn* about O is a rotation about point O through $180°$. (p. 433)

height: The *height* of a polygon or solid is the length of the altitude of the polygon or solid.

hypotenuse of a right triangle: The side opposite the right angle. (p. 98)

hypothesis: *See* conditional.

identity transformation: The mapping which maps every point to itself. (p. 445)

if-then statement: A statement expressed in the form *If p, then q,* where p and q are statements. (p. 66)

image: *See* mapping.

indirect proof: A proof in which you suppose that what you wish to prove is not true; then reason logically until you reach a contradiction of a known fact; and, finally, conclude that what you supposed to be true is false and that what you wish to prove true is true. (p. 147)

inductive reasoning: The process of reasoning from past observations to conclusions. (p. 112)

inscribed angle: An angle whose vertex is on a circle and whose sides contain chords of the circle. (p. 264)

inscribed circle: *See* circumscribed polygon.

inscribed polygon: A polygon is *inscribed in a circle* and the circle is *circumscribed about the polygon* when each vertex of the polygon lies on the circle. (p. 246)

intercepted arc: In the diagrams, the angles are said to *intercept* the arcs shown in color. Note that an angle can intercept two arcs. (p. 264)

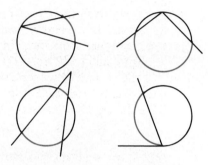

intersection of two figures: The set of points that are in both figures. (p. 2)

inverse of a conditional: The *inverse* of the statement *If p, then q* is the statement *If not p, then not q.* (p. 70)

inverse of a transformation: The *inverse* of a transformation *T* is the transformation *S* such that *T ∘ S* is the identity transformation. (p. 445)

isometry: A transformation which does *not* distort distances. (p. 424)

isosceles trapezoid: A trapezoid with equal legs. (p. 137)

isosceles triangle: A triangle with at least two sides equal. (p. 56)

lateral area of a solid: The sum of the areas of its lateral faces. (p. 360)

lateral edge of a prism: *See* prism.

lateral edge of a pyramid: *See* pyramid.

lateral face of a prism: *See* prism.

lateral face of a pyramid: *See* pyramid.

legs of an isosceles triangle: The equal sides of the isosceles triangle. (p. 94)

legs of a right triangle: The sides of the right triangle other than the hypotenuse. (p. 98)

legs of a trapezoid: The nonparallel sides of the trapezoid. (p. 137)

line parallel to a plane: A line and a plane are *parallel* if they do not intersect. (p. 42)

line perpendicular to a plane: A line and a plane are *perpendicular* if they intersect and the line is perpendicular to all lines in the plane that pass through the point of intersection. (p. 89)

line symmetry: A figure has *line symmetry* if there is a *symmetry line* such that a reflection in the line maps the figure into itself. (p. 437)

locus: A figure that is the set of all points, and only those points, that satisfy one or more conditions. (p. 303)

logically equivalent statements: Statements which must be both true or else both false. (p. 70)

major arc of a circle: *See* arc of a circle.

mapping: A correspondence between points. Each point *P* in a given set is *mapped* to some point *P'* in the same or a different set. *P'* is called the *image* of *P*, and *P* is called the *preimage* of *P'*. (p. 423)

means of a proportion: *See* proportion.

measure of an arc: The *measure of a minor arc* is the measure of its central angle. The *measure of a semicircle* is 180°. The *measure of a major arc* is the difference between 360° and the measure of the minor arc associated with the major arc. (pp. 253, 254)

median of a trapezoid: The segment that joins the midpoints of the legs. (p. 137)

median of a triangle: A segment from a vertex to the midpoint of the opposite side. (p. 103)

midpoint of an arc: M is called the *midpoint* of $\overset{\frown}{JMK}$ if $\overset{\frown}{JM} = \overset{\frown}{MK}$. (p. 258)

midpoint of a segment: The point that divides the segment into two equal segments. (p. 6)

minor arc of a circle: *See* arc of a circle.

mirror of a reflection: *See* reflection.

oblique cone: *See* cone.

oblique cylinder: *See* cylinder.

oblique prism: *See* prism.

oblique pyramid: *See* pyramid.

obtuse angle: An angle with measure between 90° and 180°. (p. 10)

obtuse triangle: A triangle with an obtuse angle. (p. 56)

opposite rays: $\overrightarrow{SR}$ and $\overrightarrow{ST}$ are called opposite rays if S is in $\overleftrightarrow{RT}$ and between R and T. (p. 5)

ordered pair of numbers: A pair of numbers is an *ordered pair* when the order in which the numbers are named has significance. (p. 387)

origin: The point of intersection of the x-axis and y-axis in the coordinate plane. (p. 387)

parallel line and plane: A line and a plane are *parallel* if they do not intersect. (p. 42)

parallel lines: Lines that do not intersect and are coplanar. (p. 41)

parallel planes: Planes that do not intersect. (p. 42)

parallelogram: A quadrilateral with both pairs of opposite sides parallel. (p. 123)

perimeter of a figure: The sum of the lengths of the sides of the figure. (p. 176)

perpendicular bisector of a segment: A line that is perpendicular to the segment and bisects the segment. (p. 18)

perpendicular line and plane: A line and a plane are *perpendicular* if they intersect and the line is perpendicular to all lines in the plane that pass through the point of intersection. (p. 89)

perpendicular lines: Two lines that form equal adjacent (90°) angles. (p. 17)

point of tangency: The point where a circle and a line tangent to the circle meet. (p. 249)

point symmetry: A figure has *point symmetry* if there is a *symmetry point* such that a half-turn about the point maps the figure into itself. (p. 437)

polygon: A figure formed by coplanar segments (called *sides*) such that:
(1) Each segment intersects exactly two other segments, one at each endpoint.
(2) No two segments with a common endpoint are in the same line. (p. 61)

postulate: A statement that is accepted without proof. (p. 13)

preimage: *See* mapping.

prism: The solids shown are *prisms.* The shaded faces are *bases*, congruent polygons lying in parallel planes. The other faces, called *lateral faces*, are parallelograms. Adjacent lateral faces intersect in parallel segments called *lateral edges.* An *altitude* of a prism is a segment joining the two base planes and perpendicular to both. The length of an altitude is the *height h* of the prism. Prism A, in which the lateral faces are rectangles, is a *right prism.* Prism B is an *oblique prism.* (pp. 359, 360)

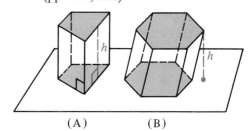

(A)   (B)

proportion: An equation stating that two ratios are equal. We can write a proportion in either of these forms:

$$\frac{a}{b} = \frac{c}{d} \qquad a:b = c:d$$

The numbers *a* and *d* are the *extremes* of the proportion. The numbers *b* and *c* are the *means*. *a* is called the first *term* of the proportion. *b, c,* and *d* are the second, third, and fourth *terms*. (pp. 168, 171)

pyramid: The solids shown are *pyramids*. Point *V* is the *vertex* of pyramid A, and pentagon *ABCDE* is the *base*. The perpendicular segment from the vertex to the base is the *altitude* and its length is the *height h* of the pyramid. The five triangular faces with *V* in common, such as △ *VAB*, are *lateral faces*, which intersect in segments called *lateral edges*. Pyramid A is an *oblique pyramid*. Pyramid B is a *regular pyramid*, which has these properties:

(1) The base is a regular polygon.
(2) The altitude meets the base at its center.
(3) All lateral edges are equal.
(4) All lateral faces are congruent isosceles triangles. The height of a lateral face is called the *slant height l* of the pyramid. (p. 364)

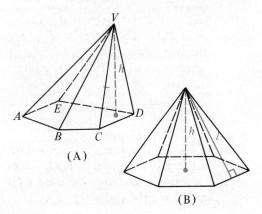

(A)

(B)

radical: The symbol $\sqrt{}$ will always indicate the positive square root of a number. The *radical* $\sqrt{45}$ has the *radicand* 45. (p. 203)

radicand: *See* radical.

radius of a circle: *See* circle.

radius of a regular polygon: The distance from the center to a vertex. (p. 334)

ratio: The *ratio* of one number to another is the quotient when the first number is divided by the second. (p. 167)

ray: *Ray AB* is the part of $\overleftrightarrow{AB}$ which starts at point *A* and extends without ending through point *B*. The *endpoint* of $\overrightarrow{AB}$ is *A*, the point named first. (p. 5)

rectangle: A quadrilateral with four right angles. (p. 132)

rectangular solid: The *rectangular solid* shown has six *faces,* each of which is a rectangle. One face is rectangle *ABCD*. The faces intersect in segments called *edges*. One edge is $\overline{EF}$. A *diagonal* is a segment whose endpoints are two vertices not in the same face. One *diagonal* is $\overline{AG}$. (pp. 360, 363)

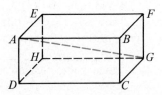

reflection: A *reflection* in a line, where the line is called the *mirror* of the reflection, maps every point *P* to a point *P'* such that:

(1) If *P* is not on the mirror, then the mirror is the perpendicular bisector of $\overline{PP'}$.
(2) If *P* is on the mirror, then $P' = P$. (p. 427)

regular polygon: A polygon which is both equiangular and equilateral. (p. 63)

regular pyramid: *See* pyramid.

remote interior angles of a triangle: *See* exterior angle of a triangle.

rhombus: A quadrilateral with four equal sides. (p. 132)

right angle: An angle with measure 90°. (p. 10)

right cone: *See* cone.

right cylinder: *See* cylinder.

right prism: *See* prism.

right triangle: A triangle with a right angle. (p. 56)

rotation: A *rotation* about O through $x°$ is a transformation such that:
(1) If point P is different from O, then $OP' = OP$ and $\angle POP' = x°$.
(2) If point P is the same as O, then $P' = O$. (p. 433)

rotational symmetry: A figure has *rotational symmetry* if there is a rotation that maps the figure into itself. (p. 437)

same-side interior angles: There are two pairs in the diagram, $\angle 3$ and $\angle 5$, and $\angle 4$ and $\angle 6$. (p. 43)

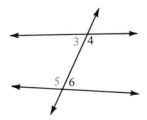

scale factor: The ratio of two corresponding sides of similar polygons. (p. 174)

scalene triangle: A triangle with no sides equal. (p. 56)

secant of a circle: A line that intersects the circle in two points. (p. 245)

sector of a circle: A region bounded by two radii and an arc of the circle. (p. 344)

segment: *Segment AB* consists of points A and B and all points in $\overleftrightarrow{AB}$ which are between A and B. Points A and B are called the *endpoints* of $\overline{AB}$. (p. 5)

semicircle: *See* arc of a circle.

sides of an angle: The rays that form the angle. (p. 8)

side of a triangle: *See* triangle.

similar figures: Figures which have the same shape. (p. 174)

similar polygons: Two polygons are *similar* if their vertices can be paired so that:
(1) Corresponding angles are equal.
(2) Corresponding sides are in proportion (have the same ratio). (p. 174)

similar solids: Solids which have the same shape but not necessarily the same size. (p. 376)

sine ratio: The *sine* of an acute angle of a right triangle is the ratio of the length of the side *opposite* the angle to the length of the hypotenuse. (p. 226)

skew lines: Lines that do not intersect and are not coplanar. (p. 41)

slant height of a regular pyramid: *See* pyramid.

slant height of a right cone: *See* cone.

slope of a line: The slope m of a line through $P_1(x_1, y_1)$ and $P_2(x_2, y_2)$, where $x_1 \neq x_2$, is defined as follows:

$$m = \frac{y_2 - y_1}{x_2 - x_1}. \text{ (p. 398)}$$

space: The set of all points. (p. 5)

sphere: The set of all points that are a given distance from a given point. (p. 246)

square: A quadrilateral with four equal sides and four right angles. (p. 132)

straight angle: An angle with measure 180°. (p. 10)

supplementary angles: Two angles whose measures have the sum 180°. Each angle is a *supplement* of the other. (p. 18)

tangent to a circle: A line that lies in the plane of a circle and meets the circle in exactly one point. (p. 249)

tangent circles: Two circles are *tangent to each other* when they are coplanar and are tangent to the same line at the same point. (p. 250)

tangent ratio: The *tangent* of an acute angle of a right triangle is the ratio of the length of the leg *opposite* the angle to the length of the leg *adjacent* to the angle. (p. 220)

term of a proportion: *See* proportion.

theorem: A statement that is proved. (p. 22)

total area of a solid: The sum of the areas of all its faces. (p. 360)

transformation of the plane: A mapping with two properties:
(1) Every point of the plane has exactly one image.
(2) Every point of the plane has exactly one preimage. (p. 423)

translation: A transformation that maps any two points P and Q to P' and Q' in such a way that $PP' = QQ'$ and $PQ = P'Q'$. (p. 425)

translational symmetry: A pattern has *translational symmetry* if there is a translation that maps the pattern into itself. (p. 437)

transversal: A line that intersects two or more coplanar lines in different points. (p. 42)

trapezoid: A quadrilateral with exactly one pair of parallel sides. (p. 137)

triangle: The figure formed by three segments joining three noncollinear points. Each of the three points is a *vertex* of the triangle (plural: *vertices*). The segments are the *sides* of the triangle. $\triangle ABC$ shown has vertices A, B, and C and sides $\overline{AB}$, $\overline{BC}$, and $\overline{CA}$. $\angle A$, $\angle B$, and $\angle C$ are the angles of $\triangle ABC$. (p. 56)

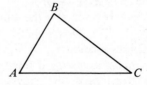

vertex of an angle: The common endpoint of the two rays that form the angle. (p. 8)

vertex angle of an isosceles triangle: The angle opposite the base. (p. 94)

vertex of a pyramid: *See* pyramid.

vertex of a triangle: *See* triangle.

vertical angles: Two angles whose sides form two pairs of opposite rays. When two lines intersect, they form two pairs of vertical angles. (p. 18)

Index

AA Similarity Postulate, 180
AAS Theorem, 98
Acute angle, 10, 57
Acute triangle, 56, 103, 213
Addition Postulate, 13
Adjacent angles, 17
Agnesi, Maria Gaetana, 107
Algebra, postulates and properties from, 13–14, 151
Alternate exterior angles, 43
Alternate interior angles, 43, 46, 50
Altitude(s)
  of a cone, 367
  of a cylinder, 367
  of a parallelogram, 322
  of a prism, 359
  of a pyramid, 364
  of a rectangle, 322
  of a trapezoid, 330
  of a triangle, 103–105, 204
    intersection of, 143, 418
Angle(s), 8–10
  acute, 10, 57
  adjacent, 17
  base, 94, 137
  bisector of, 17, 33, 99, 105, 132, 191
    construction of, 285
    intersection of, 142
  central, 253–254, 334
  in a circle, 253–254, 264–265, 268–269
  complementary, 18, 22, 57
  construction of, 284–286
  corresponding, 43, 46, 50
  of depression, 229
  dihedral, 488
  of elevation, 229
  equal, 9, 22–23, 46, 50, 57, 90, 94, 254, 265
  exterior, 42–43
  inscribed, 264–265
  interior, 42–43, 46, 50
  measure of, 9–10, 32
  obtuse, 10, 57
  of a polygon, 61–63
  plane, 488
  right, 10, 17, 57, 265
  straight, 10
  supplementary, 18, 22, 46, 50, 265
  of a triangle, 56–58, 82, 94, 152, 156–157
    exterior, 58
    remote interior, 58
    sum of, 57
  vertex, 94
  vertex of, 8

vertical, 18, 23
Apothem, 334–335
Application
  Center of Gravity, 145–146
  Electrical Circuits, 484–486
  Exercises, 34, 64–65, 88, 113, 131, 145, 170, 176, 184, 189, 194, 209, 219, 228, 230–231, 257, 306, 310, 325, 328, 341, 346, 368–369, 374, 379, 427, 430
  Mirrors, 431–432
  Scale Drawings, 178–179
  Technical Drawing, 292–293
  Vectors, 404–406
Arc(s), 253–254, 257–258
  bisector of, 258
  of a chord, 257
  equal, 254, 257
  intercepted, 264–265, 268–269
  length of, 344
  major, 253–254
  measure of, 253–254
  midpoint of, 258
  minor, 253–254, 257
  semicircle, 253–254
Archimedes, 277
Area(s), 321
  base, 360, 377
  of a circle, 338–339, 371
  of a cone, 367–368
  of congruent figures, 322
  of a cylinder, 367
  lateral, 360, 364–365, 367–368, 376–377
  of a parallelogram, 326
  of a polygonal region, 321–322
  of a rectangle, 323, 382–383
  of a region, 452–453
  of a regular polygon, 335
  of a rhombus, 326
  of a sector, 344
  of similar figures, 347–348
  of similar solids, 376–377
  of a sphere, 371, 373
  of a square, 322
  total, 360, 364–365, 367–368, 377
  of a trapezoid, 330
  of a triangle, 326–327, 332–333
Area Addition Postulate, 322
Area Congruence Postulate, 322
Arithmetic mean, 206
ASA Postulate, 86
Auxiliary line, 94
Axes (axis), 387
Axioms, 22

PHOTO CREDITS

Mechanical Art, ANCO
Title page, Cambridge Seven Associates
Page xii, Ellis Herwig, STOCK, BOSTON
Page 21, Historical Pictures Service, Chicago
Page 40, Nick Sapieha, STOCK, BOSTON
Page 55, Peter Southwick, STOCK, BOSTON
Page 80, Tom Nebbia, BLACK STAR
Page 107, The Bettmann Archive
Page 122, Barbara Marshall
Page 166, The Anaconda Company, Brass Division
Page 185, Rapho Guilumette, PHOTO RESEARCHERS
Page 202, Arley Rinehart Associates, Denver
Page 211, Bryn Mawr College Archives
Page 224, left (top), Ellis Herwig, STOCK, BOSTON; left (bottom), International
  Business Machines Corporation; right, Honeywell
Page 244, Mike Mazzaschi, STOCK, BOSTON
Page 262, top, Harvard College Observatory; bottom, Gene Daniels, BLACK STAR
Page 282, Terry McKoy
Page 298, left, CE Maguire, Inc.; right, The Foxboro Company
Page 320, Peter Vandermark, STOCK, BOSTON
Page 343, left, Bradford F. Herzog, Affiliated Hospitals Center, Boston Hospital for
  Women Division; right, Elizabeth Hamlin, STOCK, BOSTON
Page 358, Barbara Marshall
Page 381, left, U.S. Forest Service; right, Elizabeth Hamlin, STOCK, BOSTON
Page 386, Peter Southwick, STOCK, BOSTON
Page 422, Rene Burri, MAGNUM